P-ADIC ANALYSIS AND MATHEMATICAL PHYSICS

Series on Soviet & East European Mathematics – Vol. 1

P-ADIC ANALYSIS AND MATHEMATICAL PHYSICS

V. S. Vladimirov, I. V. Volovich and E. I. Zelenov

Steklov Mathematical Institute
Russia

World Scientific
Singapore • New Jersey • London • Hong Kong

Published by

World Scientific Publishing Co. Pte. Ltd.
5 Toh Tuck Link, Singapore 596224
USA office: 27 Warren Street, Suite 401-402, Hackensack, NJ 07601
UK office: 57 Shelton Street, Covent Garden, London WC2H 9HE

British Library Cataloguing-in-Publication Data
A catalogue record for this book is available from the British Library.

P-ADIC ANALYSIS AND MATHEMATICAL PHYSICS

ISBN-13 978-981-02-0880-6
ISBN-10 981-02-0880-4

CONTENTS

INTRODUCTION

Since the Newton and Leibnitz time differential equations over the real number field have been used in mathematical physics. It is not customary to discuss why exactly real numbers should be used. Why has this happened? The point is that physical processes take place in space and time and space-time coordinates are usually considered as real numbers. Since the Euclid time the three-dimensional Euclidean space $\mathbb{R}^3$ has been treated as the physical space. As it is known an important development of this point of view has been done by Riemann and Einstein using the Riemannian geometry, but basically up to now $\mathbb{R}^3$ is a mathematical model for space and $\mathbb{R}^4$ is a model for space-time.

These ideas have become so common, that $\mathbb{R}^3$ is perceived as the true physical space. But in fact *the Euclidean space* $\mathbb{R}^3$ *is not more than a mathematical model for the real physical space.* In order to convince ourselves that the Euclidean space is a good model for the physical space geometrical axioms from elementary geometry should be checked in practice. To this end lengths of segments, angles etc. should be measured precisely. However in quantum gravity and in string theory it was proved that there is the following obstacle to perform such measurements . If Δx is an uncertainty in a length measurement, then the inequality

$$\Delta x \geq l_{pl} = \sqrt{\frac{hG}{c^3}} \tag{1}$$

takes place. This inequality is stronger than the Heisenberg uncertainty principle. Here h is the Planck constant, c is the velocity of light and G is the gravitational constant. l_{pl} is called the Planck length and it is an extremely small quantity, approximately 10^{-33} cm. We are not going to discuss here a derivation of the fundamental inequality (1) which has a long history. Let us emphasize that by virtue of (1) a *measurement of distances smaller than the Planck length is impossible.*

Let us turn our attention to axioms of Euclidean geometry. In a list of axioms there exists the so-called Archimedean axiom, which was at first pointed out and analyzed by Veroneze and Hilbert. According to the Archimedean axiom any given large segment on a straight line can be surpassed by successive addition of small segments along the same line. Really, this is a physical axiom which concern the process of measurement. Two different scales are compared in this axiom. It means that we can measure distances as small as we want. But as we just discussed, the Planck length is the smallest possible distance that can in principle be measured. So *a suggestion emerges to abandon the Archimedean axiom at very small distances.* This leads to a non-Euclidean and non-Riemanian geometry of space at small distances.

How can one construct a physical theory corresponding to a non-Archimedean geometry? As it is well known there is an analytical description of geometry. One uses coordinates to describe a geometrical picture.

There are two equivalent approaches

$$\text{geometry} \longleftrightarrow \text{number system.}$$

The usual Euclidean geometry is described by means of real numbers. If we want to abandon the standard geometry for description of small distance in physical space-time we have to abandon real numbers. What should be used instead of real numbers?

In computations in everyday life, in scientific experiments and on computers we are dealing with integers and fractions, that is with rational numbers and we never have dealings with irrational numbers - infinite non-periodic decimals. Results of any practical action we can express only in terms of rational numbers which are considered to have been given us by God. Certainly, there exists generally accepted confidence that if we carry out measurements more and more precisely, then in principle we can get any large number of decimal digits and interpret a result as a real number. However this is an idealization and as it follows from the previous discussion we should be careful with such statements. Thus, *let us take as our starting point the field Q of rational numbers.* A geometric notion of distance corresponds to a notion of a norm on $\mathbb{Q}$. Norm is a real valued function $|x|$ with the following properties

1) $|x| \geq 0$, $|x| = 0 \Longleftrightarrow x = 0$,

2) $|xy| = |x||y|$,

3) $|x+y| \leq |x| + |y|$

for any rational numbers x, y. What norms do exist on $\mathbb{Q}$? There is a remarkable Ostrowski theorem describing all norms on $\mathbb{Q}$. According to this *theorem any nontrivial norm on Q is equivalent to either ordinary absolute value or p-adic norm for some prime number p.* p-Adic norm $|x|_p$ is defined by the following. Let us fix a prime number $p = 2, 3, 5, \ldots$. Any rational number x can be represented in the form $x = p^\nu \frac{m}{n}$, where ν is an integer and m and n are integers which are not divisible by p. This representation is unique. Then by definition

$$|x|_p = 1/p^\nu \ . \tag{2}$$

At first sight this definition looks artificial but according to the Ostrowski theorem there are no others nonequivalent norms on $\mathbb{Q}$. It is easy to verify the validity of the properties 1)–3) for $|x|_p$. In fact instead of triangle inequality a stronger inequality takes place

3′) $|x+y|_p \leq \max\{|x|_p, |y|_p\}$.

A norm that satisfies 3′) is called *non-Archimedean.*

The completion of the field $\mathbb{Q}$ of rational numbers with respect to usual absolute value leads to the field $\mathbb{R}$ of real numbers. Analogously, the completion of the field $\mathbb{Q}$ with respect to p-adic norm $|x|_p$ leads to the field $\mathbb{Q}_p$ of p-adic numbers for any prime p.

Any real number can be represented as a decimal, that is a series

$$\pm 10^\nu \sum_{0 \leq n < \infty} b_n 10^{-n} \ . \tag{3}$$

Analogously, *any p-adic number can be represented as a series*

$$x = p^\nu \sum_{0 \leq n < \infty} a_n p^n \ , \tag{4}$$

where a_n are integers, $0 \leq a_n \leq p-1$. If $a_0 \neq 0$, then the representation (4) is unique. Note that in contrast to (3) where we have a power series in the small parameter 1/10, in (4) we have a power series in the prime number p. The series (4) converges in p-adic norm.

The expression (4) can be considered as a definition of p-adic number. Addition, multiplication etc. of p-adic numbers of the form (4) are carried out as for power series. Any rational number including negative numbers, can be represented in the form (4). For example,

$$-1 = (p+1) + (p-1)p + (p-1)p^2 + \ldots$$

p-Adic numbers were introduced by K.Hensel at the end of XIX century. They form an integral part of number theory, algebraic geometry, representation theory and other branches of modern mathematics.

Geometry of the space $\mathbb{Q}_p$ is surprisingly unlike the geometry of the space $\mathbb{R}$, that follows from the property 3′) of the norm. In particular the Archimedean axiom is not true in $\mathbb{Q}_p$ ($\mathbb{Q}_p$ is a non-Archimedean space). For two different balls in non-Archimedean geometry one has that either one of them is contained in another or they have no common points. It resembles the behavior of two drops of mercury. *The field of p-adic numbers has a natural hierarchical structure*: every disc consists of the finite number of disjoint discs of smaller radius; the field $\mathbb{Q}_p$ is homeomorphic to a Cantor set on the real line.

Thus we come to the following picture. Fundamental physical theory should be based on rational numbers, at very small distances an important role have to play p-adic numbers and at large distances — real numbers. It leads us to the necessity of reconsidering of the basis of mathematical and theoretical physics beginning from the classical mechanics and right up to the string theory using number theory, p-adic analysis and algebraic geometry. It is a huge volume of material and in this book we cannot involve all of its aspects. The purpose of this book is to give an introduction to those parts of p-adic analysis which are proved to be the most important in applications and to describe some results of p-adic mathematical physics.

There is a natural hierarchical and fractal-like structure in the field of p-adic numbers. Therefore *p-adic analysis and non-Archimedean geometry can be used not only for the description of geometry at small distances, but also for describing chaotic behavior of complicated systems such as spin glasses and fractals in the framework of traditional theoretical and mathematical physics.*

We are only at the beginning of the p-adic mathematical physics. We hope that p-adic numbers would find applications in such fields as the theory of turbulence, biology, dynamical systems, computers, problems of an information transference, cryptography, economy etc. and also in other natural sciences which study systems with chaotic fractal behavior and hierarchical structures.

The book starts from an account of p-adic analysis. Mainly two types of analysis are considered. One is for p-adic valued functions of the type $f : \mathbb{Q}_p \to \mathbb{Q}_p$ and the other is that of complex valued functions, $f : \mathbb{Q}_p \to \mathbb{C}$.

The analysis of functions from $\mathbb{Q}_p$ to $\mathbb{Q}_p$ is constructed in many respects by analogy to ordinary real analysis. For example, the exponential function is defined by the formula

$$e^x = \sum_{0 \le n < \infty} \frac{x^n}{n!} ,$$

where $x \in \mathbb{Q}_p$ and the series converges in p-adic norm for $|x|_p < 1$ if $p \neq 2$. Derivatives are defined in a standard manner, but with the help of p-adic norm. They have standard properties. One can write out ordinary and partial differential equations and study properties of their solutions. In particular equations of p-adic classical mechanics for harmonic oscillator have the form

$$\dot{X} = \frac{\partial H}{\partial P}, \qquad \dot{P} = -\frac{\partial H}{\partial X} ,$$

where the Hamiltonian is

$$H = \frac{P^2}{2} + \frac{X^2}{2} . \tag{5}$$

Here time t, coordinate $X = X(t)$ and momentum $P = P(t)$ are p-adic variables. For p-adic valued functions there is no natural analogy of the Lebesgue integral.

On the contrary for functions $f : \mathbb{Q}_p \to \mathbb{C}$ there exists a natural integral calculus, because on $\mathbb{Q}_p$ as on any locally-compact group there exists the Haar measure. But problems of differentiation of such functions appeared.

Complex valued exponential function (additive character) is defined by the expression

$$\chi_p(x) = \exp 2\pi i\{x\} ,$$

where $\{x\}$ is the rational part of the series (4). Using the character $\chi_p(x)$ and the Haar measure dx one constructs the Fourier-transform for complex valued functions $f(x)$,

$$\tilde{f}(\xi) = \int_{\mathbb{Q}_p} f(x)\chi_p(\xi x)dx .$$

A complex valued function $f(x)$ cannot be differentiated on p-adic argument x in the usual sense. In the theory of generalized functions over $\mathbb{Q}_p$

the role of differentiability of test functions is played by a condition of local constancy of $f(x)$. An analogy of the operator *of differentiation* is a pseudodifferential operator

$$Df(x) = \int\limits_{\mathbb{Q}_p} |\xi|_p \tilde{f}(\xi)\chi(-\xi x)d\xi = \frac{p^2}{p+1}\int\limits_{\mathbb{Q}_p} \frac{f(x)-f(y)}{|x-y|_p^2}dy \,. \tag{6}$$

How to compare results of p-adic theory with a theory over real numbers? Let us remind, that only rational numbers are directly observed. Consequently, one should look into rational points of solutions of our equations. For example, for harmonic oscillator (5) let us consider rational solutions of the trajectory in phase space

$$P^2 + X^2 = 1 \,. \tag{7}$$

As it is known, rational solutions of the equation (7) are given by the formulas:

$$P = \frac{m^2 - n^2}{m^2 + n^2}, \qquad X = \frac{2mn}{m^2 + n^2} \,,$$

where m and n are integers. Thus, we have an infinite set of rational points on the algebraic curve (7) and *the form of the phase trajectory of p-adic harmonic oscillator cannot be experimentally distinguished from the phase trajectory of harmonic oscillator over real numbers.* Of course, for other systems the result will be different as the number of rational points on an algebraic curve essentially depends on the type of the curve.

Quantum mechanics is defined by a triple

$$\{L^2(\mathbb{Q}_p), W(z), U(t)\} \,, \tag{8}$$

where $L^2(\mathbb{Q}_p)$ is the Hilbert space of complex valued functions on $\mathbb{Q}_p$, $W(z)$ is a unitary representation of the Heisenberg-Weyl group in $L^2(\mathbb{Q}_p)$, that is the representation of canonical commutation relation and $U(t)$ is a unitary evolution operator. For harmonic oscillator $U(t)$ is a restriction of unitary representation of metaplectic group on an abelian subgroup. In the book we study the spectral properties of the evolution operator. To this end we investigate an expansion of the representation $U(t)$ over irreducible representations.

How p-adic quantum mechanics correlates with ordinary one? The following formula takes place:

$$\exp[2\pi i(k^2t - kx)] = \prod_p \chi_p(k^2t - kx) , \tag{9}$$

for rational numbers k, t and x. Thus the wave-function of free particle, the so-called plane wave can be represented as a product of plane waves of p-adic particles. It can be interpreted by saying that an *ordinary free particle consists of p-adic ones*, like an elementary particle consists of quarks.

Formulas like (9) are called Euler or adelic products. A well known example of such formulas is the Euler representation for the zeta function

$$\zeta(s) = \prod_p \frac{1}{1 - p^{-s}}$$

An analogous representation takes place for propagator in the field theory:

$$\frac{1}{|k^2|} \prod_p \frac{1}{|k^2|_p} = 1, \qquad k \in \mathbb{Q} ,$$

and also in p-adic string theory.

Elaboration of the formalism of mathematical physics over p-adic number field is an interesting enterprise apart from possible applications, as it promotes deeper understanding of the formalism of standard mathematical physics. One can think that there is the following principle. *Fundamental physical laws should admit a formulation invariant under a choice of a number field.* Thus we include into consideration not only rational, real and p-adic number fields but also other fields.

There are a number of other formulations of quantum mechanics and field theory which are equivalent over the real number field, but they are different over $\mathbb{Q}_p$.

In Euclidean formulation of p-adic quantum mechanics one uses an action

$$S = \int_{\mathbb{Q}_p} [\varphi(x) D^2 \varphi(x) + m^2 \varphi^2(x)] dx . \tag{10}$$

Here φ is a real valued function of p-adic argument and the operator D is defined in (6). It is not clear at the moment in which sense one has an equivalence of the quantum mechanics (8) and the quantum mechanics with the action (10). One of the possible actions for p-adic field theory has the form

$$S = \int_{\mathbb{Q}_p^n} \tilde{\varphi}(k)a(k)\tilde{\varphi}(-k)dk + \int_{\mathbb{Q}_p^n} V(\varphi(x))dx$$

where $a(k)$ is a function on k and $V(\varphi)$ describes an interaction. For this action the quantum field theory is developed using ordinary methods of theoretical physics, such as classical equations, path integrals etc.

The modern string theory began as it is known from the Veneziano amplitude describing scattering of four particles

$$A(a,b) = \int_0^1 x^{a-1}(1-x)^{b-1}dx = \frac{\Gamma(a)\Gamma(b)}{\Gamma(a+b)}\,. \tag{11}$$

Here a and b are parameters depending on momenta of colliding particles. The function x^a is a multiplicative character on the real line. One can interpret (11) as a convolution of two characters. Then a generalization of the Veneziano amplitude will be of the following expression

$$A(\gamma_a, \gamma_b) = \int_K \gamma_a(x)\gamma_b(1-x)dx, \tag{12}$$

where K is a field, γ_a is a character and dx is a measure on K. For different K, in particular for K equals to $\mathbb{R}, \mathbb{C}, \mathbb{Q}_p$ or the Galois field $\mathbb{F}_p$ one gets formulas important in string theory and number theory. The Veneziano amplitude is a special case of more general Koba-Nielsen amplitude which in abstract form looks like

$$A_n = \int \prod_{i<j} \gamma_{a_{ij}}(x_i - x_j)dx_1 \ldots dx_n\,.$$

If we take $K = \mathbb{Q}_p$ we come to a theory of p-adic dual amplitudes and strings. In the development of p-adic mathematical physics the adelic formula

$$\prod_{p=2}^{\infty} \int_{\mathbb{Q}_p} |x|_p^a |1-x|_p^b dx = 1$$

discovered by Freund and Witten is important.

It is possible to extend the formula (11) in another way using the p-adic valued gamma function. In this way there appeared interesting connections of p-adic strings with the Jacobi sums and l-adic cohomology of the Fermat curves over the Galois fields .

There exists an unexpected connection of p-adic analysis with q-analysis and quantum groups and thus with noncommutative geometry. q-analysis is sort of a q-deformation of ordinary analysis, it was spared by such mathematicians as Euler, Gauss and others. Spherical functions on quantum groups are q-special functions. The Haar measure on the quantum group $SU_q(2)$ can be expressed in terms of q-integral

$$\int_0^1 f(x)d_qx = (1-q)\sum_{n=0}^{\infty} f(q^n)q^n \ . \tag{13}$$

Here $0 < q < 1$ and $f(x)$ is a real function. On the other hand the integral with respect to the Haar measure on $\mathbb{Q}_p$ has the form

$$\int_{|x|_p \leq 1} f(|x|_p)dx = \left(1 - \frac{1}{p}\right)\sum_{n=0}^{\infty} f(p^{-n})p^{-n} \ . \tag{14}$$

We see that for $q = \frac{1}{p}$ the expressions (13) and (14) are equal. There exists a number of other surprising relations between q-deformations and p-adic analysis.

Let us summarize in conclusion general directions of applications of p-adic analysis in mathematical physics

-geometry of space-time at small distances

-extensions of the formalism of theoretical physics to other number fields

-classical and quantum chaos; investigation of complicated systems such as spin glasses and fractals

-adelic formulas giving a decomposition of certain physical systems on more simple parts

-stochastic processes and probability theory

-connections with q-analysis and quantum groups.

In concluding the Introduction let us give a short guide for reader. First of all we would like to emphasize that no knowledge of p-adic analysis is assumed. It is not necessary to read this book section by section. The interdependence table is very simple. After reading the first two subsections (1.1) and (1.2) containing the definitions of p-adic norms and p-adic numbers, one can read any other section.

* * *

We are very grateful to all our friends and colleagues for the encouragement they gave and numerous enthusiastic discussions of exciting topics covered in this book.

P-ADIC ANALYSIS AND MATHEMATICAL PHYSICS

Chapter 1

ANALYSIS ON THE FIELD OF p-ADIC NUMBERS

I. The Field of p-Adic Numbers

In this section the definition of p-adic norm and p-adic numbers is given and basic properties of the field $\mathbb{Q}_p$ of p-adic numbers are discussed.

1. *p-Adic Norm*

Let $\mathbb{Q}$ be the field of rational numbers.

The absolute value $|x|$ of any $x \in \mathbb{Q}$ satisfies the following well-known properties

(i) $|x| \geq 0$, $|x| = 0 \Leftrightarrow x = 0$,
(ii) $|xy| = |x||y|$,
(iii) $|x + y| \leq |x| + |y|$.

Any function on $\mathbb{Q}$ with properties (i)–(iii) is called a *norm*.

One can consider another norm on the field Q of rational numbers. Let p be a prime number, $p = 2, 3, 5, \ldots, 137, \ldots$. We introduce a norm $|x|_p$ on the field $\mathbb{Q}$ by the rule

$$|0|_p = 0, \qquad |x|_p = p^{-\gamma}\,, \tag{1.1}$$

where an integer $\gamma = \gamma(x)$ is defined from the representation

$$x = p^{\gamma}\frac{m}{n}\,, \tag{1.2}$$

integers m and n are not divisible by p. The norm $|x|_p$ is called the *p-adic norm*.

Examples.

$$|6|_3 = |15|_3 = \tfrac{1}{3}, \qquad |\tfrac{1}{4}|_2 = |\tfrac{3}{4}|_2 = 4, \qquad |137|_2 = 1 .$$

The p-adic norm possesses the characteristic properties (i)–(iii) of a norm even in a stronger form, namely:

1) $|x|_p \geq 0$, $|x|_p = 0 \Leftrightarrow x = 0$,
2) $|xy|_p = |x|_p |y|_p$,
3) $|x + y|_p \leq \max(|x|_p, |y|_p)$.
In the case when $|x|_p \neq |y|_p$ we have equality
3′) $|x + y| = \max(|x|, |y|)$.

For $p = 2$ we also have
3″) $|x + y|_2 \leq 1/2|x|_2$, if $|x|_2 = |y|_2$.

The properties 1) and 2) are obvious. Let us prove property 3). If x and $y = p^{\gamma'}\frac{m'}{n'}$ are represented in the form (1.2) then by (1.1) $|x|_p = p^{-\gamma'}$ and $|y|_p = p^{-\gamma'}$. Let $\min(\gamma, \gamma') = \gamma$ for definiteness. Then

$$x + y = p^{\gamma}\frac{m}{n} + p^{\gamma'}\frac{m'}{n'} = p^{\gamma}\frac{mn' + nm'p^{\gamma'-\gamma}}{nn'} . \tag{1.3}$$

The integer nn' is not divided by p but the integer $mn' + nm'p^{\gamma'-\gamma}$ may be divided by p. Therefore $\gamma(x + y) \geq \gamma = \min(\gamma, \gamma')$ and thus $|x + y|_p = p^{-\gamma(x+y)} \leq \max(p^{-\gamma}, p^{-\gamma'}) = \max(|x|_p, |y|_p)$ and inequality (3) is proved. If $\gamma' > \gamma$ then in (1.3) the integer $mn'+nm'p^{\gamma'-\gamma}$ is not divisible by p. Therefore $\gamma(x + y) = \gamma$ and

$$|x + y|_p = p^{-\gamma(x+y)} = p^{-\gamma} = \max(p^{-\gamma}, p^{-\gamma'}) = \max(|x|_p, |y|_p) ,$$

and equality 3′) is proved. For $p = 2$ and $|x|_2 = |y|_2 = 2^{\gamma}$ in (1.3) numbers m, n, m', n' are odd and thus the number $mn' + nm'$ is even, and also the number nn' is odd, so $\gamma(x + y) \geq \gamma + 1$. Therefore

$$|x + y|_2 \leq 2^{-\gamma-1} = \tfrac{1}{2}|x|_2$$

and inequality 3″) is proved. ■

Notice that the norm $|x|_p$ may take only countable set of values: $p^\gamma, \gamma \in \mathbb{Z}$. Here $\mathbb{Z}$ is the ring of integers, $\mathbb{Z} = \{0, \pm 1, \ldots\}$. The p-adic norm $|x|_p$ defines an ultrametric on $\mathbb{Q}_p$ (because of inequality 3)). This norm is a non-Archimedean one. In fact for any $n \in \mathbb{Z}$ $|nx|_p \leq |x|_p$, $x \in \mathbb{Q}$.

The following Ostrowski theorem is valid.

Theorem. *The norms $|x|$ and $|x|_p$, $p = 2, 3, \ldots$ exhaust all nonequivalent norms on the field of rational numbers $\mathbb{Q}$.*

For the proof see for example [37,186].

We denote the standard absolute value by $|x|_\infty = |x|$, $x \in \mathbb{Q}$. The following (adelic) formula takes place.

$$\prod_{2 \leq p \leq \infty} |x|_p = 1, \qquad x \in \mathbb{Q}_p, \quad x \neq 0 \ . \tag{1.4}$$

■ Expanding a rational number x by prime factors

$$x = \varepsilon p_1^{\alpha_1} p_2^{\alpha_2} \ldots p_n^{\alpha_n}, \qquad \varepsilon = \pm 1 \ ,$$

where p_j are different prime numbers and $\alpha_j \in \mathbb{Z}$, $j = 1, 2, \ldots, n$, owing to (1.1) and (1.2) we obtain

$$|x|_{p_j} = p_j^{-\alpha_j}; \quad |x|_p = 1, \quad p \neq p_j; \quad |x|_\infty = p_1^{\alpha_1} \ldots p_n^{\alpha_n} \ .$$

The formula (1.4) follows immediately. ■

2. *p-Adic Numbers*

The field $\mathbb{Q}_p$ of p-adic numbers is defined as the completion of $\mathbb{Q}$ with respect to the p-adic metric determined by the p-adic norm $|\cdot|_p$. Thus, $\mathbb{Q}_p$ is obtained from the p-adic norm $|\cdot|_p$ in the same way as the real number field $\mathbb{R}$ is obtained from the usual absolute value: as the completion of $\mathbb{Q}$.

Any p-adic number $x \neq 0$ is uniquely represented in the **canonical form**

$$x = p^\gamma (x_0 + x_1 p + x_2 p^2 + \ldots) \ , \tag{2.1}$$

where $\gamma = \gamma(x) \in \mathbb{Z}$ and x_j are integers such that $0 \leq x_j \leq p - 1$, $x_0 > 0$, $j = 0, 1, \ldots$.

Note that the series in (2.1) converges with respect to the norm $|x|_p$ because one has

$$|p^\gamma x_j p^j|_p = p^{-\gamma-j}, \qquad j = 0, 1, \ldots$$

The representation (2.1) is similar to expansion of any real number x in infinite decimal

$$x = \pm 10^\gamma (x_0 + x_1 10^{-1} + x_2 10^{-2} + \ldots), \quad \varphi \in \mathbb{Z}, \quad x_j = 0, 1, \ldots, 9, x_0 \neq 0\,,$$

and it can be proved analogously.

The representation (2.1) means that any p-adic number x is a limit (with respect to p-adic norm) of a sequence $\{x^{(n)}, n \to \infty\}$ of rational numbers

$$x^{(n)} = p^\gamma (x_0 + x_1 p + \ldots + x_n p^n)$$

If $x \in \mathbb{Q}_p$ is represented in the form (2.1) then $|x|_p = p^{-\gamma}$ and properties 1)–3) for p-adic norm are fulfilled. The representation (2.1) gives rational numbers if and only if the numbers $(x_j,\ j = 0, 1, \ldots)$ beginning from some number form a periodic sequence.

Example.

$$-1 = p - 1 + (p-1)p + (p-1)p^2 + \ldots$$

By means of representation (2.1) one defines a fractional part $\{x\}_p$ of a number $x \in \mathbb{Q}_p$:

$$\{x\}_p = \begin{cases} 0 & \text{if } \gamma(x) \geq 0 \text{ or } x = 0\,, \\ p^\gamma (x_0 + x_1 p + \ldots + x_{|\gamma|-1} p^{|\gamma|-1}) & \text{if } \gamma(x) < 0\,. \end{cases} \tag{2.2}$$

It is easy to see that

$$p^\gamma \leq \{x\}_p \leq 1 - p^\gamma, \quad \text{if} \quad \gamma(x) < 0\,. \tag{2.3}$$

A sum of two p-adic numbers x of the form (2.1) and

$$y = p^{\gamma(y)}(y_0 + y_1 p + y_2 p^2 + \ldots), \quad 0 \leq y_j \leq p-1, y_0 > 0$$

is represented in canonical form

$$x + y = p^{\gamma(x+y)}(c_0 + c_1 p + c_2 p^2 + \ldots), \qquad 0 \leq c_j \leq p-1, c_0 > 0\,,$$

where numbers $\gamma(x+y)$ and C_j are uniquely determined from the equation

$$p^{\gamma(x)}(x_0 + x_1 p + \dots) + p^{\gamma(y)}(y_0 + y_1 p + \dots) = p^{\gamma(x+y)}(c_0 + c_1 p + \dots)$$

by the method of indefinite coefficients modulo p.

The equation $a + x = 0$ is uniquely solvable in $\mathbb{Q}_p$ for any $a \in \mathbb{Q}_p, a \neq 0$, and $x = (-1)a = -a$ is its solution. The equation $ax = 1$ also is uniquely solvable in $\mathbb{Q}_p$ for any $a \in \mathbb{Q}_p, a \neq 0$, and $x = 1/a$. (For determination of a canonical form of number $1/a$ one uses the method of indefinite coefficients modulo p for equation $ax = 1$.) Thus in $\mathbb{Q}_p$ we have ordinary arithmetic operations: addition, substraction, multiplication and division. This means that $\mathbb{Q}_p$ is a field.

The field $\mathbb{Q}_p$ is a commutative and associative group with respect to addition. $\mathbb{Q}_p^* = \mathbb{Q}_p \backslash \{0\}$ is a commutative and associative group with respect to multiplication; $\mathbb{Q}_p^*$ is called the *multiplicative group* of $\mathbb{Q}_p$.

p-Adic numbers for which $|x|_p \leq 1$ (i.e. $\gamma(x) \geq 0$ or $\{x\}_p = 0$), are called *integers p-adic numbers* and their set is denoted by $\mathbb{Z}_p$. $\mathbb{Z}_p$ is a subring of the ring $\mathbb{Q}_p$; the set of natural numbers $\mathbb{Z}_+ = \{1, 2, \dots\}$ is dense in $\mathbb{Z}_p$. Integers $x \in \mathbb{Z}_p$ for which $|x|_p = 1$ are called *units* in $\mathbb{Q}_p$.

The set of $x \in \mathbb{Z}_p$ for which $|x|_p < 1$ (i.e. $\gamma(x) \geq 1$ or $|x|_P \leq 1/p$) forms the principal ideal of the ring $\mathbb{Z}_p$. Obviously this ideal has the form $p\mathbb{Z}_p$. The residue field $\mathbb{Z}_p/p\mathbb{Z}_p$ consists of p elements. In multiplicative group of the field $\mathbb{Z}_p/p\mathbb{Z}_p$ there exists a unity $\eta \neq 1$ (for $p \neq 2$; for $p = 2$ $\eta = 1$) of order $p-1$ such that the elements $0, \eta, \eta^2, \dots, \eta^{p-1} = 1$ form a complete set of representatives of residue classes of the field $\mathbb{Z}_p/p\mathbb{Z}_p$ (see [128,121]).

Examples. For $p = 3$, $\eta = -1$; for $p = 5$, $\eta = 2$. As numbers $0, 1, \dots, p-1$ form also a complete set of representatives of residue classes of the field $\mathbb{Z}_p/p\mathbb{Z}_p$ then from representation (2.1) it follows the second canonical form of any p-adic number $x \neq 0$;

$$\begin{aligned} x &= p^{\gamma(x)}(x'_0 + x'_1 p + x'_2 p^2 + \dots) , \\ x'_j &= 0, 1, \eta, \dots, \eta^{p-2}, \qquad x'_0 \neq 0, \quad (j = 0, 1, \dots) . \end{aligned} \tag{2.4}$$

3. *Non-Archimedean Topology of the Field $\mathbb{Q}_p$ of p-Adic Numbers*

Owing to the inequality 3) of Sec. 1.1 the norm on the field $\mathbb{Q}_p$ satisfies the triangle inequality

$$|x + y|_p \leq \max(|x|_p, |y|_p) \leq |x|_p + |y|_p, \quad x, y \in \mathbb{Q}_p .$$

Therefore in $\mathbb{Q}_p$ one possible introduces the metric $\rho(x,y) = |x-y|_p$, and $\mathbb{Q}_p$ becomes a complete metric space. From representation (2.1) it follows that $\mathbb{Q}_p$ is a separable space.

Denote by $B_\gamma(a)$ the disc of radius p^γ with center at a point $a \in \mathbb{Q}_p$ and by $S_\gamma(a)$ its boundary (circle):

$$B_\gamma(a) = [x : |x-a|_p \leq p^\gamma] \,,$$
$$S_\gamma(a) = [x : |x-a|_p = p^\gamma], \qquad \gamma \in \mathbb{Z}$$

It is clear that $B_\gamma(a)$ is an abelian additive group and

$$\begin{aligned} &[x : |x-a|_p < p^\gamma] = B_{\gamma-1}(a) \subset B_\gamma(a), \\ &S_\gamma(a) = B_\gamma(a) \backslash B_{\gamma-1}(a), B_\gamma(a) \subset B_{\gamma'}(a), \gamma < \gamma'; \\ &B_\gamma(a) = \bigcup_{\gamma' \leq \gamma} S_{\gamma'}(a), \bigcap_{\gamma \in \mathbb{Z}} B_\gamma(a) = \{a\}, \qquad \bigcup_{\gamma \in \mathbb{Z}} B_\gamma(a) = \bigcup_{\gamma \in \mathbb{Z}} S_\gamma(a) = \mathbb{Q}_p \,. \end{aligned} \tag{3.1}$$

For $a = 0$ we denote by $B_\gamma(0) = B_\gamma$ and $S_\gamma(0) = S_\gamma$.

Lemma 1. *If $b \in B_\gamma(a)$ then $B_\gamma(b) = B_\gamma(a)$.*

■ *Let $x \in B_\gamma(b)$. Then*

$$|x-a|_p = |x-b+b-a|_p \leq \max(|x-b|_p, |b-a|_p) \leq p^\gamma \,,$$

i.e. $x \in B_\gamma(a)$, so $B_\gamma(b) \subset B_\gamma(a)$. As $a \in B_\gamma(b)$ then as we have just proved $B_\gamma(a) \subset B_\gamma(b)$, and hence $B_\gamma(a) = B_\gamma(b)$. ■

Corollaries. 1. *The disc $B_\gamma(a)$ and the circle $S_\gamma(a)$are both open and closed sets in $\mathbb{Q}_p$.*

A closed and open set we shall call a *clopen set.*

2. *Every point of the disc $B_\gamma(a)$ is its center.*

3. *Any two discs in $\mathbb{Q}_p$ either disjoint or one is contained in another.*

4. *Every open set in $\mathbb{Q}_p$ is a union at most of a countable set of disjoint discs.*

Lemma 2. *If a set $M \subset \mathbb{Q}_p$ contains two points a and b, $a \neq b$, then it can be represented as a union of disjoint clopen (in M) sets M_1 and M_2 such that $a \in M_1, b \in M_2$.*

■ We consider three possible cases.

1) $a = 0$, $|b|_p = p^\gamma$. As M_1 and M_2 we can take sets $M_1 = M \cap B_{\gamma-1}$ and $M_2 = M \cap (\mathbb{Q}_p \backslash B_{\gamma-1})$.

2) $|a|_p = p^\gamma, |b|_p = p^{\gamma'}, \gamma' > \gamma$; then $M_1 = M \cap B_{\gamma'}$ $M_2 = M \cap (\mathbb{Q}_p \backslash B_\gamma)$.

3) $|a|_p = p^\gamma = |b|_p$. Let

$$a = p^{-\gamma}(a_0 + a_1 p + a_2 p^2 + \ldots), \qquad b = p^{-\gamma}(b_0 + b_1 p + b_2 p^2 + \ldots),$$

where $a_0 = b_0$, $a_1 = b_1, \ldots, a_{k-1} = b_{k-1}$, $a_k \neq b_k$, $|a - b|_p = p^{\gamma - k}$. Then $M_1 = M \cap B_{\gamma-k-1}(a)$, $M_2 = M \subset (\mathbb{Q}_p \backslash B_{\gamma-k-1}(a))$. ■

Lemma 2 asserts that any set of the space $\mathbb{Q}_p$ which consists of more then one points is disconnected (see [174,186]. In other words, a connected component of any point coincides with this point. Thus $\mathbb{Q}_p$ *is a totally disconnected space.*

By following the proof of Lemma 2, for the case when the set M consists only of two points, we can see that there exist disjoint neighborhoods of these points. It means, that *the space* $\mathbb{Q}_p$ *is Hausdorff.*

Lemma 3. *A set* $K \subset \mathbb{Q}_p$ *is compact in* $\mathbb{Q}_p$ *if and only if it is closed and bounded in* $\mathbb{Q}_p$.

■ **Necessity** of conditions is obvious. We will prove their **sufficiency**. As $\mathbb{Q}_p$ is a complete metric space then it is sufficient to prove countably compactness of any bounded closed (infinite) set K (see [237]), i.e. that every infinite set $M \subset K$ contains at least one limit point. Let $x \in M$, then $|x|_p = p^{-\gamma(x)} \leq C$ (M is bounded), so $\gamma(x)$ is bounded from below.

Let us consider two cases. 1) $\gamma(x)$ is not bounded from above on M. Then there exists a sequence $\{x_k, k \to \infty\} \subset M$ such that $\gamma(x_k) \to \infty$, $k \to \infty$. It means that $|x|_p = p^{-\gamma_k} \to 0$, $k \to \infty$, i.e. $x_k \to 0$, $k \to \infty$ in $\mathbb{Q}_p$ and $0 \in K$. 2) $\gamma(x)$ is bounded from above on M. Then there exists such number γ_0 that M contains an infinite set of points of the form

$$p^{\gamma_0}(x_0 + x_1 p + \ldots), 0 \leq x_j \leq p-1, x_0 \neq 0 .$$

As x_0 takes only $p-1$ values then there exists an integer $a_0, 1 \leq a_0 \leq p-1$, such that M contains infinite set of points of the form $p^{\gamma_0}(a_0 + x_1 p + \ldots)$, and so on. As a result we obtain a sequence $\{a_j, \quad j = 0, 1, \ldots\}$, $0 \leq a_j \leq$

$p-1$, $a_0 \neq 0$. The desired limit point is $p^{\gamma_0}(a_0 + a_1 p + a_2 p^2 + \ldots) \in K$ (K is closed). ■

Corollaries. 1. *Every disc $B_\gamma(a)$ and circle $S_\gamma(a)$ are compact.*

2. *The space $\mathbb{Q}_p$ is locally-compact.*

3. *Every compact in $\mathbb{Q}_p$ can be covered by a finite number of disjoint discs of a fixed radius* (see Corollary 3 from the Lemma 1).

4. *In space $\mathbb{Q}_p$ the Heine-Borel Lemma is valid: from every infinite covering of a compact K it is possible to choose a finite covering of K.*

Example 1. The circle S_γ can be covered by $(p-1)p^{\gamma-\gamma'-1}$ disjoint discs $B_{\gamma'}(a)$, $\gamma > \gamma'$, with the centers

$$a = p^{-\gamma}(a_0 + a_1 p + \ldots + a_{\gamma-\gamma'-1}p^{\gamma-\gamma'-1}), \qquad 0 \leq a_j \leq p-1, \quad a_0 \neq 0 . \tag{3.2}$$

■Any point $x = p^{-\gamma}(x_0 + x_1 p + \ldots) \in S_\gamma$ can be uniquely represented in the form $x = a + x'$ where a is of the form (3.2) and $x' \in B_{\gamma'}$. Therefore

$$S_\gamma = \bigcup_a \{a + B_{\gamma'}\} = \bigcup_a B_{\gamma'}(a) .$$

We notice now that discs $B_{\gamma'}(a)$ are disjoint as their centers $\{a\}$, owing to (3.2), remove from each other on distance $\geq p^{\gamma'+1}$ (see Corollary 3 from the Lemma 1). The number of centers is equal to $(p-1)p^{\gamma-\gamma'-1}$. ■

Example 2. The disc B_γ can be covered by $p^{\gamma-\gamma'}$ disjoint discs $B_{\gamma'}(a), \gamma > \gamma'$ with the centers

$$a = p^{-r}(a_0 + a_1 p + \ldots + a_{r-\gamma'-1}p^{r-\gamma'-1}), \qquad r = \gamma, \gamma-1, \ldots, \gamma'+1 ,$$
$$0 \leq a_j \leq p-1, \qquad a_0 \neq 0 \tag{3.3}$$

■ It follows from Example 1, if one notices, that

$$B_\gamma = B_{\gamma'} \bigcup \bigcup_{r=\gamma'+1}^{\gamma} S_r$$

and the sets $B_{\gamma'}$ and $S_{r'}$, $r = \gamma' + 1, \dots, \gamma$ are disjoint. The number of discs is

$$1 + (p-1) \sum_{r=\gamma'+1}^{\gamma} p^{r-\gamma'-1} = 1 + (p-1)\frac{1-p^{\gamma-\gamma'}}{1-p} = p^{\gamma-\gamma'} . \qquad \blacksquare$$

We call coverings of Examples 1 and 2 *canonical coverings* of the circle S_γ and the disc B_γ respectively.

Dimension of a complete metric space X is defined as the smallest integer n such that for every covering of the space X there exists a refined subcovering of multiplicity $n+1$. (Multiplicity of a covering is the largest integer m such that in this covering there exist m sets with nonempty intersection.) For example, $\dim \mathbb{R}^n = n$.

Theorem. *Dimension of the space $\mathbb{Q}_p$ is equal to 0.*

■The space $\mathbb{Q}_p$ is complete and a metric one. From every covering of $\mathbb{Q}_p$ we can find a subcovering of $\mathbb{Q}_p$ by disjoint discs (see Corollary 4 from the Lemma 1). It means that $n = 0$ in the definition of dimension of $\mathbb{Q}_p$. Thus $\dim \mathbb{Q}_p = 0$. ■

4. *Quadratic Extensions of the Field $\mathbb{Q}_p$*

Let ε be such a p-adic number which is not a square of any p-adic number, $\varepsilon \notin \mathbb{Q}_p^{*2}$. Let us join to the field $\mathbb{Q}_p$ a number (symbol) $\sqrt{\varepsilon}$; all elements of the form $z = x + y\sqrt{\varepsilon}$, $x, y \in \mathbb{Q}_p$ we shall call *quadratic extension* of the field $\mathbb{Q}_p$. We denote it by $\mathbb{Q}_p(\sqrt{\varepsilon})$. Elements $x + y\sqrt{\varepsilon}$ can be added and multiplied by the usual rules under additional condition $(\sqrt{\varepsilon})^2 = \varepsilon$.

Obviously, $x + y\sqrt{\varepsilon} = 0$ if and only if $x = y = 0$.

The equation $(x + y\sqrt{\varepsilon})z = 1$, $x + y\sqrt{\varepsilon} \neq 0$ has a unique solution in $\mathbb{Q}_p(\sqrt{\varepsilon})$ which is equal to

$$z = \frac{x}{x^2 - \varepsilon y^2} - \frac{y}{x^2 - \varepsilon y^2}\sqrt{\varepsilon} . \tag{4.1}$$

(Notice that the denominator in (4.1) $x^2 - \varepsilon y^2 \neq 0$, otherwise ε would be a square of a p-adic number.)

Thus the quadratic extension $\mathbb{Q}_p(\sqrt{\varepsilon})$ *forms a field.* Denote by $\mathbb{Q}_p^*(\sqrt{\varepsilon})$ the multiplicative group of the field $\mathbb{Q}_p(\sqrt{\varepsilon})$.

Now we find out which p-adic numbers are no square of any p-adic numbers, and for which non isomorphic quadratic extensions exist.

Be remind that an integer $a \in \mathbb{Z}$ is called a *quadratic residue modulo* p if the equation $x^2 \equiv a(\text{mod } p)$ has a solution $x \in \mathbb{Z}$; otherwise a is called *quadratic* non-*residue modulo* p. For notion of these affirmations one uses the *Legendre symbol*:

$$\left(\frac{a}{p}\right) = \begin{cases} 1 \text{ if } a \text{ is quadratic residue modulo } p, \\ -1 \text{ if } a \text{ is quadratic non-residue modulop}\,. \end{cases}$$

Lemma. *In order that the equation*

$$x^2 = a, \qquad 0 \neq a = p^{\gamma(a)}(a_0 + a_1 p + \dots)\,, \\ 0 \leq a_j \leq p-1, \qquad a_0 \neq 0 \tag{4.2}$$

has a solution $x \in \mathbb{Q}_p$, *it is necessary and sufficient that the following conditions are fulfilled* :

1) $\gamma(a)$ *is even,*
2) $(\frac{a_0}{p}) = 1$ *if* $p \neq 2$, $a_1 = a_2 = 0$ *if* $p = 2$.

■ **Necessity.** If the equation (4.2) has a solution $x = p^{\gamma(x)}(x_0 + x_1 p + \dots)$, so

$$p^{2\gamma(x)}(x_0 + x_1 p + \dots)^2 = p^{\gamma(a)}(a_0 + a_1 p + \dots) \tag{4.3}$$

whence it follows that $\gamma(a) = 2\gamma(x)$ is even and $x_0^2 \equiv a_0(\text{mod } p)$, if $p \neq 2$, i.e. $\left(\frac{a_0}{p}\right) = 1$. For $p = 2$ we have

$$2^{2\gamma(x)}(1 + x_1 2 + \dots)^2 = 2^{2\gamma(x)}\left[1 + \left(\frac{x_1 + x_1^2}{2} + x_2\right) 2^3 + \dots\right] \\ = 2^{2\gamma(a)}(1 + a_1 2 + a_2 2^2 + \dots) \tag{4.4}$$

and thus $a_1 = a_2 = 0$.

Sufficiency. Let a satisfy the conditions 1) and 2). Let us construct a solution of Eq. (4.2). We put $\gamma(x) = (1/2)\gamma(a)$.

Let $p \neq 2$. From (4.3) it follows that a number x_0 has to satisfy the conditions

$$x_0^2 \equiv a_0 \pmod{p}, \qquad 1 \leq x_0 \leq p-1 \ .$$

Such x_0 exists as $1 \leq a_0 \leq p-1$, $\left(\frac{a_0}{p}\right) = 1$. From (4.3) it follows also that numbers x_j, $j = 1, 2, \ldots$ have to satisfy the conditions

$$2x_0 x_j \equiv a_j + N_j \pmod{p}, \qquad 0 \leq x_j \leq p-1 \tag{4.5}$$

where integers N_j depend only on $x_0, x_1, \ldots, x_{j-1}$. Therefore numbers x_j are successively defined (uniquely) from Eq. (4.5) as $2x_0$ is not divisible by p.

Let $p = 2$. From (4.4.) it follows the equation

$$a_3 \equiv \frac{x_1(1+x_1)}{2} + x_2 \pmod{2}$$

which is always solvable for $a_3 = 0, 1$. From (4.4) it follows also that integers x_j, $j = 3, 4, \ldots$ have to satisfy the conditions

$$x_j \equiv a_{j+1} + N_j \pmod{2}, \qquad x_j = 0, 1 \tag{4.6}$$

where integers N_j depend only on $x_1, x_2, \ldots, x_{j-1}$. Therefore numbers x_j are successively defined (uniquely) from Eq. (4.6). ■

Let η be unity which is not a square of any p-adic number, i.e. $\left(\frac{\eta_0}{p}\right) = 1$. This fact we shall write as $\eta \notin \mathbb{Q}_p^{*2}$. (For $p \neq 2$ it is possible to take as η the unity introduced in Sec. 1.2.)

Corollaries. 1. *For $p \neq 2$ numbers $\varepsilon_1 = \eta$, $\varepsilon_2 = \rho$, $\varepsilon_3 = p\eta$ are not squares of any p-adic numbers.*

2. *Every p-adic number x can be represented in one of the four following forms: $x = \varepsilon_j y^2$ where $y \in \mathbb{Q}_p$ and $\varepsilon_0 = 1$, $\varepsilon_1 = \eta$, $\varepsilon_2 = p$, $\varepsilon_3 = p\eta$ $(p \neq 2)$.*

3. *There exists only three non-isomorphic quadratic extensions of the field $\mathbb{Q}_p : \mathbb{Q}_p(\sqrt{\varepsilon_j}), j = 1, 2, 3$ $(p \neq 2)$.*

4. *For $p = 2$ every 2-adic number x can be represented in one of the eight following forms: $x = \varepsilon_j y^2$ where $y \in \mathbb{Q}_2$ and $\varepsilon_0 = 1$, $\varepsilon_1 = 1 + 2 = 3$, $\varepsilon_2 = 1 + 4 = 5$, $\varepsilon_3 = 1 + 2 + 4 = 7$, $\varepsilon_4 = 2$, $\varepsilon_5 = 2(1+2) = 6$, $\varepsilon_6 = 2(1+4) = 10$, $\varepsilon_7 = 2(1+2+4) = 14$ (or equivalently $\varepsilon_j = \pm 1, \pm 2, \pm 3, \pm 6$).*

5. *There exists only seven non-isomorphic quadratic extensions of the field $\mathbb{Q}_2$: $\mathbb{Q}_2(\sqrt{\varepsilon_j}), j = 1, 2, ..., 7$.*

6. *The quotient group $\mathbb{Q}_p^* \backslash \mathbb{Q}_p^{*2}$ consists of four elements $\varepsilon_j, j = 0, 1, 2, 3$ for $p \neq 2$ and of eight elements $\varepsilon_j, j = 0, 1, \ldots, 7$ for $p = 2$.*

Note that for $p \equiv 3 \pmod 4$ as a number η can be taken -1 because (see [204])

$$\left(\frac{-1}{p}\right) = (-1)^{\frac{p-1}{2}} = -1 \; ; \tag{4.7}$$

for $p \equiv 3 \pmod 8$ or $p \equiv 5 \pmod 8$ as a number η can be taken 2 because $(\frac{2}{p}) = -1$ owing to the formula

$$\left(\frac{2}{p}\right) = (-1)^{\frac{p^2-1}{8}} .$$

7. *In order that the equation $x^2 = -1$ has a solution in $\mathbb{Q}_p$, it is necessary and sufficient that $p \equiv 1 \pmod 4$; in addition there exist two solutions of this equation which we denote by $\pm\tau$.*

8. *For $p \equiv 3 \pmod 4$ the equality*

$$|x^2 + y^2|_p = \max(|x|_p^2, |y|_p^2) \tag{4.8}$$

is valid. In particular, from $x^2 + y^2 = 0$, it follows that $x = y = 0$.

■ It is sufficient to verify Eq. (4.8) only for the case $|x|_p = |y|_p$. If it would be $|x^2+y^2|_p < |x^2|_p = |y|_p^2$, then the congruence $x_0^2 = -y_0^2 \pmod p$ would be solvable and thus -1 would be quadratic residue modulo p which contradicts to the formula (4.7). ■

5. Polar Coordinates and Circles in the Field $\mathbb{Q}_p(\sqrt{\varepsilon})$

Any element of the field $\mathbb{Q}_p(\sqrt{\varepsilon}), \varepsilon \notin \mathbb{Q}_p^{*2}$, is uniquely represented in the form $z = x + \sqrt{\varepsilon}y, x, y \in \mathbb{Q}_p$ (see Sec. 1.4). The numbers (x, y) are called the *Cartesian coordinates* of the element z; the element $\bar{z} = x - \sqrt{\varepsilon}y$ is called *conjugate* to z; the set of elements $z \in \mathbb{Q}_p^*(\sqrt{\varepsilon})$, which satisfy the equation

$$z\bar{z} = x^2 - \varepsilon y^2 = c, \qquad c \neq 0 \tag{5.1}$$

is called "circle" in $\mathbb{Q}_p(\sqrt{\varepsilon})$ (with the center at 0).

It is clear that the equality $z\bar{z} = 0$ is equivalent to $z = 0$.

Denote by $\mathbb{Q}_{p,\varepsilon}^*$ a subgroup of the group $\mathbb{Q}_p^*$ which consists of numbers c of the form (5.1).

By the Lemma of Sec. 1.4 the number c can be represented in the form: either $c = r^2$ or $c = \kappa r^2$ where $r \in \mathbb{Q}_p^*$, and $\kappa \notin \mathbb{Q}_p^{*2}$, hence $\kappa = cr^{-2} \in \mathbb{Q}_{p,\varepsilon}^*$ i.e. $\kappa = \nu\bar{\nu}, \nu \in \mathbb{Q}_p^*(\sqrt{\varepsilon})$.

A pair (ρ, σ) with either $(\rho = r, \sigma = r^{-1}z)$ or $(\rho = \nu r, \sigma = \nu^{-1}r^{-1}z)$ is called polar coordinates of the point $z \in \mathbb{Q}_p^*(\sqrt{\varepsilon})$, $z = \rho\sigma$, $\sigma\bar{\sigma} = 1$.

It is clear that $(-\rho, -\sigma)$ is also polar coordinates of the point z.

The unit "circle" in $\mathbb{Q}_p(\sqrt{\varepsilon})$ $z\bar{z} = 1$, which is defined by the equation (5.1) for $c = 1$ play a special role. Elements of this "circle" form a multiplicative group which we denote by C_ε.

Let us find a parametric representation of the "circle" C_ε,

$$x^2 - \varepsilon y^2 = 1 \ . \tag{5.2}$$

By introducing the parameter $t = \frac{y}{1+x}$ from the equation (5.2) we obtain

$$x = \frac{1 + \varepsilon t^2}{1 - \varepsilon t^2}, \qquad y = \frac{2t}{1 - \varepsilon t^2}, \qquad t \in \mathbb{Q}_p \tag{5.3}$$

C_ε is compact in $\mathbb{Q}_p(\sqrt{\varepsilon})$.

■ C_ε is obviously closed, we shall prove its boundedness (see Sec. 1.3). For $|\varepsilon t^2|_p > 1$ we have $|1 \pm \varepsilon t^2|_p = |\varepsilon t^2|_p$ and owing to (5.3) $|x|_p = 1$; for $|\varepsilon t^2|_p < 1$ we have $|1 \pm \varepsilon t^2|_p = 1$ and owing to (5.3) $|x|_p = 1$; for $|\varepsilon t^2|_p = 1$ we shall prove that $|1 - \varepsilon t^2|_p = 1$ and owing to (5.3) $|x|_p = |1 + \varepsilon t^2|_p \leq 1$. In all three cases $|x|_p \leq 1$, and thus from the equation (5.2) it follows that $|y|_p^2 \leq |\varepsilon|_p^{-1}$, and C_ε is bounded.

It remains to prove that from $|\varepsilon t^2|_p = 1$ it follows that $|1 - \varepsilon t^2|_p = 1$. Let $p \neq 2$. In Sec 1.4 it was proved that as ε can be taken the numbers η, p and ηp. But $\varepsilon = p$ or $\varepsilon = \eta p$ contradict to the equality $|\varepsilon t^2|_p = 1$. Therefore $\varepsilon = \eta$. But the equation $1 - \eta_0 t_0 \equiv 0 \pmod{p}$ is not solvable otherwise η_0 would be a quadratic residue modulo p what contradict to the Lemma of Sec 1.4.Thus $|1 - \eta t^2|_p = 1$. For $p = 2$ a proof is similar, but more complicated. ■

6. $\mathbb{Q}_p$ *and* $\mathbb{R}$

We shall construct a one-to-one continuous mapping φ from the set of p-adic numbers $\mathbb{Q}_p$ to some Cantor-like subset $\varphi(\mathbb{Q}_p)$ of real numbers $\mathbb{R}$.

Let us define the function $\varphi : \mathbb{Q}_p \to \mathbb{R}_+$ by the formula

$$\varphi(x) = |x|_p \sum_{0\le k<\infty} x_k p^{-2k} , \tag{6.1}$$

where numbers x_k, $0 \le x_k \le p-1$, $x_0 \ne 0$ are determined by the canonical representation (2.1) of x.

We put in order numbers of the set $\mathbb{Q}_p$ by the following way: x precedes y, $x < y$, if either $|x|_p < |y|_p$, or when $|x|_p = |y|_p$ there exists such integer $j \ge 0$ that $y_0 = x_0, y_1 = x_1, \dots, y_{j-1} = x_{j-1}$, $x_j < y_j$. It is clear that the transitive axiom is fulfilled: if $y < x$ and $x < z$ then $y < z$. Thus $\mathbb{Q}_p$ *is a completely ordered set*, and also $0 < x, x \in \mathbb{Q}_p \backslash \{0\}$.

Lemma. $\varphi(x) > \varphi(y)$ *if and only if* $x > y$.

■ **Sufficiency.** Let $x > y$ and $|x|_p > |y|_p \ne 0$, i.e. $|x|_p \ge p|y|_p$. We have

$$\varphi(y) = |y|_p \sum_{0\le k<\infty} y_k p^{-2k} \le (p-1)|y|_p \sum_{k=0}^{\infty} p^{-2k} .$$

Therefore

$$\varphi(x) - \varphi(y) \ge |x|_p - \frac{p^2}{p+1}|y|_p \ge p|y|_p \left(1 - \frac{p}{p+1}\right) = \frac{p}{p+1}|y|_p > 0 .$$

Let now $|x|_p = |y|_p$, $x_0 = y_0$, $x_1 = y_1, \dots, x_{j-1} = y_{j-1}$, $x_j > y_j$ for some $j \ge 0$. Then

$$\varphi(x) \ge |x|_p \sum_{0\le k\le j} x_k p^{-2k} ,$$

$$\varphi(y) \le |y|_p \sum_{0\le k\le j} y_k p^{-2k} + |y|_p \frac{p-1}{1-p^{-2}} p^{-2(j+1)}$$

$$= |x|_p \sum_{0\le k\le j} x_k p^{-2k} + |x|_p y_j p^{-2j} + |x|_p p^{-2j} \frac{1}{p+1} p ;$$

Therefore

$$\varphi(x) - \varphi(y) \ge |x|_p p^{-2j} \left(x_j - y_j - \frac{1}{p+1}\right) > 0 .$$

Necessity. If $\varphi(x) > \varphi(y)$ then either $x = y$ or $x < y$ is impossible, which has been proved. ■

It follows from the lemma that the *mapping* φ *is one-to-one.*

The function φ *is continuous.*

■ It follows from the estimate

$$|\varphi(x) - \varphi(y)| \leq p|x - y|_p, \qquad x, y \in \mathbb{Q}_p$$

which can be proved in a similar way to the Lemma. ■

Let us note $K = \varphi(\mathbb{Q}_p)$. The structure of the set K is given by the following

Theorem. *The set* K *is a countable union of disjoint perfect nowhere dense sets of the Lebesgue measure zero.*

■ As $\mathbb{Q}_p = \bigcup_{\gamma \in Z} S_\gamma$ (see Sec. 1.3). Then

$$K = \bigcup_{\gamma \in Z} K_\gamma, \qquad K_\gamma = \varphi(S_\gamma), \qquad K_\gamma \cap K_{\gamma'} = \phi, \quad \gamma \neq \gamma' . \tag{6.3}$$

Let us study a structure of the set K_0 (structure of the remaining sets K_γ is similar). As $1 \leq \varphi(x) < p, x \in S_0$ then $K_0 \subset [1, p)$. Let us consider the following system of disjoint intervals on $[1, p)$:

$$I_n^\varepsilon = \left(\sum_{0 \leq j \leq n} \varepsilon_j p^{-2j} + \frac{p^{-2n}}{p+1}, \quad \sum_{0 \leq j \leq n} \varepsilon_j p^{-2j} + p^{-2n} \right),$$

$$n = 0, 1, 2, \dots ,$$

$$\varepsilon = (\varepsilon_0, \varepsilon_1, \dots, \varepsilon_n), \quad 0 \leq \varepsilon_j \leq p - 1 ,$$

$$j = 1, 2, \dots, n - 1, \quad \varepsilon_0 \neq 0, \quad \varepsilon_n \neq p - 1 .$$

Denote

$$I_n = \bigcup_\varepsilon I_n^\varepsilon, \quad I = \bigcup_{0 \leq n < \infty} I_n .$$

The Lebesgue measure of the set I is equal to

$$\begin{aligned}\mu(I) &= \sum_{\varepsilon}^{\infty} \mu(I_n^{\varepsilon}) \\ &= (p-1)\left(1-\frac{1}{p+1}\right) + (p-1)^2 \sum_{1\le n<\infty} p^{n-1}p^{-2n}\left(1-\frac{1}{p+1}\right) \\ &= \frac{p(p-1)}{p+1} + \frac{(p-1)^2}{p+1} \sum_{1\le n<\infty} p^{-n} = p-1\,. \end{aligned} \tag{6.4}$$

By using the Lemma one can prove that $I \cap K_0 = \phi$, i.e. there are no $x \in S_0$ and $n = 0, 1, \ldots$ for which the inequalities

$$\sum_{0\le j\le n} \varepsilon_j p^{-2j} + \frac{p^{-2n}}{p+1} < \varphi(x) < \sum_{0\le j\le n} \varepsilon_j p^{-2j} + p^{-2n}$$

take place. Besides, using the Lemma on imbedded segments we establish that $K_0 \cup I = [1, p)$. From here and from (6.4) it follows that the Lebesgue measure of the set K_0 is equal to 0. The process of construction of the set $K_0 = [1,p)\backslash I$ coincides, within unessential details, with the known process of construction of the Cantor canonical perfect set on the segment $[0, 1]$. Therefore proofs of the remaining properties of the set K_0 are similar to the proofs of corresponding properties of the Cantor set. ■

7. *Space* $\mathbb{Q}_p^n$

$\mathbb{Q}_p^n = \mathbb{Q}_p \times \mathbb{Q}_p \times \ldots \times \mathbb{Q}_p$ consists of points $x = (x_1, x_2, \ldots, x_n)$, $x_j \in \mathbb{Q}_p$, $j = 1, 2, \ldots, n$. The norm on $\mathbb{Q}_p^n$ is

$$|x|_p = \max_{1\le j\le n} |x_j|_p, \; x \in \mathbb{Q}_p^n\,. \tag{7.1}$$

This norm is a non-Archimedean one as

$$|x+y|_p \le \max(|x|_p, |y|_p), \quad x, y \in \mathbb{Q}_p^n \tag{7.2}$$

■ $$\begin{aligned}\max_{1\le j\le n} |x_j + y_j|_p &\le \max_{1\le j\le n} \max(|x_j|_p, |y_j|_p) \\ &= \max(\max_{1\le j\le n} |x_j|_p, \max_{1\le j\le n} |y_j|_p) = \max(|x|_p, |y|_p)\end{aligned}$$ ■

The space $\mathbb{Q}_p^n$ is complete metric locally-compact and totally disconnected space.

We introduce the inner product

$$(x, y) = x_1 y_1 + x_1 y_1 + \ldots + x_n y_n, x, y \in \mathbb{Q}_p^n .$$

The following inequality is valid

$$|(x, y)|_p \leq |x|_p |y|_p, x, y \in \mathbb{Q}_p^n . \tag{7.3}$$

Denote by $B_\gamma(a)$ the ball of radius p^γ with center at the point $a \in \mathbb{Q}_p^n$ and by $S_\gamma(a)$ its boundary (sphere); $B_\gamma(0) = B_\gamma$ and $S_\gamma(0) = S_\gamma$, $\gamma \in \mathbb{Z}$ (cf. 1.3). $B_\gamma(a)$ and $S_\gamma(a)$ are closed-open sets in $\mathbb{Q}_p^n$.

It is easy to verify: if $a = (a_1, a_2, \ldots, a_n)$ then $B_\gamma(a) = B_\gamma(a_1) \times B_\gamma(a_2) \times \ldots \times B_\gamma(a_n)$.

II. Analytic Functions

In this section we consider analytic functions in the field of p-adic numbers. The bases of this theory are presented in books by H.Koblitz [121], J.-P. Serre [190] and W.H. Schikhof [186].

1. *Power series*

First we consider a numerical series in the field of p-adic numbers

$$\sum_{0 \leq k < \infty} a_k, \qquad a_k \in \mathbb{Q}_p . \tag{1.1}$$

Denote by S_n the n-th partial sum of the series (1.1)

$$S_n = \sum_{0 \leq k < n} a_k, \qquad n = 0, 1, \ldots .$$

The convergence of the series (1.1) to a p-adic number S means that $|S_n - S_p| \to 0$, $n \to \infty$; we call S its sum and

$$S = \sum_{0 \leq k < \infty} a_k .$$

Some properties of series in the field $\mathbb{Q}_p$ essentially differ from those in the field of real (or complex) numbers. For example, there is only absolute convergence of series. More precisely the following Lemma is valid.

Lemma 1. *The series* (1.1) *converges if and only if*

$$|a_k|_p \to 0, \qquad k \to \infty.$$

■ **Necessity** of the condition is obvious:

$$|a_k|_p = |S_k - S_{k-1}|_p \longrightarrow 0, k \to \infty.$$

To prove the sufficiency we use the Cauchy criterion. As $|a_k|_p \to 0$, $k \to \infty$ then for any $\varepsilon > 0$ there exists a $N = N_\varepsilon$ such that for all $k > N$ the inequality $|a_k|_p < \varepsilon$ is fulfilled. Hence for any integers $n > N$ and $m > N$

$$|S_m - S_n|_p = \left| \sum_{n \le k \le m} a_k \right|_p \le \max_{n \le k \le m} |a_k|_p < \varepsilon .$$

Thus the sequence $\{S_n, n \to \infty\}$ of partial sums converges and hence the series (1.1) converges. ■

From the Lemma 1 it follows that the sum of series (1.1) does not depend on the order of summation.

Now we examine a p-adic power series

$$f(x) = \sum_{0 \le k < \infty} f_k x^k, \quad f_k \in \mathbb{Q}_p \tag{1.2}$$

which define a p-adic valued function $f(x)$ for those $x \in \mathbb{Q}_p$ for which it converges.

Definition. A number $R = R(f)$ is called the *radius of convergence* of the series (1.2) if it converges for all $|x|_p \le R$ and diverges for $|x|_p > R$.

We note that R may take values 0 and p^γ, $\gamma \in \mathbb{Z}$. In the last case the series (1.2) converges uniformly in (open-closed) disc B_γ as by the Lemma 1

$$\left| \sum_{n \le k \le m} f_k x^k \right|_p \le \max_{m \le k \le n} |f_k R^k|_p \longrightarrow 0, \quad m, n \to \infty,$$

and defines a continuous function $f(x)$ in B_γ.

For determination of the radius of convergence of the series (1.2) like in the real case we introduce a number $r = r(f)$ by the formula

$$\frac{1}{r} = \overline{\lim_{k\to\infty}} |f_k|_p^{1/k} \tag{1.3}$$

or denoting $r = p^\sigma$

$$\sigma = \frac{-1}{\ln p} \overline{\lim_{k\to\infty}} \frac{1}{k} \ln |f_k|_p. \tag{1.3'}$$

Lemma 2. *The series* (1.2) *converges for all* $|x|_p < r$ *and diverges for* $|x|_p > r$.

∎ Let $|x|_p < r$. Then $|x|_p = (1-2\varepsilon)r$ where $0 < \varepsilon \leq 1/2$. From (1.3) it follows that there exists $N = N_\varepsilon$ such that for all $k > N$ the inequality

$$|f_k|_p^{1/k} < \frac{1}{r(1-\varepsilon)}$$

is valid, and hence

$$|f_k x^k|_p = (|x|_p |f_k|_p^{1/k})^k \left(\frac{1-2\varepsilon}{1-\varepsilon}\right)^k \longrightarrow 0, \quad k \to \infty.$$

By Lemma 1 the series (1.2) converges at the point x.

Let now $|x|_p > r$. Then $|x|_p = (1+2\varepsilon)r$ where $\varepsilon > 0$. From (1.3) it follows that there exists a subsequence $\{n_k, k \to \infty\}$ such that

$$\lim_{k\to\infty} |f_{n_k}|_p^{1/n_k} = \frac{1}{r}.$$

Therefore the inequality

$$|f_{n_k}|_p^{1/n_k} > \frac{1}{r(1+\varepsilon)}, \quad k > N$$

is valid for some $N = N_\varepsilon$. Hence

$$|f_{n_k} x^{n_k}|_p = (|x|_p |f_{n_k}|_p^{1/n_k})^{n_k} > \left(\frac{1+2\varepsilon}{1+\varepsilon}\right)^{n_k} \longrightarrow \infty, \quad k \to \infty,$$

and by Lemma 1 the series (1.2) diverges. ■

Note that r is not necessarily an integer power of p, i.e. the number σ in (1.3′) is not always an integer (see examples in Sec. 2.4).

Relations between numbers $R(f)$ and $r(f)$ are given by the following lemma which follows from Lemmas 1 and 2.

Lemma 3. *$R(f) \leq r(f)$, moreover if $p^\gamma < r(f) < p^{\gamma+1}$, $\gamma \in \mathbb{Z}$ then $R(f) = p^\gamma$; if $r(f) = p^\gamma$ then either $R(f) = p^\gamma$ or $R(f) = p^{\gamma-1}$.*

Note that in the case $r(f) = p^\gamma$, $\gamma \in \mathbb{Z}$ convergence of the series (1.2) on circumference $|x|_p = p^\gamma$ requires a special investigation like in the real case.

2. *Analytic functions*

Definition. A function $f(x)$ is called *analytic* in a disc B_γ if it can be represented by a power series convergent in B_γ. (Obviously, one can always assume $R(f) = p^\gamma$, see Sec. 2.1.)

Analytic functions in a disc possess some usual properties, for instance, they form a ring with respect to ordinary operations of addition and multiplication. However there are some differences, for instance, superposition of analytic functions may turn out to be a non-analytic function (see [186]).

We introduce the series for $n = 0, 1, \ldots$

$$f^{(n)}(x) = \sum_{n \leq k < \infty} k(k-1)\ldots(k-n+1) f_k x^{k-n}, \tag{2.1'}$$

$$f^{(-n)}(x) = \sum_{0 \leq k < \infty} \frac{1}{(k+1)(k+2)\ldots(k+n)} f_n x^{k+n}, \tag{2.1''}$$

which are obtained from the series (2.1) by termwise differentiation and integration respectively; $f(x) = f^{(0)}(x)$.

These functions are called *derivative* and *primitive* of order n respectively.

It is clear that radii of convergence of the series (2.1) satisfy inequalities

$$R(f^{(-n)}) \leq R(f) \leq R(f^{(n)}), n = 1, 2, \ldots \tag{2.2}$$

To obtain a more detailed information about radii of convergence of the series (2.1) we shall prove

$$|k|_p \geq \frac{1}{k}, \quad k \in \mathbb{Z}_+, \qquad \lim_{k\to\infty, k\in Z_+} |k|_p^{1/k} = 1. \tag{2.3}$$

■ In fact, let $k \in \mathbb{Z}_+$, $k = p^m k_0$, $|k|_p = p^{-m}$ where m and k_0 are integers, $m \geq 0$, $1 \leq k_0 \leq p-1$. Then

$$m = \frac{\ln k - \ln k_0}{\ln p} \geq \frac{\ln k}{\ln p}$$

and thus

$$|k|_p = p^{-m} \geq p^{-\frac{\ln k}{\ln p}} = \frac{1}{k}, \lim_{k\to\infty, k\in Z_+} |k|_p^{1/k}$$
$$= \lim_{k\to\infty, k\in Z_+} p^{-\frac{m}{k}} = \lim_{k\to\infty, k\in Z_+} p^{-\frac{\ln k - \ln k_0}{k \ln p}} = 1\,. \qquad ■$$

By using the formula (1.3) to the series (2.1) and the relation (2.3) we obtain the equalities

$$r(f^{(n)}) = r(f) = r(f^{(-n)}), \quad n = 0, 1, \ldots\,. \tag{2.4}$$

From here and from Lemma 3 of Sec. 2.1 it follows that if $p^\gamma < r(f) < p^{\gamma+1}$ then $R(f) = p^\gamma$ and

$$R(f^{(n)}) = R(f) = R(f^{(-n)}), \quad n = 0, 1, \ldots\,; \tag{2.5}$$

if $r(f) = p^\gamma$ then two cases are possible: 1) $R(f) = p^\gamma$ then either the equalities (2.5) are valid or

$$pR(f^{(-n)}) = R(f) = R(f^{(n)}), \quad n \geq n_0 \tag{2.5$'$}$$

for some $n_0 \geq 1$; 2) $R(f) = p^{\gamma-1}$ then either the equalities (2.5) are valid or

$$R(f^{(-n)}) = R(f) = \frac{1}{p} R(f^{(n)}), \quad n \geq n_0 \tag{2.5$''$}$$

for some $n_0 \geq 1$.

The formulas (2.5) can be interpreted by the following way. It is possible to differentiate and to integrate an analytic termwise any times; by differentiation a radius of convergence may increase in p times, but by integration this radius may decrease in p times.

As we see the situation somewhat differs from the case of real numbers.

3. *Algebra of Analytic Functions*

We denote by A a set of analytic functions in the unit disc B_0. Such functions are the only ones defined by the series (1.2) for which the condition $|f_k|_p \longrightarrow 0$, $k \to \infty$ is fulfilled (see Sec. 2.1).

The set A is linear over the field $\mathbb{Q}_p$. On the set A we introduce norm $||f||$ by the formula

$$||f|| = \max_{0 \le k < \infty} |f_k|_p, \quad f \in A. \tag{3.1}$$

The functional (3.1) *is in fact a norm, besides the non-Archimedean one.*

■ Let $||f|| = 0$, $f \in A$, i.e. $\max\limits_{0 \le k < \infty} |f_k|_p = 0$ then $f_k = 0$, $k = 0, 1, \ldots$ and hence $f = 0$. Let $\alpha \in \mathbb{Q}_p$, $\alpha \neq 0$. Then

$$||\alpha f|| = \max_k |\alpha f_k|_p = |\alpha|_p \max_k |f_k|_p = |\alpha|_p ||f||.$$

Finally, if $f, g \in A$ then

$$||f + g|| = \max_k |f_k + g_k|_p \le \max_k \max\left[|f_k|_p, |g_k|_p\right] \le \max[||f||, ||g||]. \quad ■$$

Theorem. *The space A is a Banach algebra.*

■ Prove completeness of A. Let a sequence $\{f^n, n \to \infty\}$, $f^n \in A$ be fundamental. As

$$||f^n - f^m|| = \max_{0 \le k < \infty} |f_k^n - f_k^m|_p$$

then the sequences $\{f_k^n, n \to \infty\}$ are fundamental for every $k = 0, 1, \ldots,$ thus they converge to some $f_k \in \mathbb{Q}_p$ uniformly with respect to k (see Sec. 1.3) and hence

$$\lim_{k \to \infty} f_k = \lim_{k \to \infty} \lim_{n \to \infty} f_k^n = \lim_{n \to \infty} \lim_{k \to \infty} f_k^n = 0.$$

Therefore the function

$$f(x) = \sum_{0 \le k < \infty} f_k x^k$$

belongs to A and

$$\lim_{n\to\infty} \|f^n - f\| = \lim_{n\to\infty} \max_{0\le k<\infty} .$$

Let $f, g \in A$ and $h = fg$. Then $h \in A$ (see Sec. 2.2) and

$$h_k = \sum_{0 \le j \le k} f_j g_{k-j} .$$

Hence

$$\|h\| = \max_k |h_k|_p = \max_k \left| \sum_{0\le j\le k} f_j g_{k-j} \right|_p$$

$$\le \max_k \max_{0\le j\le k} |f_j|_p |g_{k-j}|_p \le \max_j |f_j|_p \max_k |g_k|_p = \|f\|\,\|g\| . \qquad \blacksquare$$

Remark. The algebra A is called *algebra of bounded power series* and is a special case of Tate's algebras.

4. *Functions* e^x, $\ln(1+x)$, $\sin x$, $\cos x$

These functions like in real case are defined by the series

$$e^x = \sum_{0\le k<\infty} \frac{x^k}{k!}, \tag{4.1}$$

$$\ln(1+x) = \sum_{1\le k<\infty} \frac{(-1)^{k+1}}{k} x^k, \tag{4.2}$$

$$\sin x = \sum_{0\le k<\infty} \frac{(-1)^k}{(2k+1)!} x^{2k+1}, \tag{4.3}$$

$$\cos x = \sum_{0\le k<\infty} \frac{(-1)^k}{(2k)!} x^{2k}. \tag{4.4}$$

To study convergence of these series we need to estimate $|n!|_p$ for any $n \in \mathbb{Z}_+$.

Let $n \in \mathbb{Z}_+$, $n \leq p^{s+1} - 1$, $s \in \mathbb{Z}_+$. Then n can be uniquely represented in the form

$$n = n_0 + n_1 p + \ldots + n_s p^s, \quad 0 \leq n_j \leq p - 1, \quad j = 0, 1, \ldots, s. \tag{4.5}$$

Denote

$$s_n = \sum_{0 \leq j < s} n_j. \tag{4.6}$$

From (4.5) it follows that

$$\lim_{n \to \infty, n \in Z_+} \frac{s_n}{n} = 0. \tag{4.7}$$

Now we shall prove *the equality*

$$|n!|_p = p^{-\frac{n - s_n}{p-1}}, \quad n \in \mathbb{Z}_+. \tag{4.8}$$

■ A power $M(n)$ in which factor p enters in $n!$ is equal to

$$\left[\frac{n}{p}\right] + \left[\frac{n}{p^2}\right] + \ldots + \left[\frac{n}{p^s}\right] = M(n).$$

Here $[a]$ is the entire part of a number a.

By using the representation (4.5) and the notation (4.6) we get

$$\begin{aligned} M(n) &= \frac{n - n_0}{p} + \frac{n - n_0 - n_1 p}{p^2} + \ldots + \frac{n - n_0 - n_1 p - \ldots - n_s p^s}{p^{s+1}} \\ &= n\frac{1 - p^{-s-1}}{p-1} - n_0 \frac{1 - p^{-s-1}}{p-1} - n_1 \frac{1 - p^{-s}}{p-1} - \ldots - n_s \frac{1 - p^{-1}}{p-1} \\ &= \frac{n - s_n}{p-1} - \frac{n}{p-1} p^{-s-1} + (n_0 + n_1 p + \ldots + n_s p^s) \frac{p^{-s-1}}{p-1} = \frac{n - s_n}{p-1} \end{aligned}$$

from which the equality (4.8) follows. ■

Function e^x is defined by the series (4.1) from which it follows $(e^x)' = e^x$. Taking into account the equalities (4.7) and (4.8) for the radius $r(e^x)$ (see Sec. 2.2) we get

$$r(e^x) = \overline{\lim_{k \to \infty}} \, |k!|_p^{1/k} = \lim_{k \to \infty} p^{-\frac{k - s_k}{k(p-1)}} = p^{-\frac{1}{p-1}}. \tag{4.9}$$

For $p \neq 2$ we obtain from (4.9)

$$\frac{1}{p} < r(e^x) < 1.$$

Therefore

$$R(e^x) = R((e^x)^{(n)}) = \frac{1}{p}, n \in \mathbb{Z}. \tag{4.10}$$

For $p = 2$ we get from (4.9) $r(e^x) = 2^{-1}$. Let us investigate the convergence of the series (4.1) on the circle $|x|_2 = 2^{-1}$. Let $x = 2$ and $k = 2^n$. Then $s_k = 1$ and owing to (4.3)

$$\left|\frac{2^k}{k!}\right|_2 = 2^{-k}2^{k-1} = 2^{-1} \not\to 0, \quad k \to \infty.$$

Hence on the circle $|x|_2 = 2^{-1}$ the series (4.1) diverges, and by Lemma 3 of Sec. 2.1 $R(e^x) = 2^{-2}$, and then (see Sec. 2.2)

$$R(e^x) = R((e^x)^{(n)}) = 2^{-2}, \quad n \in \mathbb{Z}. \tag{4.11}$$

So for any p the series (4.1) for e^x can be termwise integrated and differentiated any times in the same radius of convergence.

Denote by G_p the disc of convergence of the series (4.1); G_p is the additive group $|x|_p \leq p^{-1}$ for $p \neq 2$ and $|x|_2 \leq 2^{-2}$ for $p = 2$.

By using the series (4.1) we verify the relation

$$e^x e^y = e^{x+y}, \quad x, y \in G_p\,. \tag{4.12}$$

Now we prove the equalities

$$|e^x - 1|_p = |x|_p, \quad |e^x|_p = 1, \quad x \in G_p. \tag{4.13}$$

■ The second one follows from the first one. To prove it we establish beforehand the inequality

$$\left|\frac{x^k}{k!}\right|_p \leq |x|_p p^{(1-k)\varepsilon_p}, \quad x \in G_p, k \in \mathbb{Z}_+, \tag{4.14}$$

where $\varepsilon_p = \frac{p-2}{p-1}$, $p \neq 2$ and $\varepsilon_2 = 1$. In fact, it follows from the equality (4.8) that

$$\left|\frac{x^k}{k!}\right| = |x|_p \frac{|x^{k-1}|_p}{|k!|_p} \leq |x|_p p^{1-k} p^{\frac{k-s_k}{p-1}} \leq |x|_p p^{(1-k)\varepsilon_p}, \quad p \neq 2$$

as owing to (4.6) $s_k \geq 1$; for $p = 2$ the proof is similar. ■

Now by using the inequality (4.14) we obtain (4.13) :

$$|e^x - 1|_p = \left|\sum_{1 \leq k < \infty} \frac{x^k}{k!}\right| = |x|_p \max_{k \in \mathbb{Z}_+} \left|\frac{x^{k-1}}{k!}\right|_p = |x|_p$$

as

$$\left|\frac{x^{k-1}}{k!}\right|_p \leq p^{(1-k)\varepsilon_p} < 1, \quad k = 2, 3, \ldots, x \in G_p.$$

Function $\ln(1+x)$ is defined by the series (4.2). Its radius r owing to (2.3) is equal to

$$r(\ln(1+x)) = \lim_{k \to \infty} |k|_p^{1/k} = 1.$$

But at the point $x = 1$ the series (4.2) is divergent as $|k^{-1}|_p \geq 1$, $k \in \mathbb{Z}_+$. Therefore

$$R(\ln(1+x)) = \frac{1}{p}. \tag{4.15}$$

The function $\ln(1+x)$ possesses the property

$$\ln[(1+x)(1+y)] = \ln(1+x) + \ln(1+y), \quad |x|_p \leq \frac{1}{p}, |y|_p \leq \frac{1}{p}, \tag{4.16}$$

which can be verified with the help of the series (4.2).

The following equality is valid

$$|\ln(1+x)|_p = |x|_p, \quad x \in G_p. \tag{4.17}$$

■ Taking into account the equality (2.3) $|k^{-1}|_p \leq k$ we have (4.17)

$$|\ln(1+x)|_p = \left|\sum_{1 \leq k < \infty} (-1)^{k+1} \frac{x^k}{k}\right|_p = |x|_p \max_{k \in Z_+} |x|_p^{k-1} |k^{-1}|_p = |x|_p$$

as

$$|x|_p^{k-1}|k^{-1}|_p \leq \begin{cases} p^{1-k}k, & p \neq 2 \\ 2^{2-2k}k, & p = 2 \end{cases} < 1, \quad k = 2, 3, \ldots, x \in G_p \quad \blacksquare$$

Denote by

$$J_p = [x \in \mathbb{Q}_p, 1 - x \in G_p].$$

J_p is a multiplicative group where it follows from the identity

$$1 - xy = 1 - x + 1 - y + (1 - x)(1 - y).$$

From the relations (4.12), (4.13), (4.16) and (4.17) it follows that *the function e^x realizes an isomorphism of the additive group G_p onto the multiplicative group J_p. The inverse isomorphism is realized by the function* $\ln x = \ln(1 + x - 1)$ *and the equalities are valid*

$$\ln e^x = x, \quad x \in G, \quad e^{\ln x}, \quad x \in J_p, \tag{4.18}$$

which can be verified directly with the help of the series (4.1) and (4.2).

Functions $\sin x$ and $\cos x$ are defined by the series (4.3) and (4.4). These series like in the case of the function e^x converge in G_p, and the following relations are valid

$$|\sin x|_p = |x|_p, \quad |\cos x|_p = 1, \quad x \in G_p. \tag{4.19}$$

The standard trigonometrical formulas are valid, established by the help of the series (4.3) and (4.4), in particular

$$\cos^2 x + \sin^2 x = 1, \quad e^{\tau x} = \cos x + \tau \sin x, \quad x \in \mathbb{Q}_p \tag{4.20}$$

where $\tau^2 = -1$, $\tau \in \mathbb{Q}_p$, $p \equiv 1 \pmod 4$ (see Sec 1.4).

Functions

$$e^x, \cos x, \frac{\sin x}{x}, \frac{\ln(1+x)}{x} \tag{4.21}$$

are squares of p-adic functions in G_p (for $p = 2$ functions e^x and $\frac{\ln(1+x)}{x}$ are squares in $\mathbb{Q}_p$ only for $|x|_2 \leq 2^{-3}$).

■ These assertion follow from the Lemma of 1.4. In fact, the norms of functions (4.21) are equal to 1 owing to the equations (4.13),(4.17) and (4.19). Moreover their canonical forms are

$$1 + C_1(x)p + \ldots, p \neq 2, \quad 1 + C_3(x)2^{-3} + \ldots, p = 2. \tag{4.22}$$

Let $p \neq 2$. For e^x the representation (4.22) follows from (4.13)

$$|e^x - 1|_p = |x|_p \leq \frac{1}{p}, \quad x \in G_p.$$

For other functions it follows from the similar estimates which can be obtained by using the estimates (4.14) and (2.3).

For $p = 2$ the situation is similar, more complicated. By using the estimate (4.14) we have for $|x|_2 \leq 2^{-2}$

$$|\cos x - 1|_2 \leq \max_{k \in Z_+} \left|\frac{x^{2k}}{(2k)!}\right|_2 \leq |x|_2 2^{1-2k} \leq 2^{-3},$$

$$\left|\frac{\sin x}{x} - 1\right|_2 \leq \max_{k \in Z_+} \left|\frac{x^{2k}}{(2k+1)!}\right|_2 \leq |x|_p \max_{k \in Z_+} 2^{-1-2k} \leq 2^{-3}.$$

The similar estimate for functions $e^x = (e^{x/2})^2$ and $\frac{\ln(1+x)}{x}$ are valid only for $|x|_2 \leq 2^{-3}$. Hence the representation (4.22) follows. ∎

Finally we note that *the functions* $\sin x$ *and* $\tan x = \frac{\sin x}{\cos x}$ *map one-to-one* G_p *onto* G_p. *The inverse functions are*

$$\arcsin x = \sum_{0 \leq k < \infty} \frac{(2k)!}{2^{2k}(k!)^2} x^{2k+1}, \quad x \in G_p, \tag{4.23}$$

$$\arctan x = \sum_{0 \leq k < \infty} \frac{(-1)^k}{2k+1} x^{2k+1}, \quad x \in G_p, \tag{4.24}$$

besides

$$|\arcsin x|_p = |\arctan x|_p = |x|_p. \tag{4.25}$$

5. *Theorem on Inverse Function*

Let $f(x)$ be an analytic function in the disc $B_r(a)$ and $f'(a) \neq 0$, $|f'(a)|_p = p^n$. Then there exists a disc $B_\rho(a)$, $\rho \leq r$ such that f maps it on a disc $B_{\rho+n}(b)$, $b = f(a)$, one-to-one, the inverse function $g(y)$ is analytic in the disc $B_{\rho+n}(b)$ and the equality

$$g'(b) = \frac{1}{f'(a)}. \tag{5.1}$$

is valid.

■ By using the classical Theorem on inverse function we conclude that there exists a disc $B_\rho(a)$ of sufficiently small radius p^ρ, $\rho \leq r$ which the function f maps one-to-one onto some neighborhood $U(b)$ of the point b and the inverse function $g(y)$ is analytic in $U(b)$ and the equality (5.1) holds. It remains to prove that for a sufficiently small p^ρ $U(b) = B_{\rho+n}(b)$.

As the function $f(x)$ is analytic in the disc $B_r(a)$ then it can be represented in the form of a series (see Sec. 2.2)

$$f(x) = f(a) + f'(a)(x-a) + (x-a)^2 \sum_{2\leq k<\infty} a_k (x-a)^{k-2} \tag{5.2}$$

besides

$$|a_k|_p p^{rk} \longrightarrow 0, \quad k \to \infty.$$

Denote

$$\max_{k=2,3,\ldots} |a_k|_p p^{r(k-2)} = p^s, \quad s \in \mathbb{Z}.$$

Then choosing $\rho < n - s$ ($\rho \leq r$) from (5.2) we have

$$|f(x) - b|_p = |f'(a)|_p |x-a|_p, \quad x \in B_\rho(a). \tag{5.3}$$

From (5.3) it follows that

$$f(B_\rho(a)) = U(b) \subset B_{\rho+n}(b) \tag{5.4}$$

and there exists a point $x' \in S_\rho(a)$ such that

$$|f(x') - b|_p = |f'(a)|_p |x' - a|_p = p^{n+\rho}.$$

Hence $f(x') = y' \in S_{\rho+n}(b) \cap U(b)$. Therefore the power series for the inverse function

$$g(y) = a + g'(b)(y-b) + \ldots$$

converges at the point $y' \in S_{\rho+n}(b)$ and thus in the disc $B_{\rho+n}(b)$. Reducing if necessary the radius $p^{\rho+n}$ we achieve that the function $g(y)$ also will satisfy the equality (5.3)

$$|g(y) - a|_p = \frac{1}{|f'(a)|_p} |y-b|_p, \quad y \in B_{\rho+n}.$$

From here it follows that the inclusion

$$g(B_{\rho+n}(b)) \subset B_\rho(a) = g(U(b))$$

which together with (5.4) gives $U(b) = B_{\rho+n}(b)$. ■

III. Additive and Multiplicative Characters

The field $\mathbb{Q}_p$ is an additive group. We denote by $\mathbb{Q}_p^* = \mathbb{Q}_p \backslash \{0\}$ its multiplicative group.

1. *Additive Characters of the Field* $\mathbb{Q}_p$

Additive character of the field $\mathbb{Q}_p$ is called a character of additive group $\mathbb{Q}_p$, i.e. a continuous (complex-valued) function $\chi(x)$ defined on $\mathbb{Q}_p$ and satisfied the conditions $|\chi(x)| = 1$,

$$\chi(x+y) = \chi(x)\chi(y), \qquad x, y \in \mathbb{Q}_p \ . \tag{1.1}$$

Analogously one defines (additive) characters of the field $\mathbb{Q}_p(\sqrt{\varepsilon})$ and of subgroup B_γ, $\gamma \in \mathbb{Z}$, of the group $\mathbb{Q}_p$.

It is clear that every additive character of the field $\mathbb{Q}_p$ is a character of any group B_γ.

The function

$$\chi_p(\xi x) = \exp(2\pi i \{\xi x\}_p) \tag{1.2}$$

for every fixed $\xi \in \mathbb{Q}_p$ is an additive character of the field $\mathbb{Q}_p$ and the group B_γ. It follows from the relation for fractional parts (see Sec. 1.2)

$$\{x+y\}_p = \{x\}_p + \{y\}_p - N, \qquad N = 0, 1 \ .$$

Our goal is to prove *that the formula* (1.2) *gives a general representation of additive characters of the field* $\mathbb{Q}_p$ *and the group* B_γ. Let $\chi(x)$ be an arbitrary additive character. From (1.1) we have the relations

$$\chi(0) = 1, \quad \chi(-x) = \overline{\chi(x)} = \chi^{-1}(x), \quad \chi(nx) = \{\chi(x)\}^n, \qquad n \in \mathbb{Z} \tag{1.3}$$

At first we investigate characters of the group B_γ.

Let $\chi \not\equiv 1$ *be such a character. Prove that there exists* $k \in \mathbb{Z}$ *such that*

$$\chi(x) \equiv 1, \qquad x \in B_k \ . \tag{1.4}$$

■ By virtue of the conditions $\chi(0) = 1$, $|\chi(x)| = 1$ and $\chi(x)$ is a continuous function on B_γ it is possible to choose such branch of the function

$\ln\chi(x) = i\arg\chi(x)$ that it will be continuous at 0 and $\arg\chi(0) = 0$. In particular there exists $k \in \mathbb{Z}$ such that $|\arg\chi(x)| < 1$ for all $x \in B_k$.

Taking into account that $nx \in B_k$ for all $x \in B_k$ and $n \in \mathbb{Z}_+$ we conclude from (1.3) that

$$|\arg\chi(x)| = |\frac{1}{n}\arg c(nx)| < \frac{1}{n}, \qquad n \in \mathbb{Z}_+, \quad x \in B_k$$

and thus $\arg\chi(x) = 0$ and $\chi(x) \equiv 1, x \in B_k$. ■

We assume that the disc B_k in (1.4) is *maximal* so that as $\chi(x) \not\equiv 1$ in B_γ then $k < \gamma$.

Now we *prove for any integer r, $k < r \le \gamma$, the equality*

$$\chi(p^{-r}) = \exp(2\pi i m p^{-r+k}), \qquad \exists m = 1, 2, \dots, p^{\gamma-k} - 1\,, \tag{1.5}$$

where m does not depend on r.

■ For $r = \gamma$ it follows from (1.4) and (1.3) as

$$1 = \chi(p^{-k}) = \chi(p^{-\gamma+\gamma-k}) = [\chi(p^{-r})]^{p^{\gamma-k}}\,.$$

For $k < r < \gamma$

$$\chi(p^{-r}) = \chi(p^{-r+\gamma-\gamma}) = [\chi(p^{-\gamma})]^{p^{\gamma-r}} = [\exp(2\pi i m p^{-\gamma+k})]^{p^{\gamma-r}}\,. \quad ■$$

Denote $\xi = p^k m$ where $|\xi|_p = p^{-k}|m|_p > p^{-k}p^{-\gamma+k} = p^{-\gamma}$ and $|\xi|_p \le p^{-k}$. Then owing to (1.2) the representation (1.5) takes the form $\chi_p(p^{-\gamma}) = \chi_p(p^{-\gamma}\xi)$, and thus owing to (1.3) we have

$$\chi(p^{-r}) = \chi_p(p^{-r}\xi), \qquad k < r \le \gamma \tag{1.6}$$

for some $\xi \in \mathbb{Q}_p$ such that $p^{-k} \ge |\xi|_p > p^{-\gamma}$.

Let now $x \in B_\gamma \backslash B_k$. *The following representation is valid*

$$\chi(x) = \chi_p(\xi x), \qquad \exists\xi \in \mathbb{Q}_p, \qquad |\xi|_p > p^{-\gamma}\,. \tag{1.7}$$

■ Let $x \in B_\gamma \backslash B_k$. Such x can be represented in the form

$$x = x_0 p^{-r} + x_1 p^{-r+1} + \dots + x_{r-k+1} p^{-k+1} + x', \qquad x' \in B_k, \quad x_0 \ne 0$$

for some r, $k < r \leq \gamma$. By using (1.6) and (1.4) we get the representation (1.7):

$$\begin{aligned}\chi(x) &= [\chi(p^{-r})]^{x_0}[\chi(p^{-r+1})]^{x_1}\dots[\chi(p^{-k+1})]^{x_{r-k+1}}\chi(x') \\ &= [\chi_p(p^{-r}\xi)]^{x_0}[\chi_p(p^{-r+1}\xi)]^{x_0}\dots[\chi_p(p^{-k+1}\xi)]^{x_{r-k+1}}\chi_p(x'\xi) \\ &= \chi_p(x_0p^{-r}\xi + x_1p^{-r+1}\xi + \dots + x_{r-k+1}p^{-k+1}\xi + x'\xi) = \chi_p(x\xi)\ .\end{aligned}$$

The case $\xi = 0$ is impossible otherwise $\chi(x) = \chi_p(0) = 1$ in B_γ which contradicts to the definition of the number k. ■

Hence we have just proved that *any additive character of the group* B_γ *has a form* (1.2) *where either* $\xi = 0$ *or* $|\xi|_p \geq p^{-\gamma+1}$.

Now let $\chi(\xi) \not\equiv 1$ be an additive character of the field $\mathbb{Q}_p$. Then in a disc B_0 it is represented in the form

$$\chi(x) = \chi_p(\xi^{(0)}x), \qquad \xi^{(0)} \in \mathbb{Q}_p, \qquad |\xi^{(0)}|_p > 1\ . \tag{1.8}$$

We shall prove that in the disc B_1 it is represented in the form

$$\chi(x) = \chi_p(\xi^{(1)}x), \qquad \xi^{(1)} = \xi^{(0)} + \xi_0, \qquad \exists\xi_0 = 0, 1, \dots, p-1 \tag{1.9}$$

■ As $B_1 = B_0 \cup S_1$ and $B_0 \cap S_1 = \phi$ we shall prove at first the representation (1.9) in the circumference S_1. Any point $x \in S_1$ is represented in the form

$$x = p^{-1}x_0 + x', \qquad \exists x_0 = 1, 2, \dots, p-1, \qquad x' \in B_0\ .$$

Then using (1.8) we shall have the representation (1.9) in S_1:

$$\begin{aligned}\chi(x) &= [\chi(p^{-1}]^{x_0}\chi_p(\xi^{(0)}x') = [\chi(1)]^{x_0/p}\chi_p(\xi^{(0)}x') \\ &= [\chi_p(\xi^{(0)})]^{x_0/p}\chi_p(\xi^{(0)}x') = \chi_p\left(\xi^{(0)}\frac{x_0}{p}\right)\chi_p\left(\frac{\xi_0 x_0}{p}\right) \\ \chi_p(\xi^{(0)}x' + \xi_0 x') &= \chi_p\left((\xi^{(0)} + \xi_0)\left(\frac{x_0}{p} + x'\right)\right) = \chi_p(\xi^{(1)}x)\end{aligned}$$

for some $\xi_0 = 0, 1, \dots, p-1$. The representation (1.9) is valid also in B_0 owing to (1.8). ■

Continuing this process we obtain in the disc B_2 the representation

$$\chi(x) = \chi_p(\xi^{(2)}x), \qquad \xi^{(2)} = \xi^{(0)} + \xi_0 + \xi_1 p$$

for some $\xi_1 = 0, 1, \dots, p-1$, and so on. As a result in $\mathbb{Q}_p$ we obtain the representation (1.2)

$$\chi(x) = \chi_p(\xi x), \qquad \xi = \xi^{(0)} + \xi_0 + \xi_1 p + \dots \in \mathbb{Q}_p \,.$$

Hence, any additive character of the field $\mathbb{Q}_p$ *has a form* (1.2) *for some* $\xi \in \mathbb{Q}_p$.

In other words, the mapping $\xi \to \chi_p(\xi x)$ is a homomorphism of the additive group of the field $\mathbb{Q}_p$ onto the group of additive characters. This mapping is one-to-one (i.e. from the equality $\chi_p(\xi_1 x) = \chi_p(\xi_2 x)$ for all $x \in \mathbb{Q}_p$ it follows that $\xi_1 = \xi_2$).

Now we have

Theorem. *The group of additive characters of the field* $\mathbb{Q}_p$ *is isomorphic to its additive group* $\mathbb{Q}_p$, *and the mapping* $\xi \to \chi_p(\xi x)$ *gives this isomorphism.*

Let us denote

$$\chi_\infty(x) = \exp(-2\pi i x), \qquad x \in R \,. \tag{1.10}$$

Then *the following (adelic) formula is valid*

$$\prod_{2 \le p \le \infty} \chi_p(x) = 1, \qquad x \in \mathbb{Q} \,. \tag{1.11}$$

■ Let x be an arbitrary positive rational number

$$x = N p_1^{-\alpha_1} p_2^{-\alpha_2} \dots p_n^{-\alpha_n}$$

where $\alpha_j \in \mathbb{Z}_+$, p_j are prime numbers and N is positive integer not divisible by p_j, $j = 1, 2, \dots, n$. From the theory of congruence it follows (see [204])* that the number x is represented in the form

$$x = \frac{N_1}{p^{\alpha_1}} + \frac{N_2}{p^{\alpha_2}} \dots + \frac{N_n}{p^{\alpha_n}} + M \tag{1.12}$$

---

* We use the following result: if natural numbers c and d are relatively prime then the congruence $cx + dy \equiv 1 \pmod{cd}$ is solvable; indeed, $dy \equiv 1 \pmod{c}$, $x \equiv \frac{1-dy}{c} \pmod{d}$.

for some $N_j \in \mathbb{Z}_+$, $1 \leq N_j < p_j^{\alpha_j} - 1$ and $M \in \mathbb{Z}$. From (1.12) it follows that

$$\{x\}_{p_j} = \frac{N_j}{p_j^{\alpha_j}}, \qquad \{x\}_p = 0, \qquad p \neq p_j, \quad j = 1, 2, \ldots, n$$

and hence the equality (1.11) is valid

$$\prod_{2 \leq p < \infty} \chi_p(x) = \prod_{1 \leq j \leq n} \chi_{p_j}(x) = \prod_{1 \leq j \leq n} \exp(2\pi i \{x\}_{p_j})$$
$$= \exp(2\pi i \sum_{1 \leq j \leq n} \{x\}_{p_j}) = \exp(2\pi i x) . \qquad \blacksquare$$

At last we note that any additive character χ of the field $\mathbb{Q}_p(\sqrt{\varepsilon})$ (see Sec. 1.4) has the form

$$\chi(x + i\sqrt{y}) = \chi_p(\xi_1 x)\chi_p(\xi_2 y), \qquad \xi_1, \xi_2 \in \mathbb{Q}_p \tag{1.13}$$

2. *Multiplicative Characters of the Field* $\mathbb{Q}_p$

Multiplicative character of the field $\mathbb{Q}_p$ is called a character of the multiplicative group $\mathbb{Q}_p^*$, i.e. a continuous (complex valued) function $\pi(x)$, defined on $\mathbb{Q}_p^*$ and satisfied the condition

$$\pi(xy) = \pi(x)\pi(y), \qquad x, y \in \mathbb{Q}_p^* .$$

Analogously one defines (multiplicative) characters of the field $\mathbb{Q}_p(\sqrt{\varepsilon})$, of subgroup S_ε of the group $\mathbb{Q}_p^*(\sqrt{\varepsilon})$ and of subgroup S_0 of the group $\mathbb{Q}_p^*$.

Any multiplicative character π of the field $\mathbb{Q}_p$ has the form

$$\pi(x) = |x|_p^{\alpha - 1}\pi_0(|x|_p x), \qquad |\pi_0(x')|_p = 1, \qquad x' \in S_0, \quad \alpha \in C \tag{2.1}$$

where π_0 is a character of the group S_0; conversely, any multiplicative character of the group S_0 is extended up to a multiplicative character of the field $\mathbb{Q}_p$ by the formula (2.1).

$\blacksquare$ Let π be a multiplicative character of the field $\mathbb{Q}_p$. Any element $x \in \mathbb{Q}_p^*$ is represented in the form (see Sec. 1.2)

$$x = |x|_p^{-1} x', \qquad x' = |x|_p x \in S_0 ,$$

and therefore the representation (2.1) is valid

$$\pi(x) = \pi(|x|_p^{-1}x') = \pi(|x|_p^{-1})\pi(x') = \pi(p^N)\pi_0(x')$$
$$= [\pi(p)]^N \pi_0(x') = p^{(1-\alpha)N}\pi_0(x') = |x|_p^{1-\alpha}\pi_0(x')$$

where it is denoted $|x|_p = p^{-N}$, $\pi(p) = p^{1-\alpha}$. ■

We notice that the set of multiplicative characters of the compact group S_0 is discrete (see [174], ch.VI).

From (2.1) it follows that $\pi_0(1) = 1$.

Let $\pi_0(x) \not\equiv 1$ be a multiplicative character of the group S_0. Then there exists such $k \in \mathbb{Z}_+$ that

$$\pi_0(x) \equiv 1, \qquad x \in B_{-k}(1) = [x \in \mathbb{Q}_p : |x-1|_p \leq p^{-k}] . \tag{2.2}$$

■ A proof follows from the conditions $\pi_0(1) = 1$, $|\pi_0(x)| = 1$ and $\pi_0(x)$ is a continuous function on S_0, and it is similar to the proof of the corresponding statement for an additive character of the group B_γ (see Sec. 4.1). ■

The lowest $k \in \mathbb{Z}_+ \cup \{0\}$ for which the equality (2.2) is fulfilled is called a *rank* of the character $\pi_0(x)$.

There exists only one character of rank $0 : \pi_0(x) \equiv 1$, as $S_0 \subset B_0(1)$.

Let the rank of a character $\pi_0(x)$ be positive. Then

$$\pi_0(x) = \pi_0(x_0 + x_1 p + \ldots + x_{k-1}p^{k-1}) \tag{2.3}$$

So $\pi_0(x)$ depends only on $x_0, x_1, \ldots, x_{k-1}$.

■ In fact a number $x \in S_0$ is represented in the form (see Sec. 1.2)

$$x = x_0 + x_1 p + \ldots = (x_0 + x_1 p + \ldots + x_{k-1}p^{k-1})(1 + tp^k)$$

where $|t|_p \leq 1$ so $1 + tp^k \in B_{-k}(1)$. Then owing to (2.2) $\pi_0(1 + tp^x) = 1$ and the equality (2.3) follows. ■

Now we prove the equality: *if the rank* $k \geq 2$ then

$$\sum_{0 \leq x_{k-1} \leq p-1} \pi_0(x_0 + x_1 p + \ldots + x_{k-1}p^{k-1}) = 0 . \tag{2.4}$$

■ There exists a number $t = 0, 1, \ldots, p-1$ such that

$$\rho = \pi_0(1 + tp^{k-1}) \neq 1 .$$

(Otherwise the rank of the character $\pi_{(x)}$ would be less then k.) Therefore using (2.4) we get the equalities

$$\begin{aligned}
&\sum_{0 \le x_{k-1} \le p-1} \pi_0(x_0 + x_1 p + \ldots + x_{k-1} p^{k-1}) \\
&= \frac{1}{\rho} \sum_{0 \le x_{k-1} \le p-1} \pi_0(1 + tp^{k-1}) \pi_0(x_0 + x_1 p + \ldots + x_{k-1} p^{k-1}) \\
&= \frac{1}{\rho} \sum_{0 \le x_{k-1} \le p-1} \pi_0[(x_0 + x_1 p + \ldots + x_{k-1} p^{k-1})(1 + tp^{k-1})] \\
&= \sum_{0 \le x_{k-1} \le p-1} \pi_0(x_0 + x_1 p + \ldots + (tx_0 + x_{k-1}) p^{k-1}) \\
&= \frac{1}{\rho} \sum_{0 \le x'_{k-1} \le p-1} \pi_0(x_0 + x_1 p + \ldots + x'_{k-1} p^{k-1})
\end{aligned}$$

from which the equality (2.4) follows. ■

Examples of multiplicative characters.

1) $\pi(x) = |x|_p^{\alpha-1}$, $\pi_0(x') = 1$

2) $\pi(x) = \mathrm{sgn}_\varepsilon x = \begin{cases} 1, & \text{if } x \in \mathbb{Q}^*_{p,\varepsilon}, \\ -1, & \text{if } x \notin \mathbb{Q}^*_{p,\varepsilon}, \quad p \neq 2, \end{cases}$

where $\varepsilon \notin \mathbb{Q}_p^{*2}$ (see Sec. 1.4) and $\mathbb{Q}^*_{p,\varepsilon}$ is defined in Sec. 1.5:

$$\mathbb{Q}^*_{p,\varepsilon} = [x \in \mathbb{Q}^*_p : x = a^2 - \varepsilon b^2, \quad a, b \in \mathbb{Q}_p] . \tag{2.5}$$

To be convinced that the function $\mathrm{sgn}_\varepsilon x$ defines a multiplicative character of the field $\mathbb{Q}_p$ it is sufficient to prove that

$$\mathrm{rank}\,(\mathbb{Q}^*_p/\mathbb{Q}^*_{p,\varepsilon}) = 2, \qquad \mathrm{rank}\,(\mathbb{Q}^*_{p,\varepsilon}/\mathbb{Q}^{*2}_p) = 2 . \tag{2.6}$$

■ Let $p \neq 2$. In Sec. 1.4 it was shown that the rank of the group $\mathbb{Q}^*_p/\mathbb{Q}^{*2}_p$ is equal to 4. Therefore owing to inclusions $\mathbb{Q}^{*2}_p \subset \mathbb{Q}^*_{p,\varepsilon} \subset \mathbb{Q}^*_p$ it remains to prove that

$$\mathbb{Q}^{*2}_p \neq \mathbb{Q}^*_{p,\varepsilon} \neq \mathbb{Q}^*_p . \tag{2.7}$$

It is necessary to consider three cases: $\varepsilon = \eta, \eta p, p$, where $\eta \notin \mathbb{Q}_p^{*2}$ (see Sec. 1.4).

We shall prove

$$p\eta \notin \mathbb{Q}_{p,\eta}^*, \qquad \eta \notin \mathbb{Q}_{p,p\eta}^*, \qquad \eta \notin \mathbb{Q}_{p,p}^* . \tag{2.8}$$

which means that $\mathbb{Q}_{p,\varepsilon}^* \neq \mathbb{Q}_p^*$.

Let conversely $p\eta \in \mathbb{Q}_{p,\eta}^*$ i.e. $p\eta = a^2 - \eta b^2$ for some $a, b \in \mathbb{Q}_p$, $b \neq 0$. But the last equality is impossible for none of a and $b \neq 0$. Indeed rewriting it in the form

$$p\eta\left(1 + \frac{b^2}{p}\right) = a^2$$

we see that for 1) $|\frac{b^2}{p}|_p < 1$ the number $1 + \frac{b^2}{p}$ is a square of a p-adic number (see Sec. 1.4) and then $p\eta$ would be a square, 2) $|\frac{b^2}{p}|_p > 1$ the number $1 + \frac{p}{b^2}$ is a square of a p-adic number and then η would be a square, 3) $|\frac{b^2}{p}|_p = 1$ is impossible. Other statements (2.8) are proved similarly.

Now we prove that $\mathbb{Q}_{p,\varepsilon}^* \neq \mathbb{Q}_p^{*^2}$. Let $\varepsilon = \eta$. Then $\eta \in \mathbb{Q}_{p,\eta}^*$ by the Lemma of Sec. 8.2. Let $\varepsilon = p, p\eta$. Then $-\varepsilon \in \mathbb{Q}_{p,\varepsilon}^*$ as it is represented in the form (2.5) with $a = 0$ and $b = 1$. ■

Remark. For $p = 2$ (2.6) takes the form

$$\text{rank}\,(\mathbb{Q}_2^*/\mathbb{Q}_{2,\varepsilon}^*) = 2, \qquad \text{rank}\,(\mathbb{Q}_{2,\varepsilon}^*/\mathbb{Q}_2^{*2}) = 4 ,$$

see [37] Sec. 6.

3. *Multiplicative Characters of the Field* $\mathbb{Q}_p(\sqrt{\varepsilon})$

According to Sec. 1.5 every element z of the field $\mathbb{Q}_p(\sqrt{\varepsilon})$ is represented in the form either $z = r\sigma$ or $z = \nu r\sigma$ where $r \in \mathbb{Q}_p$, $\nu\bar{\nu} \notin \mathbb{Q}_p^{*^2}$, $\nu \in \mathbb{Q}_p^*(\sqrt{\varepsilon})$ and $\sigma\bar{\sigma} = 1$ (i.e. $\sigma \in C_\varepsilon$).

Let $\pi(z)$ be a multiplicative character of the field $\mathbb{Q}_p(\sqrt{\varepsilon})$. Let us denote by π_1 and π_2 its restrictions on the field $\mathbb{Q}_p$ and on the "circle" C_ε respectively. Then we shall have

$$\pi(r\sigma) = \pi_1(r)\pi_2(\sigma), \qquad \pi(\nu r\sigma) = \pi(\nu)\pi(r\sigma) . \tag{3.1}$$

From the equality $\rho\sigma = (-\rho)(-\sigma)$ it follows the condition

$$\pi_1(-1)\pi_2(-1) = 1 . \tag{3.2}$$

Further as

$$\nu^2 = \nu\bar{\nu}\frac{\nu}{\bar{\nu}}, \qquad \nu\bar{\nu} \in \mathbb{Q}_p^*, \qquad \frac{\nu}{\bar{\nu}} \in C_\varepsilon$$

we have the equality

$$\pi^2(\nu) = \pi_1(\nu\bar{\nu})\pi_2(\nu/\bar{\nu}) \,. \tag{3.3}$$

Conversely, *let* π_1 *and* π_2 *be arbitrary multiplicative characters of the field* $\mathbb{Q}_p$ *and the "circle"* C_ε *respectively satisfied the condition* (3.2). *Then the function* $\pi(z)$ *defined on the multiplicative group* $\mathbb{Q}_p^*(\sqrt{\varepsilon})$ *by the formulas* (3.1) *and* (3.3) *for some* $\nu \in \mathbb{Q}_p^*(\sqrt{\varepsilon})$, $\nu\bar{\nu} \notin \mathbb{Q}_p^{*2}$ *is a multiplicative character of the field* $\mathbb{Q}_p(\sqrt{\varepsilon})$.

Remark. The set of multiplicative characters π_2 of the compact group C_ε (see Sec. 1.5) is discrete.

IV. Integration Theory

In this section the theory of integration of complex valued functions of p-adic arguments will be presented and some examples of calculations of specific integrals will be given.

1. *Invariant Measure on the Field* $\mathbb{Q}_p$

As the field $\mathbb{Q}_p$ is a locally-compact commutative group with respect to addition then in $\mathbb{Q}_p$ there exists the Haar measure, a positive measure dx which is invariant with respect to shifts, $d(x+a) = dx$ (see [163]). We normalize the measure dx such that

$$\int_{|x|_p \le 1} dx = 1 \,. \tag{1.1}$$

Under such agreement the measure dx is unique.

For any compact $K \subset \mathbb{Q}_p$ the measure dx defines a positive linear continuous functional on $C(K)$ by the formula $\int_K f(x)dx$, $f \in C(K)$. Here $C(K)$ is the space of continuous functions on K with the norm

$$\|f\|_{C(K)} = \max_{x\in K} |f(x)|$$

A function $f \in L^1_{loc}$ is called *integrable on* $\mathbb{Q}_p$ (improper integral) if there exists

$$\lim_{N\to\infty} \int_{B_N} f(x)dx = \lim_{N\to\infty} \sum_{-\infty<\gamma\le N} \int_{S_\gamma} f(x)dx . \tag{1.2}$$

This limit is called an *integral* (improper) of the function f on $\mathbb{Q}_p$, and it is denoted by $\int\limits_{\mathbb{Q}_p} f(x)dx$ so that

$$\int_{\mathbb{Q}_p} f(x)dx = \sum_{-\infty<\gamma<\infty} \int_{S_\gamma} f(x)dx . \tag{1.3}$$

Analogously one defines the improper integral with respect to a point $a \in \mathbb{Q}_p$: if $f \in L^1_{loc}(\mathbb{Q}_p\backslash\{a\})$ then

$$\int_{\mathbb{Q}_p} f(x)dx = \lim_{\substack{N\to+\infty \\ M\to-\infty}} \sum_{M\le\gamma\le N} \int_{S_\gamma(a)} f(x)dx . \tag{1.4}$$

2. *Change of Variables in Integrals*

At first we shall prove the formula

$$d(xa) = |a|_p dx, \qquad a \in \mathbb{Q}_p^* . \tag{2.1}$$

■ For any $a \in \mathbb{Q}_p^*$ the measure $d(xa)$ is invariant with respect to shifts, and therefore it differs from the measure dx by a factor $C(a) > 0$, $d(xa) = C(a)dx$. From here it follows that $C(a)$ is a continuous function satisfying the condition $C(ab) = C(a)C(b)$ i.e. C is a multiplicative character of the group $\mathbb{Q}_p^*$, and hence it has a form (2.1) of Sec. (3.2).

$$C(a) = |a|_p^{\alpha-1}\pi_0(a'), \quad |\pi_0(a')| = 1, \qquad a' \in S_0, \quad \alpha \in C .$$

But $C(a) > 0$ and therefore $\pi_0(a') = 1$. Hence $C(a) = |a|_p^{\alpha-1}$, $\alpha \in C$. We shall find a number α. As B_0 is a union of disjoint sets $pB_0 + k = B_{-1}(k)$, $k = 0, 1, \ldots, p-1$ whose measures are equal then measure $B_0 = p$ measure B_{-1} and hence $d(xp) = \frac{1}{p}dx$ i.e. $C(p) = \frac{1}{p} = |p|_p$. Thus $\alpha = 2$, $C(a) = |a|_p$ and (2.1) is valid. ■

By using the formula (2.1) to an integral we get

$$\int_K f(x)dx = |a|_p \int_{\frac{k-b}{a}} f(ay+b)dy, \quad a \neq 0 . \tag{2.2}$$

Let us show how to use this formula to do simple integrals.

Example 1.

$$\int_{B_\gamma} dx = p^\gamma, \qquad \gamma \in \mathbb{Z} . \tag{2.3}$$

It follows from the formulas (2.1) and (2.2) that

$$\int_{|x|_p \leq p^\gamma} dx = \int_{|y|_p \leq 1} d(p^{-\gamma} y) = |p^{-\gamma}|_p \int_{B_0} dy = p^\gamma .$$

Example 2.

$$\int_{S_\gamma} dx = p^\gamma (1 - \frac{1}{p}), \qquad \gamma \in \mathbb{Z} . \tag{2.4}$$

It follows from the formula (2.3)

$$\int_{S_\gamma} dx = \int_{B_\gamma} dx - \int_{B_{\gamma-1}} dx = p^\gamma - p^{\gamma-1} .$$

Thus the "area" of a circle in $\mathbb{Q}_p$ is positive!

Example 3. From (1.3) and (2.4) it follows that

$$\int_{\mathbb{Q}_p} f(|x|_p)dx = \left(1 - \frac{1}{p}\right) \sum_{-\infty<\gamma<\infty} f(p^\gamma)p^\gamma .$$

Example 4. In particular,

$$\int_B |x|_p^{\alpha-1} dx = \frac{1-p^{-1}}{1-p^{-\alpha}}, \quad \text{Re}\, \alpha > 0 . \tag{2.5}$$

Example 5.

$$\int_{B_0} \ln |x|_p dx = -\frac{\ln p}{p-1} \,. \tag{2.6}$$

■ $$\int_{|x|_p \le 1} \ln |x|_p dx = -\left(1 - \frac{1}{p}\right) \ln p \sum_{0 \le \gamma < \infty} \gamma p^{-\gamma} = -\frac{\ln p}{p-1} \,.$$

Here we have used the equality

$$\sum_{0 \le \gamma < \infty}^{\infty} \gamma p^{-\gamma} = \frac{p}{(p-1)^2} \,. \quad ■ \tag{2.7}$$

Remark. An invariant measure dx^* of the multiplicative group $\mathbb{Q}_p^*$ of the field $\mathbb{Q}_p$ has the form

$$d^* x = |x|_p^{-1} dx \,.$$

■ It follows from (2.1)

$$d^*(ax) = |ax|_p^{-1} d(ax) = |x|_p^{-1} dx = d^* x, \qquad a \in \mathbb{Q}_p^* \,. \quad ■$$

General change of variables in integrals. *Let $\sigma(y)$ be an analytic function which maps homeomorphically a clopen set $K_1 \subset \mathbb{Q}_p$ onto a clopen set $K \subset \mathbb{Q}_p$ and also $\sigma'(y) \ne 0$, $y \in K_1$. Then for any $f \in C(K)$ the formula (2.8) is valid,*

$$\int_K f(x) dx = \int_{K_1} |\sigma'(y)|_p f(\sigma(y)) dy \,. \tag{2.8}$$

■ As an integral is a linear continuous functional on $C(K)$ (see Sec. 4.1) then it is sufficient to prove the formula (2.8) for locally-constant functions on K, the set of which is dense in $C(K)$ (see Sec. 6.2). Hence it is sufficient to prove the equality (2.8) for $f(x) \equiv 1$, $x \in K$ i.e.

$$\int_K dx = \int_{K_1} |\sigma'(y)|_p dy \,. \tag{2.9}$$

We cover compact K_1 by a finite number of disjoint discs $B_\rho(y_k)$ (see Sec. 1.3) of a sufficiently small radius p^ρ such that

$$|\sigma'(y)|_p = |\sigma'(y_k)|_p = p^{r_k}, \qquad y \in B_\rho(y_k)$$

and the disc $B_\rho(y_k)$ is mapped onto the disc $B_{\rho+r_k}(x_k)$, $x_k = \sigma(y_k)$ according to the Theorem on inverse function of Sec. 2.5. From here using the formula (2.3) we get the equality (2.9)

$$\int_K dx = \sum_k \int_{B_{\rho+r_k}(x_k)} dx = \sum_k p^{\rho+r_k} = \sum_k \int_{B_\rho(y_k)} p^{r_k} dy$$

$$= \sum_k \int_{B_\rho(y_k)} |\sigma'(y_k)| dy = \sum_k \int_{B_\rho(y_k)} |\sigma'(y)|_p dy = \int_{K_1} |\sigma(y)| dy . \quad \blacksquare$$

Examples. If a function f is integrable on $\mathbb{Q}_p$ then (see (1.4))

$$\int_{\mathbb{Q}_p} f(x)dx = \int_{\mathbb{Q}_p} \frac{1}{|y|_p^2} f\left(\frac{1}{y}\right) dy \ldots \tag{2.10}$$

The following formulas are valid (see Sec. 2.4)

$$\int_{G_p} f(x)dx = \int_{G_p} f(\sin y)dy = \int_{G_p} f(\tan y)dy . \tag{2.11}$$

$$\int_{G_p} f(x)dx = \int_{J_p} f(\ln y)dy, \qquad \int_{J_p} f(x)dx = \int_{G_p} f(e^y)dy . \tag{2.12}$$

3. *Some Examples of Calculation of Integrals*

In addition to Sec. 4.2 we shall list some examples of calculation of integrals in explicit form (see [206]).

Example 6.

$$\int_{B_\gamma} \chi_p(\xi x)dx = \begin{cases} p^\gamma, & |\xi|_p \le p^{-\gamma}, \\ 0, & |\xi|_p \ge p^{-\gamma+1}, \quad \gamma \in \mathbb{Z} \end{cases} \tag{3.1}$$

■ For $|\xi|_p \le p^{-\gamma}$ we have $|\xi x|_p \le 1$ and therefore $\chi_p(\xi x) = 1$. Hence owing to (2.3)

$$\int_{B_\gamma} \chi_p(\xi x)dx = \int_{B_\gamma} dx = p^\gamma .$$

If $|\xi|_p \ge p^{-\gamma+1}$ then for some $x' \in S_\gamma$ we have $|\xi x'|_p = |\xi|_p |x'|_p \ge p$ and therefore $\chi_p(\xi x') \not\equiv 1$. Then performing the change of variable $x = y - x'$ we get that the desired integral is equal to zero:

$$\int_{B_\gamma} \chi_p(\xi x)dx = \int_{B_\gamma(x')} \chi_p(\xi(y - x'))dy = \chi_p(-\xi x') \int_{B_\gamma} \chi_p(\xi y)dy .$$

Here we have used the fact that $B_\gamma(x') = B_\gamma$ if $|x'|_p \le p^\gamma$ (see Sec. 1.3). ■

Example 7.

$$\int_{S_\gamma} \chi_p(\xi x)dx = \begin{cases} p^\gamma(1 - \frac{1}{p}), & |\xi|_p \le p^\gamma, \\ -p^{\gamma-1}, & |\xi|_p = p^{-\gamma+1}, \\ 0, & |\xi|_p \ge p^{-\gamma+2}, \qquad \gamma \in \mathbb{Z} . \end{cases} \tag{3.2}$$

■ It follows from the formulas (3.1) as

$$\int_{S_\gamma} \chi_p(\xi x)dx = \int_{B_\gamma} \chi_p(\xi x)dx - \int_{B_{\gamma-1}} \chi_p(\xi x)dx .$$

Example 8.* If $\sum_{0 \le \gamma < \infty} |f(p^{-\gamma})| p^{-\gamma} < \infty$ then for $\xi \ne 0$

$$\int_{\mathbb{Q}_p} f(|x|_p)\chi_p(\xi x)dx$$

$$= \left(1 - \frac{1}{p}\right) |\xi|_p^{-1} \sum_{0 \le \gamma < \infty} p^{-\gamma} f(p^{-\gamma}|\xi|_p^{-1}) - |\xi|_p^{-1} f(p|\xi|_p^{-1}) . \tag{3.3}$$

* About improper integrals see Sec. 4.1

■ It follows from the formulas (1.3) and(3.2). Denoting $|\xi|_p = p^N$ we have

$$\int_{\mathbb{Q}_p} f(|x|_p)\chi_p(\xi x)dx = \sum_{-\infty<\gamma<\infty} f(p^\gamma)\int_{S_\gamma}\chi_p(\xi x)dx$$
$$= (1-\frac{1}{p})\sum_{-\infty<\gamma\leq -N} f(p^\gamma)p^\gamma - f(p^{-N+1})p^{-N}$$
$$= (1-\frac{1}{p})p^{-N}\sum_{0\leq\gamma<\infty} f(p^{-\gamma-N})p^{-\gamma} - p^{-N}f(p^{-N+1}).$$

■

Example 9.

$$\int_{\mathbb{Q}_p}\chi_p(\xi x)dx = 0, \qquad \xi\neq 0\,. \tag{3.4}$$

It follows from (3.3) for $f\equiv 1$.

Example 10.

$$\int_{\mathbb{Q}_p}|x|_p^{\alpha-1}\chi_p(\xi x)dx = \frac{1-p^{\alpha-1}}{1-p^{-\alpha}}|\xi|_p^{-\alpha}, \qquad \Re\alpha>0\,. \tag{3.5}$$

It follows from (3.3) for $f = |x|_p^{\alpha-1}$:

$$\int_{\mathbb{Q}_p}|x|_p^{\alpha-1}\chi_p(\xi x)dx = (1-\frac{1}{p})|\xi|_p^{-1}\sum_{0\leq\gamma<\infty} p^{-\gamma-(\alpha-1)\gamma}|\xi|_p^{-\alpha+1}$$
$$-|\xi|_p^{-1}p^{\alpha-1}|\xi|_p^{-\alpha+1} = \left(\frac{1-p^{-1}}{1-p^{-\alpha}} - p^{\alpha_1}\right)|\xi|_p^{-\alpha}\,.$$

Example 11.

$$\int_{\mathbb{Q}_p}\ln|x|_p\chi_p(\xi x)dx = -\frac{p\ln p}{p-1}|\xi|_p^{-1}, \qquad \xi\neq 0\,. \tag{3.6}$$

It follows from (3.3) for $f = \ln |x|_p$:

$$\int\limits_{\mathbb{Q}_p} \ln |x|_p \chi_p(\xi x) dx$$
$$= (1 - \frac{1}{p}) |\xi|_p^{-1} \sum_{0 \leq \gamma < \infty} p^{-\gamma} (-\gamma \ln p - \ln |\xi|_p) + |\xi|_p^{-1} (\ln |\xi|_p - \ln p)$$
$$= -|\xi|_p^{-1} \ln p \left[1 + \left(1 - \frac{1}{p}\right) \sum_{0 \leq \gamma < \infty} \gamma p^{-\gamma} \right] = -|\xi|_p^{-1} \left(1 + \frac{1}{p-1}\right) \ln p \, .$$

Here we have used the equality (2.7).

Example 12.

$$\int\limits_{\mathbb{Q}_p} \frac{\chi_p(\xi x)}{|x|_p^2 + m^2} dx = \left(1 - \frac{1}{p}\right) \frac{|\xi|_p}{p^2 + m^2 |\xi|_p^2} \sum_{0 \leq \gamma < \infty} p^{-\gamma} \frac{p^2 - p^{-2\gamma}}{p^{-2\gamma} + m^2 |\xi|_p^2} . \quad (3.7)$$

It follows from (3.3) for $f = (|x|_p^2 + m^2)^{-1}$:

$$\int\limits_{\mathbb{Q}_p} \frac{\chi_p(\xi x)}{|x|_p^2 + m^2} dx$$
$$= (1 - \frac{1}{p}) |\xi|_p^{-1} \sum_{0 \leq \gamma < \infty} \frac{p^{-\gamma}}{p^{-2\gamma} |\xi|_p^{-2} + m^2} - \frac{|\xi|_p^{-1}}{p^2 |\xi|_p^{-2} + m^2}$$
$$= \left(1 - \frac{1}{p}\right) |\xi|_p \sum_{0 \leq \gamma < \infty} p^{-\gamma} \left(\frac{1}{p^{-2\gamma} + m^2 |\xi|_p^2} - \frac{1}{p^2 + m^2 |\xi|_p^2} \right) .$$

From the formula (3.7) it follows an asymptotics

$$\int\limits_{\mathbb{Q}_p} \frac{\chi_p(\xi x)}{|x|_p^2 + m^2} dx \sim \frac{p^4 + p^3}{p^2 + p + 1} \frac{1}{m^4 |\xi|_p^3}, \qquad |\xi|_p \to \infty \, . \quad (3.8)$$

■ $$\lim_{|\xi|_p\to\infty} |\xi|_p^3 \int_{\mathbb{Q}_p} \frac{\chi_p(\xi x)}{|x|_p^2+m^2}dx$$

$$= \left(1-\frac{1}{p}\right)\frac{1}{m^4}\lim_{|\xi|_p\to\infty}\sum_{0\leq\gamma<\infty} p^{-\gamma}(p^2-p^{-2\gamma})\left(1-\frac{p^{-2\gamma}}{p^{-2\gamma}+m^2|\xi|_p^2}\right)$$

$$= \left(1-\frac{1}{p}\right)\frac{1}{m^4}\sum_{0\leq\gamma<\infty}(p^{2-\gamma}-p^{-3\gamma})$$

$$= \left(1-\frac{1}{p}\right)\frac{1}{m^4}\left(\frac{p^2}{1-p^{-1}}-\frac{1}{1-p^{-3}}\right)$$

■

Note that in the real case the similar integral is equal to

$$\int_{-\infty}^{\infty}\frac{\chi_\infty(\xi x)}{x^2+m^2}dx = \int_{-\infty}^{\infty}\frac{\exp(-2\pi i\xi x)}{x^2+m^2}dx = \frac{\pi}{m}e^{-2\pi m|\xi|} \tag{3.9}$$

and hence exponentially decreases as $|\xi|\to\infty$.

Example 13.

$$\int_{S_{\gamma},x_0=k} dx = p^{\gamma-1}, \qquad \gamma\in\mathbb{Z},\quad k=1,2,..,p-1 \tag{3.10}$$

■ By virtue of invariance of the measure dx with respect to shifts from the formula (2.4) we have

$$\int_{S_\gamma} dx = p^\gamma\left(1-\frac{1}{p}\right) = \sum_{1\leq k\leq p-1}\int_{S_\gamma,x_0=k} dx = (p-1)\int_{S_\gamma,x_0=k} dx\,,$$

and the formula (3.10) follows. ■

Example 14.

$$\int_{S_\gamma,x_0\neq k} dx = p^\gamma\left(1-\frac{2}{p}\right), \qquad \gamma\in\mathbb{Z},\quad k=1,2,\ldots,p-1\,. \tag{3.11}$$

It follows from the formulas (2.1) and (3.10):

$$\int\limits_{S_\gamma, x_0 \neq k} dx = \int\limits_{S_\gamma} dx - \int\limits_{S_\gamma, x_0 = k} dx = p^\gamma \left(1 - \frac{1}{p}\right) - p^{\gamma-1} .$$

Example 15. $l = 1, 2, \ldots$

$$\int\limits_{S, x_l = k} dx = p^{\gamma-1}\left(1 - \frac{1}{p}\right), \qquad \gamma \in \mathbb{Z}, \quad k = 0, 1, 2, \ldots, p-1, \tag{3.12}$$

Example 16. $l = 0, 1, 2, \ldots$

$$\int\limits_{S_\gamma, x_0 = k_0, x_1 = k_1, \ldots, x_l = k_l} dx = p^{\gamma-l-1}, \qquad \gamma \in \mathbb{Z}, \quad 0 \le k_j < p-1, \quad k_0 \neq 0 . \tag{3.13}$$

Example 17. $l = 0, 1, 2, \ldots$

$$\int\limits_{\substack{S_\gamma, x_0 = k_0, x_1 = k_1, \ldots, \\ x_l = k_l, x_{l+1} \neq k_{l+1}}} dx = p^{\gamma-l-1}\left(1 - \frac{1}{p}\right), \qquad \gamma \in \mathbb{Z}, 0 \le k_j \le p-1 , \quad k_0 \neq 0 . \tag{3.14}$$

Example 18.

$$\int\limits_{S_0} |1 - x|_p^{\alpha-1} dx = \frac{p - 2 + p^{-\alpha}}{p(1 - p^{-\alpha})}, \qquad \Re\alpha > 0 . \tag{3.15}$$

■ By using the formulas (3.11) and (3.14) we get (3.15):

$$\int_{S_0} |1-x|_p^{\alpha-1} dx = \int_{S_0, x_0=1} |1-x|_p^{\alpha-1} dx$$
$$+ \int_{S_0, x_0=1, x_1\neq 0} |1-x|_p^{\alpha-1} dx + \int_{S_0, x_0=1, x_1=0, x_2\neq 0} |1-x|_p^{\alpha-1} dx + \dots$$
$$= \int_{S_0, x_0\neq 1} dx + p^{1-\alpha} \int_{S_0, x_0=1, x_1\neq 0} dx + p^{2(1-\alpha)} \int_{S_0, x_0=1, x_1=0, x_2\neq 0} dx + \dots$$
$$= 1 - \frac{2}{p} + \frac{1}{p^\alpha}\left(1-\frac{1}{p}\right) + \frac{1}{p^{2\alpha}}\left(1-\frac{1}{p}\right) + \dots$$
$$= \left(1-\frac{1}{p}\right)\frac{1}{1-p^{-\alpha}} - \frac{1}{p}\,.$$
■

Example 19.

$$\int_{S_0} \ln |1-x|_p dx = -\frac{\ln p}{p-1} \tag{3.16}$$

■ Acting as in Example 18 we get

$$\int_{S_0} \ln |1-x|_p dx = \ln 1 \int_{S_0, x_0\neq 1} dx + \ln \frac{1}{p} \int_{S_0, x_0=1, x_1\neq 0} dx$$
$$+ \ln \frac{1}{p^2} \int_{S_0, x_0=1, x_1=0, x_2\neq 0} dx + \dots$$
$$= \ln \frac{1}{p}\left(1-\frac{1}{p}\right)\left(\frac{1}{p} + \frac{2}{p^2} + \dots\right) = \ln \frac{1}{p}\frac{1}{p(1-p^{-1})}$$

Here we have used the equality (2.7). ■

4. *Integration in* $\mathbb{Q}_p^n$

The invariant measure dx on $\mathbb{Q}_p$ in a standard way is extended up to an invariant measure $dx = dx_1 dx_2 \dots dx_n$ on $\mathbb{Q}_p^n$ (see Sec. 1.7), and all which was said in 4.1 for $\mathbb{Q}_p$ caries over to $\mathbb{Q}_p^n$. A reduction of multidimensional integration to a one dimentional one is given by the following theorem (see [163]).

Fubini's theorem (on a change of order of integration).
Let a function $f(x,y)$, $x \in \mathbb{Q}_p^n$, $y \in \mathbb{Q}_p^m$ be such that an iterated integral

$$\int_{\mathbb{Q}_p^n} \left[\int_{\mathbb{Q}_p^m} |f(x,y)| dy \right] dx$$

exists. Then the function f is integrable on $\mathbb{Q}_p^{n+m}$, and there exist all iterated integrals of f, and they are equal

$$\int_{\mathbb{Q}_p^n} \left[\int_{\mathbb{Q}_p^m} f(x,y) dy \right] dx = \int_{\mathbb{Q}_p^{n+m}} f(x,y) dx dy = \int_{\mathbb{Q}_p^m} \left[\int_{\mathbb{Q}_p^n} f(x,y) dx \right] dy \,. \quad (4.1)$$

Conversely, if a function f is (absolutely) integrable on $\mathbb{Q}_p^{n+m}$ then all integrals in (4.1) exist and the equality (4.1) is valid.

Change of variables in integral. *Let $x = x(y)$ (i.e. $x_i = x_i(y_1, y_2, \ldots, y_n), i = 1, 2, \ldots, n$) be a homeomorphic mapping of an open compact $K_1 \subset \mathbb{Q}_p^n$ onto (open) compact $K \subset \mathbb{Q}_p^n$, and also functions $x_i(y)$ are analytic in K_1 and*

$$\det \frac{\partial x(y)}{\partial y} = \det \left\| \frac{\partial x_i}{\partial y_j} \right\| \neq 0, \qquad y \in K_1 \,.$$

Then for any $f \in L^1(K)$ the equality (4.2) is valid,

$$\int_K f(x) dx = \int_{K_1} \left| \det \frac{\partial x(y)}{\partial y} \right|_p f(x(y)) dy \,. \quad (4.2)$$

■ A proof of the formula (4.2) carries out by induction on n like in a real case using the one dimensional formula (2.8). ■

We note also an analogy of **the Lebesgue Theorem on limiting passage under the sign of an integral** (see [163]): *If a sequence $\{f_k, k \to \infty\}$ of functions $f_k \in L^1(\mathbb{Q}_p^n)$ converges almost everywhere in $\mathbb{Q}_p^n$ (with respect to the measure dx) to a function f,*

$$f_k(x) \to f(x), \qquad k \to \infty, \quad x \in \mathbb{Q}_p^n \quad a.e. \,,$$

and there exists a function $\psi \in L^1(\mathbb{Q}_p^n)$ *such that*

$$|f_k(x)| \leq \psi(x), \qquad k = 1, 2, \ldots, \quad x \in L^1(\mathbb{Q}_p^n) \quad a.e.$$

then the equality is true

$$\lim_{k\to\infty} \int_{\mathbb{Q}_p^n} f_k(x)dx = \int_{\mathbb{Q}_p^n} f(x)dx \,. \tag{4.3}$$

Example 20. The function

$$(|x_1|_p^2 + |x_2|_p^2 + \ldots + |x_n|_p^2)^{-\alpha}$$

is locally integrable in $\mathbb{Q}_p^n$ if $2\alpha < n$.

■ By using the formula (2.4) we have

$$I = \int_{B_0}\int_{B_0} \ldots \int_{B_0} (|x_1|_p^2 + |x_2|_p^2 + \ldots + |x_n|_p^2)^{-\alpha} dx_1 dx_2 \ldots dx_n$$

$$= \left(1 - \frac{1}{p}\right)^n \sum_{-\infty < \gamma_1, \ldots, \gamma_n \leq 0} p^{\gamma_1 + \ldots + \gamma_n} (p^{2\gamma_1} + p^{2\gamma_2} + \ldots + p^{2\gamma_n})^{-\alpha}$$

$$= \left(1 - \frac{1}{p}\right)^n \sum_{0 \leq N < \infty} p^{-N}$$

$$\sum_{0 \leq \nu_1 + \ldots + \nu_n \leq N} (p^{-2\nu_1} + \ldots + p^{-2\nu_{n-1}} + p^{-2N + 2\nu_1 + \ldots + 2\nu_{n-1}})^{-\alpha} \,.$$

Note the equality

$$\min_{0 \leq \alpha_j \leq 1} \left(\alpha_1 + \ldots + \alpha_{n-1} + \frac{p^{-2N}}{\alpha_1 \ldots \alpha_{n-1}}\right)$$

$$= \min_{0 \leq \alpha \leq 1} \left[(n-1)\alpha + \frac{p^{-2N}}{\alpha^{n-1}}\right] = np^{-\frac{2N}{n}} \,.$$

Continuing our estimates we get for $0 < \alpha < 2n$

$$I \leq \left(1 - \frac{1}{p}\right)^n \sum_{0 \leq N < \infty} p^{-N} \sum_{0 \leq \nu_1 + \ldots + \nu_{n-1} \leq N} n^{-\alpha} (p^{-\frac{2N}{n}})^{-\alpha}$$

$$\leq n^{-\alpha} \sum_{0 \leq N < \infty} p^{-N(1 - \frac{2\alpha}{n})} N^{n-1} < \infty \,.$$

■

Example 21. $p \equiv 3 \pmod 4$.

$$\int_{\mathbb{Q}_p^2} f((x,x))\chi_p((\xi,x))dx_1dx_2$$

$$= \frac{1}{|(\xi,\xi)^2|_p}\left[\left(1-\frac{1}{p^2}\right)\cdot\sum_{0\le\gamma<\infty} p^{-2\gamma} f\left(\frac{p^{-2\gamma}}{|(\xi,\xi)^2|_p}\right) - f\left(\frac{p^2}{|(\xi,\xi)^2|_p}\right)\right], \tag{4.4}$$

if

$$\sum_{0\le\gamma<\infty} p^{-2\gamma}|f(p^{-2\gamma})| < \infty .$$

■ Let us denote the desired integral by $J(\xi)$. Owing to its symmetrey on $\xi = (\xi_1, \xi_2)$ it is sufficient to prove the formula (4.4) for $|\xi_1|_p \le |\xi_2|_p$. Denote $|\xi_1|_p = p^{-N}$, $|\xi_2|_p = p^{-M}$, $N \le M$. Using the equality

$$|(x,x)|_p = |x_1^2 + x_2^2|_p = \max(|x_1|_p^2, |x_2|_p^2)$$

(see Sec. 1.1) and the formulas (3.1)–(3.4) we get the equalities

$$J(\xi) = \int_{\mathbb{Q}}\left[\int_{|x_2|_p\le|x_1|_p} f(|x_1|_p^2)\chi_p(\xi_2x_2)dx_2 + \int_{|x_2|_p>|x_1|_p} f(|x_2|_p^2)\chi_p(\xi_2x_2)dx_2\right]\chi_p(\xi_1x_1)dx_1$$

$$= \int_{\mathbb{Q}} f(|x_1|_p^2)|x_1|_p\chi_p(\xi_1x_1)\Omega(|\xi_2x_1|_p)dx_1 + \int_{\mathbb{Q}_p} f(|x_2|_p^2)\chi_p(\xi_2x_2)\cdot\int_{|x_1|_p\le\frac{1}{p}|x_2|_p}\chi_p(\xi_1x_1)dx_1dx_2$$

$$= \int_{|x_1|_p\le|\xi_2|_p^{-1}} f(|x_1|_p^2)|x_1|_pdx_1 + \frac{1}{p}\int_{\mathbb{Q}_p} f(|x_2|_p^2)|x_2|_p\chi_p(\xi_2x_2)\Omega\left(\frac{1}{p}|\xi_1x_2|_p\right)dx_2$$

$$= \int\limits_{|x_1|_p \le |\xi_2|_p^{-1}} f(|x_1|_p^2)|x_1|_p dx_1 + \frac{1}{p} \int\limits_{|x_2|_p \le |\xi_2|_p^{-1}} f(|x_2|_p^2)|x_2|_p dx_2$$

$$+ \int\limits_{p|\xi_2|_p^{-1} \le |x_2|_p \le p|\xi_1|_p^{-1}} f(|x_2|_p^2)|x_2|_p \chi_p(\xi_2 x_2) dx_2$$

$$= \left(1 + \frac{1}{p}\right) \sum_{-\infty < \gamma \le N} p^{2\gamma} f(p^{2\gamma})(1 - \frac{1}{p})$$

$$+ \frac{1}{p} \sum_{N+1 \le \gamma \le M+1} p^{\gamma} f(p^{2\gamma}) \int\limits_{S_\gamma} \chi_p(\xi_2 x_2) dx_2$$

$$= \left(1 - \frac{1}{P^2}\right) \sum_{0 \le \gamma < \infty} p^{-2\gamma+2N} f(p^{-2\gamma+2N}) - \frac{1}{p} p^{2N+1} f(p^{2N+2})$$

$$= |\xi_2|_p^{-2} \left[\left(1 - \frac{1}{P^2}\right) \sum_{0 \le \gamma < \infty} p^{-2\gamma} f(p^{-2\gamma} |\xi_2|_p^{-2}) - f(p^2 |\xi_2|_p^{-2}) \right],$$

from where the formula (4.4) follows if we note that $|\xi_2|_p^2 = |\xi_1^2 + \xi_2^2|_p = |(\xi,\xi)|_p$ when $|\xi_1|_p \le |\xi_2|_p$. ■

In particular, for the propagator

$$G = (|(x,x)|_p + m^2)^{-1} \tag{4.5}$$

(cf. (3.7)) the formula (4.4) takes the form

$$\int\limits_{\mathbb{Q}_p} \frac{\chi_p((\xi,x))}{|(x,x)|_p + m^2}$$

$$= \left(1 - \frac{1}{p^2}\right) \sum_{0 \le \gamma < \infty} p^{-2\gamma} \left(\frac{1}{p^{-2\gamma} + m^2 |(\xi,\xi)|_p^2} - \frac{1}{p^2 + m^2 |(\xi,\xi)|_p^2} \right). \tag{4.6}$$

From here it follows as in the case of the propagator (3.7) that the right-hand part of the equality (4.6) is positive and have an asymptotics

$$\sim \frac{p^4}{P^2 + 1} \frac{1}{m^4 |(\xi,\xi)|_p^2}, \qquad |(\xi,\xi)|_p \to \infty . \tag{4.7}$$

By another method the integral (4.6) has been calculated in Bikulov [35].

Example 22. $p \equiv 1 \pmod 4$.

$$\int_{\mathbb{Q}} f((x,x))\chi_p((\xi,x))dx$$

$$= \frac{1}{|(\xi,\xi)|_p}\left[\left(1-\frac{1}{p}\right)^2 \sum_{0\le\gamma<\infty}\left(\gamma+\frac{p-3}{p-1}\right)p^{-\gamma}f\left(\frac{p^{-\gamma}}{|(\xi,\xi)|_p}\right)\right.$$

$$\left. -2\left(1-\frac{1}{p}\right)f\left(\frac{p}{|(\xi,\xi)|_p}\right)+f\left(\frac{p^2}{|(\xi,\xi)|_p}\right)\right] \tag{4.8}$$

if

$$\sum_{0\le\gamma<\infty} \gamma p^{-\gamma}|f(p^{-\gamma})| < \infty\,.$$

■ There exists a solution $\tau \in \mathbb{Q}_p$ of the equation $\tau^2 = -1$ (see Sec. 1.4). Change of variables in the integral

$$\begin{aligned} y_1 &= \tfrac{1}{2}(x_1 - \tau x_2), & x_1 &= y_1 + y_2, \\ y_2 &= \tfrac{1}{2}(x_1 + \tau x_2), & x_2 &= \tau(y_1 - y_2), \end{aligned} \qquad \det\left(\frac{\partial x}{\partial y}\right) = -2\tau$$

gives

$$|(x,x)|_p = |x_1^2 + x_2^2|_p = |2(1-\tau)y_1y_2| = |y_1|_p|y_2|_p\,,$$

$$(\xi,x) = \xi_1x_1 + \xi_2x_2 = \xi_1(y_1+y_2) + \xi_2\tau(y_1-y_2) = \zeta_1y_1 + \zeta_2y_2 = (\zeta,y)$$

where it is denoted

$$\zeta_1 = \xi_1 + \tau\xi_2, \qquad \zeta_2 = \xi_1 - \tau\xi_2 \qquad (\zeta_1\zeta_2 = \xi_1^2 + \xi_2^2 = (\xi,\xi))\,.$$

Therefore the desired integral $J(\xi)$ takes the form (see (4.2))

$$J(\xi) = \int_{\mathbb{Q}_p} f(|y_1|_p|y_2|_p)\chi_p((\zeta,y))dy$$

$$= \int_{\mathbb{Q}_p}\left[\int_{\mathbb{Q}_p} f(|y_1|_p|y_2|_p)\chi_p(\zeta_2y_2)dy_2\right]\chi_p(\zeta_1y_1)dy_1\,. \tag{4.9}$$

By using the formula (3.3) twice to the iterated integral in (4.9) we get the formula (4.8):

$$J(\xi) = \left(1-\frac{1}{p}\right)\sum_{\gamma'=0}^{\infty} p^{-\gamma'}\left[\left(1-\frac{1}{p}\right)\sum_{\gamma=0}^{\infty} p^{-\gamma} f(p^{-\gamma-\gamma'}|\zeta_1|_p^{-1}|\zeta_2|_p^{-1})|\zeta_2|_p^{-1}\right.$$

$$\left. - f(p^{-\gamma'+1}|\zeta_1|_p^{-1}|\zeta_2|_p^{-1})|\zeta_2|_p^{-1}\right]|\zeta_1|_p^{-1} - \left(1-\frac{1}{p}\right)\sum_{\gamma=0}^{\infty} p^{-\gamma}$$

$$\cdot f(p^{-\gamma+1}|\zeta_1|_p^{-1}|\zeta_2|_p^{-1})|\zeta_1|_p^{-1}|\zeta_2|_p^{-1} + f(p^2|\zeta_1|_p^{-1}|\zeta_2|_p^{-1})|\zeta_1|_p^{-1}|\zeta_2|_p^{-1}$$

$$= |\zeta_1\zeta_2|_p^{-1}\left[\left(1-\frac{1}{p}\right)^2\sum_{\gamma=0}^{\infty}(\gamma+1)p^{-\gamma} f(p^{-\gamma}|\zeta_1\zeta_2|_p^{-1})\right.$$

$$\left. - 2\left(1-\frac{1}{p}\right)\sum_{\gamma=0}^{\infty} p^{-\gamma} f(p^{1-\gamma}|\zeta_1\zeta_2|_p^{-1}) + f(p^2|\zeta_1\zeta_2|_p^{-1})\right]$$

$$= |\zeta_1\zeta_2|_p^{-1}\left[\left(1-\frac{1}{p^2}\right)\sum_{\gamma=0}^{\infty}\left(\gamma+\frac{p-3}{p-1}\right)p^{-\gamma} f(p^{-\gamma}|\zeta_1\zeta_2|_p^{-1})\right.$$

$$\left. - 2\left(1-\frac{1}{p}\right) f(p|\zeta_1\zeta_2|_p^{-1}) + f(p^2|\zeta_1\zeta_2|_p^{-1})\right] \qquad \blacksquare$$

V. The Gaussian Integrals

Integrals of the form $\int \chi_p(ax^2+bx)dx$ are called Gaussian ones. Here we shall consider the cases when sets of integration are the circumference S_γ, the circle B_γ and the whole space $\mathbb{Q}_p$.

In particular, the important formula will be obtained (see (3.1) of Sec. 5.3)

$$\int_{\mathbb{Q}_p} \chi_p(ax^2+bx)dx = \lambda_p(a)|a|_p^{-1/2}\chi_p\left(-\frac{b^2}{4a}\right), \qquad a \neq 0\,.$$

Here we used an important number-theoretical function $\lambda_p : \mathbb{Q}_p^* \to \mathbb{C}$ by the following way:

$$\lambda_p(a) = \begin{cases} 1, & \text{if } \gamma(a) \text{ is even}, \\ \left(\frac{a_0}{p}\right), & \text{if } \gamma(a) \text{ is odd and } p \equiv 1 \pmod 4, \\ i\left(\frac{a_0}{p}\right), & \text{if } \gamma(a) \text{ is odd and } p \equiv 3 \pmod 4; \end{cases}$$

$$\lambda_2(a) = \begin{cases} \frac{1}{\sqrt{2}}[1+(-1)^{a_1}i] & \text{if } \gamma(a) \text{ is even}, \\ \frac{1+i}{\sqrt{2}} i^{a_1}(-1)^{a_2} & \text{if } \gamma(a) \text{ is odd}. \end{cases}$$

We note the simplest properties of the function $\lambda_p(a)$:

$$|\lambda_p(a)|_p = 1, \qquad \lambda_p(a)\lambda_p(-a) = 1, \qquad a \neq 0\,,$$
$$\lambda_p(ac^2) = \lambda_p(a), \qquad a, c \neq 0\,.$$

The furthest properties of the function $\lambda_p(a)$ will be given in Sec. 5.4.

Calculation of the Gaussian integrals for $p \neq 2$ is based on the Gauss sums (see [37])

$$\sum_{0 \le k \le p-1} \exp\left(2\pi i \frac{mk^2}{p}\right) = \begin{cases} \left(\frac{m}{p}\right)\sqrt{p} & \text{if } \ p \equiv 1 \pmod 4\,, \\ i\left(\frac{m}{p}\right)\sqrt{p} & \text{if } \ p \equiv 3 \pmod 4 \end{cases} \qquad (*)$$

where m is an integer not divisible by p, $\left(\frac{m}{p}\right)$ is the Legendre symbol (see Sec. 1.4).

In terms of the function $\lambda_p(a)$ the Gauss sum $(*)$ takes the form

$$\sum_{0 \le k \le p-1} \exp\left(2\pi i \frac{mk^2}{p}\right) = \lambda_p(m)\sqrt{p}, \qquad p \neq 2\,. \qquad (*)$$

1. *The Gaussian Integrals on the Circles S_γ*

Example 1. $p \neq 2$, $|\varepsilon|_p = 1$, (see [215,218])

$$\int_{S_\gamma} \chi_p(\varepsilon(x-y)^2)dy = \begin{cases} p^\gamma\left(1-\frac{1}{p}\right)\chi_p(\varepsilon x^2), & |x|_p \le p^{-\gamma}, \quad \gamma \le 0, \\ -p^{\gamma-1}\chi_p(\varepsilon x^2), & |x|_p = p^{-\gamma+1}, \quad \gamma \le 0, \\ 0, & |x|_p \ge p^{-\gamma+2}, \quad \gamma \le 0, \\ 1, & |x|_p = p^\gamma, \quad \gamma \ge 1, \\ 0, & |x|_p \neq p^\gamma, \quad \gamma \ge 1. \end{cases} \qquad (1.1)$$

■ For $\gamma \le 0$, $|x|_P \le p^{-\gamma}$, $y \in S_\gamma$ we have $|\varepsilon(y^2 - 2xy)|_p = |\varepsilon|_p|(y^2 - 2xy)|_p \le \max(|y|_p^2, |2xy|_p) \le \max(p^{2\gamma}, p^\gamma p^{-\gamma}) = 1$ and hence

$$\chi_p(\varepsilon(x-y)^2) = \chi_p(\varepsilon(y^2 - 2xy) + \varepsilon x^2)$$
$$= \chi_p(\varepsilon(y^2 - 2xy))\chi_p(\varepsilon x^2) = \chi_p(\varepsilon x^2) .$$

The desired integral owing to (2.4) of Sec. 4.2 is equal to

$$\chi_p(\varepsilon x^2) \int_{S_\gamma} dx = p^\gamma \left(1 - \frac{1}{p}\right) \chi_p(\varepsilon x^2) .$$

For $\gamma \le 0$, $|x|_p = p^{-\gamma+1}$, $y \in S_\gamma$ we have $|\varepsilon y^2|_p = p^{2\gamma} \le 1$ and hence

$$\chi_p(\varepsilon(x-y)^2) = \chi_p(\varepsilon x^2 + \varepsilon y^2 - 2\varepsilon xy) = \chi_p(\varepsilon x^2)\chi_p(-2\varepsilon xy) .$$

The integral owing to (3.2) of Sec. 4.3 is equal to

$$\chi_p(\varepsilon x^2) \int_{S_\gamma} \chi_p(-2\varepsilon xy) dy = -p^{\gamma-1}\chi_p(\varepsilon x^2) .$$

For $\gamma \le 0$, $|x|_p = p^N$, $N \ge -\gamma + 2$, $y \in S_\gamma$ we have

$$\{\varepsilon(x-y)^2\}_p = \Big\{p^{-2N}(\varepsilon_0 + \varepsilon_1 p + \ldots)(x_0 + x_1 p + \ldots + (x_{N-\gamma} - y_0)p^{N-\gamma}$$
$$+ \ldots + (x_{2N-1} - y_{N+\gamma-1})p^{2N-1} + \ldots)^2\Big\}_p$$
$$= L(y_0, y_1, \ldots, y_{N+\gamma-2}) - \frac{2}{p}\varepsilon_0 x_0^2 y_{N+\gamma-1}$$

where L does not depend on $y_{N+\gamma-1}$. Then taking into account the formula (3.13) of Sec. 4.3 for $l = N + \gamma - 1$ we have for the integral

$$p^{-N} \sum_{1 \le y_0 \le p-1} \sum_{0 \le y_1 \le p-1} \cdots \sum_{0 \le y_{N+\gamma-2} \le p-1} \exp(2\pi i L)$$
$$\cdot \sum_{0 \le y_{N+\gamma-1} \le p-1} \exp\left(-\frac{4\pi i}{p}\varepsilon_0 x_0^2 y_{N+\gamma-1}\right) = 0 .$$

Acting similarly we are convinced that in cases $\gamma \geq 1$, $|x|_p \neq p^\gamma$ the desired integral equal to 0, and in the case $\gamma \geq 1$, $|x|_p = p^\gamma$ it is equal to

$$\int_{S_\gamma, y_0 = x_0} \chi_p(\varepsilon(x-y)^2)dy = \int_{S_\gamma, y_0 = x_0, y_1 = x_1} \chi_p(\varepsilon(x-y)^2)dy = \dots$$

$$= \int_{S_\gamma, y_0 = x_0, \dots, y_{\gamma-1} = x_{\gamma-1}} \chi_p(\varepsilon(x-y)^2)dy \int_{S_\gamma, y_0 = x_0, \dots, y_{\gamma-1} = x_{\gamma-1}} dy$$

$$= p^{\gamma-(\gamma-1)-1} = 1 \,,$$

owing to the formula (3.13) of Sec. 4.3 for $l = \gamma - 1$. ■

Example 2. $p \neq 2$, $|\varepsilon|_p = 1$, $\gamma \in Z$ (see [215,218])

$$\int_{S_\gamma} \chi_p(\varepsilon p(x-y)^2)dy = \begin{cases} p^\gamma \left(1 - \frac{1}{p}\right) \chi_p(\varepsilon p x^2), & |x|_p \leq p^{-\gamma+1}, \ \gamma \leq 0, \\ -p^{\gamma-1} \left(1 - \frac{1}{p}\right) \chi_p(\varepsilon p x^2), & |x|_p = p^{-\gamma+2}, \ \gamma \leq 0, \\ 0, & |x|_p \geq p^{-\gamma+3}, \ \gamma \leq 0, \\ \lambda_p(\varepsilon p)\sqrt{p} - \chi_p(\varepsilon p x^2), & |x|_p \leq p, \ \gamma = 1, \\ 0, & |x|_p \geq p^2, \ \gamma = 1, \\ \lambda_p(\varepsilon p)\sqrt{p}, & |x|_p = p^\gamma, \ \gamma \geq 2, \\ 0, & |x|_p \neq p^\gamma, \ \gamma \geq 2. \end{cases} \tag{1.2}$$

■ Similar to the Example 1. Some peculiar are:
For $\gamma = 1$, $|x|_p \leq 1$, $y \in S_1$ we have

$$\begin{aligned} |\varepsilon p(x^2 - 2xy)|_p &= |\varepsilon|_p |p|_p |x^2 - 2xy|_p \\ &\leq \frac{1}{p} \max(|x|_p^2, |2xy|_p) \leq \frac{1}{p} \max(1, p) = 1 \,. \end{aligned}$$

Therefore

$$\chi_p(\varepsilon p(x-y)^2) = \chi_p(\varepsilon p y^2) \chi_p[\varepsilon p(x^2 - 2xy)] = \chi_p(\varepsilon p y) \,.$$

But

$$\{\varepsilon p y^2\}_p = \left\{\frac{1}{p}(\varepsilon_0 + \varepsilon_1 p + \dots)(y_0 + y_1 p + \dots)^2\right\}_p = \frac{\varepsilon_0 y_0^2}{p} \,,$$

and owing to (3.10) of Sec. 4.3 and (*) the integral is equal to

$$\int_{S_1} \exp\left(2\pi i \frac{\varepsilon_0 y_0^2}{p}\right) dy = \sum_{1 \le y_0 \le p-1} \exp\left(2\pi i \frac{\varepsilon_0 y_0^2}{p}\right)$$

$$= \sum_{0 \le y_0 \le p-1} \exp\left(2\pi i \frac{\varepsilon_0 y_0}{p}\right) - 1 = \lambda_p(\varepsilon p)\sqrt{p} - 1 .$$

For $\gamma = 1$, $|x|_p = p$, $y \in S_1$ we have

$$\{\varepsilon p(x-y)^2\}_p = \left\{\frac{1}{p}(\varepsilon_0 + \varepsilon_1 p + \dots)[x_0 - y_0 + (x_1 - y_1)p + \dots]^2\right\}_p$$

$$= \frac{\varepsilon_0}{p}(x_0 - y_0)^2$$

and the integral is equal to

$$\int_{S_1} \exp\left(\frac{2\pi i}{p}\varepsilon_0(x_0 - y_0)^2\right) dy = \sum_{1 \le y_0 \le p-1} \exp\left(\frac{2\pi i}{p}\varepsilon_0(x_0 - y_0)^2\right)$$

$$= \sum_{0 \le y_0 \le p-1} \exp\left(\frac{2\pi i}{p}\varepsilon_0(x_0 - y_0)^2\right) - \exp\left(\frac{2\pi i}{p}\varepsilon_0 x_0^2\right)$$

$$= \sum_{0 \le k \le p-1} \exp\left(2\pi i \frac{\varepsilon_0 k^2}{p}\right) - \exp 2\pi i \left\{\varepsilon p x^2\right\}_p = \lambda_p(\varepsilon p)\sqrt{p} - \chi_p(\varepsilon p x^2) .$$

For $\gamma \ge 2$, $|x|_p = p^\gamma$, $y \in S_\gamma$ we have

$$\begin{aligned}
&\left\{\varepsilon p(x-y)^2\right\}_p \\
&= \left\{p^{-2\gamma+1}\right\}(\varepsilon_0 + \varepsilon_1 p + \dots)[x_0 - y_0 + (x_1 - y_1)p + \dots \\
&\quad + (x_{2\gamma-2} - y_{2\gamma-2})p^{2\gamma-2} + \dots]^2\}_p \\
&= L(y_0, y_1, \dots, y_{2\gamma-3}) - \frac{2\varepsilon_0}{p}(x_0 - y_0)y_{2\gamma-2}
\end{aligned}$$

As in Example 1 the integral over that part of S_γ where $x_0 \ne y_0$ is equal to 0. Therefore the integral is equal over that part of S_γ where $x_0 = y_0$, and so on. So we have

$$\int_{S_\gamma, y_0=x_0} \chi_p(\varepsilon p(x-y)^2)dy = \int_{S_\gamma, y_0=x_0, y_1=x_1} \chi_p(\varepsilon p(x-y)^2)dy = \dots$$

$$= \int_{S_\gamma, y_0=x_0, \dots, y_{\gamma-2}=x_{\gamma-2}} \chi_p(\varepsilon p(x-y)^2)dy$$

$$= \sum_{0 \le y_{\gamma-1} \le p-1} \exp\left[2\pi i \frac{\varepsilon_0}{p}(x_{\gamma-1} - y_{\gamma-1})^2\right]$$

$$= \sum_{0 \le k \le p-1} \exp\left[2\pi i \frac{\varepsilon_0}{p} k^2\right] = \lambda_p(\varepsilon p)\sqrt{p} \,.$$

Here we have used the formulas (3.13) of Sec. 4.3 for $l = \gamma - 1$ and (*) ■

Example 3. $p = 2$, $|\varepsilon|_2 = 1$, $\gamma \in Z$ (see[239,218])

$$\int_{S_\gamma} \chi_2(\varepsilon(x-y)^2)dy = \begin{cases} 2^{\gamma-1}\chi_2(\varepsilon x^2), & |x|_2 \le 2^{-\gamma+1}, \quad \gamma \le 0, \\ -2^{\gamma-1}\chi_2(\varepsilon x^2), & |x|_2 = 2^{-\gamma+2}, \quad \gamma \le 0, \\ 0, & |x|_2 \ge 2^{-\gamma+3}, \quad \gamma \le 0, \\ (-1)^{\varepsilon_1} i, & |x|_2 \le 1, \quad \gamma = 1, \\ 1, & |x|_2 = 2, \quad \gamma = 1, \\ 0, & |x|_2 \ge 4, \quad \gamma = 1, \\ \sqrt{2}\lambda_2(\varepsilon), & |x|_2 = 2^\gamma, \quad \gamma \ge 2, \\ 0, & |x|_2 \ne 2^\gamma, \quad \gamma \ge 2. \end{cases} \tag{1.3}$$

■ Similar to Examples 1 and 2. Special cases are the following. For $\gamma = 1$, $|x|_2 \le 1$, $y \in S_1$ we have

$$|\varepsilon(x^2 - 2xy)|_2 \le \max(|x|_2^2, |2xy|_2) \le \max(1, \frac{1}{2}|y|_2) = 1 \,,$$

$$\{\varepsilon y^2\}_2 = \left\{\frac{1}{4}(1 + \varepsilon_1 2 + \dots)(1 + y_1 2 + \dots)^2\right\}_2 = \frac{1}{4} + \frac{\varepsilon_1}{2} \,,$$

and the integral is equal to

$$\int_{S_1} \chi_2(\varepsilon y^2)dy = \int_{S_1} \exp\left[2\pi i \left(\frac{1}{4} + \frac{\varepsilon_1}{2}\right)\right] dy$$

$$= \exp\left[2\pi i \left(\frac{1}{4} + \frac{\varepsilon_1}{2}\right)\right] = i(-1)^{\varepsilon_1} \,.$$

For $\gamma = 1$, $|x|_2 = 2$, $y \in S_1$ we have

$$|\varepsilon(x-y)^2|_2 = |(x-y)^2|_2 = \left|\frac{1}{4}[(x_1-y_1)2+(x_2-y_2)4+\ldots]^2\right|_2 \leq 1,$$

and the integral is equal to

$$\int_{S_1} dy = 2\left(1-\frac{1}{2}\right) = 1.$$

For $\gamma \geq 2$, $|x|_2 = 2^\gamma$, $y \in S_\gamma$ we have

$$\begin{aligned}\{\varepsilon(x-y)^2\}_2 &= \{2^{-2\gamma}(1+\varepsilon_1 2+\ldots)[(x_1-y_1)2+(x_2-y_2)4+\ldots]^2\}_2 \\ &= L(y_0, y_1, \ldots, y_{2\gamma-4}) - \frac{1}{2}(x_1-y_1)y_{2\gamma-3},\end{aligned}$$

and the desired integral is equal to

$$\begin{aligned}\int_{S_\gamma, y_1=x_1} \chi_2(\varepsilon(x-y)^2)dy &= \int_{S_\gamma, y_1=x_1, y_2=x_2} \chi_2(\varepsilon(x-y)^2)dy = \ldots \\ &= \int_{S_\gamma, y_1=x_1, \ldots, y_{\gamma-2}=x_{\gamma-2}} \chi_2(\varepsilon(x-y)^2)dy.\end{aligned}$$

For the last integral we have

$$\begin{aligned}\{\varepsilon(x-y)^2\}_2 &= \left\{\frac{1}{4}(1+\varepsilon_1 2+\ldots)[(x_{\gamma-1}-y_{\gamma-1})+(x_\gamma-y_\gamma)2+\ldots]^2\right\}_2 \\ &= \left(\frac{1}{4}+\frac{\varepsilon_1}{2}\right)(x_{\gamma-1}-y_{\gamma-1})^2,\end{aligned}$$

and hence the desired integral is equal to

$$\begin{aligned}&\int_{S_\gamma, y_1=x_1, \ldots, y_{\gamma-2}=x_{\gamma-2}} \exp\left[2\pi i\left(\frac{1}{4}+\frac{\varepsilon_1}{2}\right)(x_{\gamma-1}-y_{\gamma-1})^2\right] dy \\ &= \sum_{y_{\gamma-1}=0,1} \exp\left[2\pi i\left(\frac{1}{4}+\frac{\varepsilon_1}{2}\right)(x_{\gamma-1}-y_{\gamma-1})^2\right] \\ &= \sum_{k=0,1} \exp\left[2\pi i k^2\left(\frac{1}{4}+\frac{\varepsilon_1}{2}\right)\right] = 1+\exp\left[2\pi i\left(\frac{1}{4}+\frac{\varepsilon_1}{2}\right)\right] \\ &= 1+i(-1)^{\varepsilon_1} = \sqrt{2}\lambda_2(\varepsilon).\end{aligned}$$

Here we have used the formular (3.13) of Sec. 4.3 for $l = \gamma - 2$. ∎

Example 4. $p = 2$, $|\varepsilon|_2 = 1$, $\gamma \in Z$ (see [239,218])

$$\int_{S_\gamma} \chi_2(2\varepsilon(x-y)^2)dy = \begin{cases} 2^{\gamma-1}\chi_2(2\varepsilon x^2), & |x|_2 \le 2^{-\gamma+2}, \quad \gamma \le 0, \\ -2^{\gamma-1}\chi_2(2\varepsilon x^2), & |x|_2 = 2^{-\gamma+3}, \quad \gamma \le 0, \\ 0, & |x|_2 \ge 2^{-\gamma+4}, \quad \gamma \le 0, \\ -1, & |x|_2 \le 1, \quad \gamma = 1, \\ 1, & |x|_2 = 2, \quad \gamma = 1, \\ \lambda_2(2\varepsilon), & |x|_2 = 4, \quad \gamma = 1, \\ 0, & |x|_2 \ge 8, \quad \gamma = 1, \\ 2\lambda_2(2\varepsilon), & |x|_2 \le 2, \quad \gamma = 2, \\ 0, & |x|_2 \ge 4, \quad \gamma = 2, \\ 2\lambda_2(2\varepsilon), & |x|_2 = 2^\gamma, \quad \gamma \ge 3, \\ 0, & |x|_2 \ne 2^\gamma, \quad \gamma \ge 3. \end{cases} \tag{1.4}$$

∎ Similar to Examples 1–3. Special cases are the following.
For $\gamma = 1$, $|x|_2 = 4$, $y \in S_1$ we have

$$\begin{aligned} \{2\varepsilon(x-y)^2\}_2 &= \left\{\frac{1}{8}(1+\varepsilon_1 2+\varepsilon_2 4\ldots)[1+(x_1-1)2+(x_2-y_1)4+\ldots]^2\right\}_2 \\ &= \left\{\frac{1}{8}+\frac{\varepsilon_1}{4}+\frac{\varepsilon_2}{2}+\frac{(x_1-1)^2}{2}+\frac{x_1-1}{2}\right\}_2 \\ &= \frac{1}{2}+\frac{\varepsilon_1}{4}+\frac{\varepsilon_2}{2}, \end{aligned}$$

and the integral is equal to

$$\begin{aligned} \int_{S_1} \exp\left[2\pi i\left(\frac{1}{8}+\frac{\varepsilon_1}{4}+\frac{\varepsilon_2}{2}\right)\right]dy &= \exp\left[2\pi i\left(\frac{1}{8}+\frac{\varepsilon_1}{4}+\frac{\varepsilon_2}{2}\right)\right] \\ &= \frac{1+i}{\sqrt{2}} i^{\varepsilon_1}(-1)^{\varepsilon_2} = \lambda_2(2\varepsilon). \end{aligned}$$

For $\gamma = 2$, $|x|_2 = 2^N$, $N \leq 1$, $y \in S_2$ we have

$$\begin{aligned}
&\{2\varepsilon(x-y)^2\}_2 \\
&= \left\{\frac{1}{8}(1+\varepsilon_1 2+\varepsilon_2 4\ldots)\left[-(1-y_1)2+\ldots+(1-y_{2-N})2^{2-N}+\ldots\right]^2\right\}_2 \\
&= \left\{\frac{1}{8}+\frac{\varepsilon_1}{4}+\frac{\varepsilon_2}{2}+\frac{y_1^2}{2}+\frac{y_1}{2}\right\}_2 \\
&= \frac{1}{8}+\frac{\varepsilon_1}{4}+\frac{\varepsilon_2}{2},
\end{aligned}$$

and the integral is equal to

$$\int_{S_2} \exp\left[2\pi i\left(\frac{1}{8}+\frac{\varepsilon_1}{4}+\frac{\varepsilon_2}{2}\right)\right] dy = 2\exp\left[2\pi i\left(\frac{1}{8}+\frac{\varepsilon_1}{4}+\frac{\varepsilon_2}{2}\right)\right] = 2\lambda_2(2\varepsilon)$$

For $\gamma \geq 3$, $|x|_2 = 2^\gamma$, $\gamma \in S_\gamma$ we have

$$\begin{aligned}
&\Big\{2^{-2\gamma+3}(1+\varepsilon_1 2+\ldots)[(x_1-y_1)+(x_2-y_2)2+\ldots \\
&\qquad +(x_{\gamma-3}+y_{\gamma-3})2^{\gamma-4}+\ldots+(x_{2\gamma-4}-y_{2\gamma-4})2^{2\gamma-5}+\ldots]^2\Big\}_2 \\
&\quad = L(y_0,y_1,\ldots,y_{2\gamma-5}) - \frac{1}{2}(x_1-y_1)y_{2\gamma-4},
\end{aligned}$$

and the integral is equal to

$$\int_{S_\gamma, y_1=x_1} \chi_2(2\varepsilon(x-y)^2)dy = \ldots \int_{S_\gamma, y_1=x_1,\ldots,y_{\gamma-3}=x_{\gamma-3}} \chi_2(2\varepsilon(x-y)^2)dy.$$

In the last integral we have

$$\begin{aligned}
&\{2\varepsilon(x-y)^2\}_2 \\
&= \left\{\frac{1}{8}(1+\varepsilon_1 2+\varepsilon_2 4+\ldots)\left[x_{\gamma-2}-y_{\gamma-2}+(x_{\gamma-1}-y_{\gamma-1})2+\ldots\right]^2\right\}_2 \\
&= \frac{1}{8}\left[(x_{\gamma-2}-y_{\gamma-2})^2+4(x_{\gamma-1}-y_{\gamma-1})^2+4(x_{\gamma-1}-y_{\gamma-1})(x_{\gamma-2}-y_{\gamma-2})\right] \\
&\quad +(x_{\gamma-2}-y_{\gamma-2})^2\left(\frac{\varepsilon_1}{4}+\frac{\varepsilon_2}{2}\right),
\end{aligned}$$

and the integral is equal to

$$\int\limits_{S_\gamma, y_1=x_1,\dots,y_{\gamma-3}=x_{\gamma-3}} \exp\left\{\frac{\pi i}{4}[(x_{\gamma-2}-y_{\gamma-2})^2+4(x_{\gamma-1}-y_{\gamma-1})^2\right.$$

$$\left.+4(x_{\gamma-1}-y_{\gamma-1})(x_{\gamma-2}-y_{\gamma-2})+(2\varepsilon_1+\varepsilon_2)(x_{\gamma-2}-y_{\gamma-2})^2]\right\}dy$$

$$=\sum_{y_{\gamma-2}=0,1}\sum_{y_{\gamma-1}=0,1}\exp\left\{\frac{\pi i}{4}[(x_{\gamma-2}-y_{\gamma-2})^2+4(x_{\gamma-1}-y_{\gamma-1})^2\right.$$

$$\left.+4(x_{\gamma-1}-y_{\gamma-1})(x_{\gamma-2}-y_{\gamma-2})+(2\varepsilon_1+\varepsilon_2)(x_{\gamma-2}-y_{\gamma-2})^2]\right\}$$

$$=\sum_{k=0,1}\sum_{j=0,1}\exp\left\{\frac{\pi i}{4}\left[k^2+4j^2+4kj+(2\varepsilon_1+\varepsilon_2)k^2\right]\right\}$$

$$=1+e^{\pi i}+2\exp\left[\frac{\pi i}{4}(1+2\varepsilon_1+\varepsilon_2)\right]=2\frac{1+i}{\sqrt{2}}(-1)^{\varepsilon_1}i^{\varepsilon_2}=2\lambda_2(2\varepsilon) \quad ■$$

Example 5. $p\neq 2$, $|a_p|\geq p^{2-2\gamma}$, $\gamma\in Z$

$$\int\limits_{S_\gamma}\chi_p(ax^2+bx)dx=\begin{cases}\lambda_p(a)|a|_p^{-1/2}\chi_p\left(-\frac{b^2}{4a}\right), & \left|\frac{b}{2a}\right|_p=p^\gamma,\\ 0, & \left|\frac{b}{2a}\right|_p\neq p^\gamma\end{cases} \tag{1.5}$$

■ Let $a=\sigma p^{-2N}$ where either $\sigma=\varepsilon$ or $\sigma=\varepsilon p$, $|\varepsilon|_p=1$ (see Sec. 1.4). Under the condition $|a|_p\geq p^{2-2\gamma}$, either $N\geq 1-\gamma$ (for $\sigma=\varepsilon$) or $N\geq 2-\gamma$ (for $\sigma=\varepsilon p$). Performing in the integral change of variables of integration $p^{-N}x=y$, $dx=p^{-N}dy$ we get

$$\int\limits_{S_\gamma}\chi_p(p^{-2N}\sigma x^2+bx)dx$$

$$=p^{-N}\int\limits_{S_{N+\gamma}}\chi_p(\sigma y^2+bp^N y)dy$$

$$=p^{-N}\chi_p\left(-\frac{b^2p^{2N}}{4\sigma}\right)\int\limits_{S_{N+\gamma}}\chi_p\left[\sigma\left(y+\frac{bp^N}{2\sigma}\right)^2\right]dy\,.$$

Using the formulas (1.1) for $\sigma = \varepsilon$, $N+\gamma \geq 1$ and (1.2) for $\sigma = \varepsilon p$, $N+\gamma \geq 2$ we get for the desired integral the expressions

$$p^{-N}\chi_p\left(-\frac{b^2}{4a}\right) = \begin{cases} 1 & \text{if } |\frac{bp^N}{2\sigma}|_p = p^{N+\gamma}, \sigma = \varepsilon, \\ \lambda_p(\varepsilon p)\sqrt{p} & \text{if } |\frac{bp^N}{2\sigma}|_p = p^{N+\gamma}, \sigma = \varepsilon p, \\ 0 & \text{if otherwise.} \end{cases}$$

By combining these cases and taking into account that

$$1 = \lambda_p(\varepsilon), \quad p^N = |a|_p^{1/2}, \quad p^N|\frac{p^N}{2\sigma}|_p^{-1} = |a|_p \text{ if } \sigma = \varepsilon\ ,$$

$$\lambda_p(a) = \lambda_p(\varepsilon p), \quad p^{N-1/2} = |a|_p^{1/2}, \quad p^N|\frac{p^N}{2\sigma}|_p^{-1} = |a|_p \text{ if } \sigma = \varepsilon p$$

we get the formulas (1.5). ■

Example 6. $p = 2$, $|a|_2 \geq 2^{4-2\gamma}$, $\gamma \in \mathbb{Z}$.

$$\int_{S_\gamma} \chi_2(ax^2+bx)dx = \begin{cases} \sqrt{2}\lambda_2(a)|a|_2^{-1/2}\chi_2\left(-\frac{b^2}{4a}\right), & |\frac{b}{2a}|_2 = 2^\gamma, \\ 0, & |\frac{b}{2a}|_2 \neq 2^\gamma. \end{cases} \tag{1.6}$$

■ Similar to Example 5 by using the formulas (1.3) and (1.4). ■

The formulas (1.5) and (1.6) admit unification.

Example 7. $|4a|_p \geq p^{2-2\gamma}$, $\gamma \in \mathbb{Z}$.

$$\int_{S_\gamma} \chi_p(ax^2+bx)dx = \begin{cases} \lambda_p(a)|2a|_p^{-1/2}\chi_p\left(-\frac{b^2}{4a}\right), & |\frac{b}{2a}|_p = p^\gamma, \\ 0, & |\frac{b}{2a}|_p \neq p^\gamma. \end{cases} \tag{1.7}$$

Example 8. $p \neq 2$, $|a|_p = p^{1-2\gamma}$, $\gamma \in \mathbb{Z}$.

$$\int_{S_\gamma} \chi_p(ax^2+bx)dx = \begin{cases} |a|_p^{-1/2}\left[\lambda_p(a)\chi_p\left(-\frac{b^2}{4a}\right) - \frac{1}{\sqrt{p}}\right], & |b|_p \leq p^{-\gamma+1}, \\ 0, & |b|_p \geq p^{-\gamma+2}. \end{cases} \tag{1.8}$$

■ As $a = p^{2\gamma-1}\varepsilon$, $|\varepsilon|_p = 1$ then after substitution $p^{\gamma-1}x = y$ the integral takes the form

$$p^{\gamma-1}\chi_p\left(-\frac{b^2}{4a}\right)\int_{S_1}\chi_p\left[\varepsilon p\left(y+\frac{bp^{-\gamma+1}}{2\varepsilon p}\right)^2\right]dy\,.$$

If $|b|_p \le p^{-\gamma+1}$ then $\left|\frac{bp^{-\gamma+1}}{2\varepsilon p}\right|_p \le p$, and we use the formula (1.2) for $\gamma = 1$, $|x|_p \le p$:

$$p^{\gamma-1}\chi_p\left(-\frac{b^2}{4a}\right)\left[\lambda_p(\varepsilon p)\sqrt{p} - \chi_p\left(\varepsilon p\frac{b^2p^{-2\gamma+2}}{4\varepsilon^2p^2}\right)\right]$$
$$= |a|_p^{-1/2}\left[\lambda_p(a)\chi_p\left(-\frac{b^2}{4a}\right) - \frac{1}{\sqrt{p}}\right]\,.$$

If $|b|_p \ge p^{-\gamma+2}$ then $\left|\frac{bp^{-\gamma+1}}{2\varepsilon p}\right|_p \ge p^2$, and we use the formula (1.2) for $\gamma = 1$, $|x|_p \ge p^2$, so the integral is equal to 0. ■

2. *The Gaussian integrals on the discs* B_γ

Example 9. $p \ne 2$, $\gamma \in \mathbb{Z}$

$$\int_{B_\gamma}\chi_p(ax^2+bx)dx = \begin{cases} p^\gamma\Omega(p^\gamma|b|_p), & |a|_pp^{2\gamma} \le 1, \\ \lambda_p(a)|a|_p^{-1/2}\chi_p\left(-\frac{b^2}{4a}\right)\Omega\left(p^{-\gamma}\left|\frac{b}{a}\right|_p\right), & |a|_pp^{2\gamma} > 1. \end{cases} \tag{2.1}$$

■ For $|a|_pp^{2\gamma} \le 1$, $y \in B_\gamma$ we have $|ax^2|_p = |a|_p|x^2|_p \le 1$ hence $\chi_p(ax^2) = 1$, and the formula (2.1) follows from the formula (3.1) of Sec. 4.3

$$\int_{B_\gamma}\chi_p(bx)dx = p^\gamma\Omega(p^\gamma|b|_p)\,.$$

Let now $|a|_pp^{2\gamma} > 1$, $|a|_p = p^{2N-2\gamma}$, $N = 1, 2, \ldots$, $a = \varepsilon p^{2\gamma-2N}$, $|\varepsilon|_p = 1$. Performing the change of variable of integration $x = p^{N-\gamma}y$, $dx = p^{\gamma-N}dy$, $|y|_p = p^{N-\gamma}|x|_p \le p^N$ we get

$$p^{\gamma-N}\int_{B_\gamma}\chi_p(\varepsilon y^2 + p^{N-\gamma}by)dy$$
$$= p^{\gamma-N}\left[\int_{B_0}\chi_p(p^{N-\gamma}by)dy + \sum_{1\le\gamma'\le N}\int_{S_{\gamma'}}\chi_p(\varepsilon y^2 + p^{N-\gamma}by)dy\right]\,. \tag{2.2}$$

For $|\frac{b}{a}|_p > p^\gamma$, i.e. $|b|_p > |a|_p p^\gamma = p^{2N-\gamma}$, we have $|p^{N-\gamma}b|_p = p^{\gamma-N}|b|_p > p^N$. Taking in account that in (2.2) $|\varepsilon|_p p^{2\gamma'} \geq p^2$, $\gamma' = 1, 2, \ldots, N$ we conclude owing to the formulas (3.1) of Sec. 4.3 and (1.7) that all integrals in (2.2) are equal to 0, and the formula (2.1) is proved in this case.

For $|\frac{b}{a}|_p \leq p^\gamma$, i.e. $|p^{N-\gamma}b|_p \leq p^N$, the integral in the left-hand side of (2.2) takes the form (2.1):

$$p^{\gamma-N}\chi_p\left(-\frac{p^{2N-2\gamma}b^2}{4\varepsilon}\right)\int\limits_{B_N}\chi_p\left(\varepsilon\left(y+\frac{p^{N-\gamma}b}{\varepsilon}\right)^2\right)dy$$

$$= |a|_p^{-1/2}\chi_p\left(-\frac{b^2}{4a}\right)\int\limits_{B_N}\chi_p(\varepsilon y^2)dy$$

$$= |a|_p^{-1/2}\chi_p\left(-\frac{b^2}{4a}\right)\left[\int\limits_{B_0}dx + \sum_{1\leq\gamma'\leq N}\int\limits_{S_{\gamma'}}\chi_p(\varepsilon y^2)dy\right]$$

$$= \lambda_p(a)|a|_p^{-1/2}\chi_p\left(-\frac{b^2}{4a}\right),$$

because the integrals under sign of sum are equal to 0.

The case $|a|_p p^{2\gamma} > 1$, $|a|_p = p^{2N-1-2\gamma}$, $N = 1, 2, \ldots$, $a = \varepsilon p^{2\gamma-2N+1}$, $|\varepsilon|_p = 1$ is considered analogously, owing to the formulas (3.1) of Sec. 4.3 and (1.8). ■

Example 10. $p = 2$, $\gamma \in \mathbb{Z}$,

$$\int\limits_{B_\gamma}\chi_2(ax^2+bx)dx$$

$$= \begin{cases} 2^\gamma\Omega(2^\gamma|b|_2), & |a|_2 2^{2\gamma} \leq 1, \\ \lambda_2(a)|2a|_2^{-1/2}\chi_2\left(-\frac{b^2}{4a}\right)\delta(|b|_2 - 2^{1-\gamma}), & |a|_2 2^{2\gamma} = 2, \\ \lambda_2(a)|2a|_2^{-1/2}\chi_2\left(-\frac{b^2}{4a}\right)\Omega(2^\gamma|b|_2), & |a|_2 2^{2\gamma} = 4, \\ \lambda_2(a)|2a|_2^{-1/2}\chi_2\left(-\frac{b^2}{4a}\right)\Omega\left(2^{-\gamma}\left|\frac{b}{2a}\right|_2\right), & |a|_2 2^{2\gamma} \geq 8, \end{cases} \tag{2.3}$$

where

$$\delta(|b|_p - p^\gamma) = \begin{cases} 1 \text{ if } & |b|_p = p^\gamma, \\ 0 \text{ if } & |b|_p \neq p^\gamma. \end{cases}$$

■ Similar to the Example 9. The cases $|a|_2 2^{2\gamma} = 2$ and $|a|_2 2^{2\gamma} = 4$ are considered specifically. ■

The formulas (2.1) and (2.3) admit the inification.

Example 11. $\gamma \in \mathbb{Z}$

$$\int_{B_\gamma} \chi_p(ax^2 + bx)dx$$

$$= \begin{cases} p^\gamma \Omega(p^\gamma |b|_p), & |a|_p p^{2\gamma} \le 1, \\ \lambda_p(a)|2a|_p^{-1/2} \chi_p\left(-\frac{b^2}{4a}\right) \Omega\left(p^{-\gamma}\left|\frac{b}{2a}\right|_p\right), & |4a|_p p^{2\gamma} > 1. \end{cases} \quad (2.4)$$

3. *The Gaussian integrals on* $\mathbb{Q}_p$

Example 12. $a \neq 0$

$$\int_{\mathbb{Q}} \chi_p(ax^2 + bx)dx = \lambda_p(a)|2a|_p^{-1/2} \chi_p\left(-\frac{b^2}{4a}\right). \quad (3.1)$$

It follows from the formula (2.3) by $\gamma \to \infty$.
Note that a formula similar to (3.1) is valid in the real case $\mathbb{Q}_\infty = \mathbb{R}$:

$$\int_{\mathbb{Q}_\infty} \chi_\infty(ax^2 + bx)dx = \lambda_\infty(a)|2a|_\infty^{-1/2} \chi_\infty\left(-\frac{b^2}{4a}\right), \quad a \neq 0 \quad (3.2)$$

$|a|_\infty = |a|$, $\chi_\infty(x) = \exp(-2\pi i x)$ and

$$\lambda_\infty(a) = \exp\left(-i\frac{\pi}{4}\,\mathrm{sign} a\right) = \begin{cases} \frac{1-i}{\sqrt{2}} \text{ if } & a > 0, \\ \frac{1+i}{\sqrt{2}} \text{ if } & a < 0. \end{cases} \quad (3.3)$$

■ The formula (3.2) follows from the classical formula

$$\frac{1}{\sqrt{2\pi}} \int_{-\infty}^{\infty} \exp(-iax^2 - ibx)dx = \lambda_\infty(a)|2a|^{-1/2} \exp\left(i\frac{b^2}{4a}\right), \quad a \neq 0 .$$

■ (3.4)

4. Further Properties of the Function $\lambda_p(a)$

We prove the equality

$$\lambda_p(a)\lambda_p(b) = \lambda_p(a+b)\lambda_p\left(\frac{1}{a}+\frac{1}{b}\right), \quad a,b,a+b \in \mathbb{Q}_p^* . \tag{4.1}$$

■ It follows from the formula (3.1)

$$\begin{aligned}
&\int_{\mathbb{Q}_p} \chi_p(ax^2)dx \int_{\mathbb{Q}_p} \chi_p(by^2)dy = \lambda_p(a)\lambda_p(b)|2a|^{-1/2}|2b|^{-1/2} \\
&= \int_{\mathbb{Q}_p^2} \chi_p(a(x-y)^2)\chi_p(by^2)dxdy = \int_{\mathbb{Q}_p}\int_{\mathbb{Q}_p} \chi_p[a(x-y)^2+by^2]dxdy \\
&= \int_{\mathbb{Q}_p} \chi_p(ax^2)dx \int_{\mathbb{Q}_p} \chi_p[(a+b)y^2 - 2axy]dydx \\
&= \int_{\mathbb{Q}_p} \chi_p(ax^2)\lambda_p(a+b)|2a+2b|_p^{-1/2}\chi_p\left(-\frac{a^2x^2}{a+b}\right)dx \\
&= \lambda_p(a+b)|2a+2b|_p^{-1/2}\int_{\mathbb{Q}_p} \chi_p\left(\frac{ab}{a+b}x^2\right)dx \\
&= \lambda_p(a+b)|2a+2b|_p^{-1/2}\lambda_p\left(\frac{ab}{a+b}\right)\left|\frac{ab}{a+b}\right|_p^{-1/2} \\
&= \lambda_p(a+b)\lambda_p\left(\frac{a+b}{ab}\right)|2a|^{-1/2}|2b|^{-1/2} .
\end{aligned}$$

Integrals entering in this chain of equalities extend in fact over discs of finite radii (depending on a and b) (see Sec. 4.1). Hence use of the Fubini Theorem is justified here. ■

The following adelic formula is valid

$$\prod_{2\le p\le\infty} \lambda_p(a) = 1, \quad a \in \mathbb{Q}^* . \tag{4.2}$$

■ The product (4.2) converges for all rational numbers $a \neq 0$, as only a finite number of factors in it differs from the unity. Let a rational number $a \neq 0$ have a form

$$a = \pm 2^{\alpha_0} p_1^{\alpha_1} p_2^{\alpha_2} \dots p_n^{\alpha_n}, \qquad \alpha_j \in \mathbb{Z} \quad (j = 0, 1, \dots, n)$$

where $2, p_1, p_2, \dots, p_n$ are relatively prime numbers.

Owing to the relations

$$\lambda_p(a)\lambda_p(-a) = 1, \quad \lambda_p(ac^2) = \lambda_p(a), \quad 2 \leq p \leq \infty$$

it is sufficient to prove the formula (4.2) for numbers a of the form

$$a = 2^{\alpha_0} p_1 p_2 \dots p_n, \qquad \alpha_0 = 0, 1,$$

so that $\lambda_\infty(a) = \exp(-i\frac{\pi}{4})$.

Let $\alpha_0 = 0$. Denote by $l, 0 \leq l \leq n$ a number of prime numbers of the form $4N + 3$ in the set $(p_1, p_2, \dots, p_n)$ so $n - l$ is a number of prime numbers of the form $4N + 1$ in this set. Note that the product of numbers of the form $4N + 1$ is again a number of the same form, but the product of numbers of the form $4N + 3$ is of the form $4N + 1$ if a number of factors are even and it is of the form $4N + 3$ if a number of factors is odd. Then we have

$$\lambda_2(a) = \frac{1}{\sqrt{2}}[1 + (-1)^l i], \quad \lambda_p(a) = 1 \text{ if } p \neq 2, p \neq p_j; \quad j = 1, 2, \dots, n$$

$$\lambda_{p_j}(a) = \begin{cases} \left(\dfrac{\prod\limits_{k \neq j} p_k}{p_j}\right) & \text{if } p_j \equiv 1 \pmod 4, \\ i\left(\dfrac{\prod\limits_{k \neq j} p_k}{p_j}\right) & \text{if } p_j \equiv 3 \pmod 4. \end{cases}$$

Therefore

$$\sum_{2 \leq p \leq \infty} \lambda_p(a) = \exp\left(-i\frac{\pi}{4}\right) \frac{1}{\sqrt{2}}[1 + i(-1)^l] i^l \prod_{1 \leq j \leq n} \left(\frac{\prod\limits_{k \neq j} p_k}{p_j}\right). \tag{4.3}$$

Taking into account properties of the Legendre symbol in particular the reciprocity law (see [204])

$$\left(\frac{p}{q}\right)\left(\frac{q}{p}\right) = (-1)^{\frac{p-1}{2}\frac{q-1}{2}}$$

where p and q are prime odd numbers we get

$$\prod_{1\le j\le n}\left(\frac{\prod\limits_{k\ne j} p_k}{p_j}\right) = \prod_{\substack{1\le j,k\le n\\ j\ne k}}\left(\frac{p_k}{p_j}\right) = \prod_{1\le j<k\le n}\left(\frac{p_k}{p_j}\right)\left(\frac{p_j}{p_k}\right)$$
$$= \prod_{1\le j<k\le n}(-1)^{\frac{p_k-1}{2}\frac{p_j-1}{2}} = (-1)^{\frac{l(l-1)}{2}}. \qquad (4.4)$$

From here and (4.3) the formula (4.2) follows

$$\prod_{2\le p\le\infty}\lambda_p(a) = \exp\left(-i\frac{\pi}{4}\right)\frac{1}{\sqrt{2}}[1+i(-1)^l]i^l(-1)^{\frac{l(l-1)}{2}} = 1, \quad l = 0,1,\dots.$$

Let now $\alpha_0 = 1$. Denote by l_1, l_2, l_3, $0 \le l_1 + l_2 + l_3 \le n$ a number of prime numbers of the form $8N+3$, $8N+5$, $8N+7$ in the set $(p_1, p_2, \dots, p_n)$ respectively. Note the product of numbers of the form $8N+1$ is a number of the same form, but the product of numbers of the form $8N+j (j = 3,5,7)$ is of the form $8N+1$ if a number of factors are even and is of the form $8N+j (j = 3,5,7)$ if a number of factors are odd. Therefore

$$\lambda_2(a) = \begin{cases} \frac{1+i}{\sqrt{2}} \text{ if } l_1, l_2, l_3 \text{ even or } l_1, l_2, l_3 \text{ odd}, \\ \frac{1+i}{\sqrt{2}} \text{ if } l_1 \text{ odd }, l_2, l_3 \text{ even or } l_1 \text{ even}, l_2, l_3 \text{ odd}, \\ \frac{-1+i}{\sqrt{2}} \text{ if } l_2 \text{ odd}, l_1, l_3 \text{ even or } l_2 \text{ even}, l_1, l_3 \text{ odd}, \\ \frac{1-i}{\sqrt{2}} \text{ if } l_3 \text{ odd}, l_1, l_2 \text{ even or } l_3 \text{ even}, l_1, l_2 \text{ odd}; \end{cases} \qquad (4.5)$$

$$\lambda_{p_j}(a) = \begin{cases} \left(\frac{\prod\limits_{k\ne j} 2p_k}{p_j}\right) & \text{if } p_j \equiv 1 \pmod 4, \\ i\left(\frac{\prod\limits_{k\ne j} 2p_k}{p_j}\right) & \text{if } p_j \equiv 3 \pmod 4, \end{cases}$$

$$\lambda_p(a) = 1, \quad \text{if } p \ne 2, \quad p \ne p_j, \quad j = 1,2,\dots,n.$$

Thus acting like in (4.4) and using the formula (4.7) of Sec. 1.4 we get

$$\prod_{2<p<\infty} \lambda_p(a) = i^{l_1+l_3} \prod_{j=1}^{n} \left(\frac{\prod\limits_{k\neq j} 2p_k}{p_j} \right) = i^{l_1+l_2} \prod_{j=1}^{n} \left(\frac{2}{p_j}\right) \prod_{\substack{1\le j,k\le n \\ j\neq k}} \left(\frac{p_k}{p_j}\right)$$

$$= i^{l_1+l_3} \prod_{j=1}^{n} (-1)^{\frac{p_j^2-1}{2}} \prod_{1\le j<k\le n} \left(\frac{p_k}{p_j}\right)\left(\frac{p_j}{p_k}\right)$$

$$= i^{l_1+l_3} (-1)^{l_1+l_2} (-1)^{\frac{(l_1+l_3)(l_1+l_3-1)}{2}} = (-1)^{l_1+l_2+\frac{1}{2}(l_1+l_3)^2} ,$$

From this formula using (4.5) we obtain the formula (4.2)

$$\prod_{2\le p\le\infty} \lambda_p(a) = \exp\left(-i\frac{\pi}{4}\right) \lambda_2(a) (-1)^{l_1+l_2+\frac{1}{2}(l_1+l_3)^2} = 1 . \qquad \blacksquare$$

Remark. A function similar to $\lambda_p(a)$ has been considered by A. Weil [225] for locally compact fields. Its particular expressions for the fields $\mathbb{Q}_p$ are contained in papers [215,239,3,182,151]. The function $\lambda_p(a)$ is connected with the Hilbert symbol (see [37]).

By definition the Hilbert symbol (a,b), $a,b\in\mathbb{Q}_p^*$ is equal to $+1$ or -1 subject to if the form $ax^2+by^2-z^2$ represents 0 in the field $\mathbb{Q}_p$ or not.

The Hilbert symbol obeys the properties (see [37]):

$$(a,b) = (b,a), \qquad (a\alpha^2, b\beta^2) = (a,b),$$
$$(a,-a) = 1, \qquad (a_1a_2, b) = (a_1,b)(a_2,b)$$

and for $p\neq 2$

$$(\varepsilon,\varepsilon_1) = 1, \qquad (p,\varepsilon) = \left(\frac{\varepsilon_0}{p}\right), \qquad |\varepsilon|_p = |\varepsilon_1|_p = 1 ,$$
$$(a,-\varepsilon) = \operatorname{sgn}_\varepsilon a, \qquad \varepsilon\notin\mathbb{Q}_p^{*2} \text{ see (2.4) of Sec. 3.2}.$$

The following formula is valid:

$$\lambda_p(a)\lambda_p(b) = (a,b)\lambda_p(ab), \qquad a,b\in\mathbb{Q}_p^* . \tag{4.6}$$

■ Let $p\neq 2$. Owing to the properties of the Hilbert symbol and the function $\lambda_p(a)$, it is sufficient to prove the equality (4.6) for the following cases (see Sec. 1.2):

1) $a = b = \varepsilon$, 4) $a = \varepsilon$, $b = p$,
2) $a = b = p$, 5) $a = \varepsilon$, $b = \varepsilon p$,
3) $a = b = \varepsilon p$, 6) $a = p$, $b = \varepsilon p$, $|\varepsilon|_p = 1$,

that can be done by means of elementary calculations.

From the formula (4.6) for $b = -\varepsilon a$ it follows the formula

$$\lambda_p(a)\lambda_p(-\varepsilon a) = \operatorname{sgn}_\varepsilon a \lambda_p(-\varepsilon), \tag{4.7}$$

obtained in paper [182].

5. *Example 13.* $p \neq 2$ (*see* [206])

$$\int_{\mathbb{Q}_p} e^{-|y|_p^2}\chi_p(a(x-y)^2)dy$$

$$= \left(1-\frac{1}{p}\right)|a|_p^{-1/2}S\left(|a|_p^{-1},\frac{1}{p}\right), \gamma(a) \text{ even}, \qquad (5.1_1)$$

$$= \left(1-\frac{1}{p}\right)\frac{1}{\sqrt{p}}|a|_p^{-1/2}S\left(\frac{1}{p}|a|_p^{-1},\frac{1}{p}\right) + |a|_p^{-1/2}\left[\lambda_p(a) - \frac{1}{\sqrt{p}}\right]e^{-p|a|_p^{-1}},$$

$\gamma(a)$ odd, $\qquad (5.1_2)$

if $|x|_p \leq |a|_p^{-1/2}$;

$$= \lambda_p(a)|a|_p^{-1/2}e^{-|x|_p^2} + |ax|_p^{-1}\chi_p(ax^2)\left[\left(1-\frac{1}{p}\right)S\left(|ax|_p^{-2},\frac{1}{p}\right) - e^{-p^2|ax|_p^{-2}}\right]$$

if $|x|_p > |a|_p^{-1/2}$. $\qquad (5.1_3)$

Here the function $S(\alpha, q)$ is defined by the formula

$$S(\alpha,q) = \sum_{0\leq k<\infty}\frac{(-\alpha)^k}{k!(1-q^{2k+1})} = \sum_{0\leq \nu<\infty} q^\nu e^{-\alpha q^{2\nu}}, \qquad |q| < 1. \tag{5.2}$$

From the equality (5.1$_3$) it follows the asymptotes

$$\int_{\mathbb{Q}_p} e^{-|y|_p^2}\chi_p(a(x-y)^2)dy$$

$$\sim \frac{p^4+p^3}{p^2+p+1}|ax|_p^{-3}\chi_p(ax^2) + O(|x|_p^{-5}), \quad |x|_p \to \infty. \tag{5.3}$$

■ Let $\gamma(a)$ be even, $|a|_p = p^{2N}$, $N \in \mathbb{Z}$. Then $|a|_p \geq p^{2-2\gamma}$ if $\gamma \geq -N+1$. The integral is equal to

$$\begin{aligned}
&\sum_{-\infty<\gamma<\infty} e^{-p^{2\gamma}} \int_{S_\gamma} \chi_p(a(x-y)^2)dy \\
&= \chi_p(ax^2) \sum_{-\infty<\gamma\leq -N} e^{-p^{2\gamma}} \int_{S_\gamma} \chi_p(-2axy)dy \\
&\quad + \sum_{1-N\leq\gamma<\infty} e^{-p^{2\gamma}} \int_{S_\gamma} \chi_p(ay^2 - 2axy)dy \,. \qquad (5.4)
\end{aligned}$$

For $|x|_p \leq |a|_p^{-1/2} = p^{-N}$ we have $\chi_p(ax^2) = 1$, $\chi_p(-2axy) = 1$, $y \in S_\gamma$, $\gamma \leq -N$, $|\frac{2ax}{2a}|_p = |x|_p \neq p^\gamma$, $\gamma \geq -N+1$, and by virtue of the formulas (5.4) and (1.7) the integral is expressed by the formula (5.1_1):

$$\begin{aligned}
&\sum_{-\infty<\gamma\leq -N} e^{-p^{2\gamma}} \int_{S_\gamma} dy = \left(1-\frac{1}{p}\right) \sum_{-\infty<\gamma\leq -N} e^{-p^{2\gamma}} p^\gamma \\
&= \left(1-\frac{1}{p}\right) \sum_{N\leq\gamma<\infty} e^{-p^{-2\gamma}} p^{-\gamma} = \left(1-\frac{1}{p}\right) p^{-N} \sum_{0\leq\gamma<\infty} e^{-p^{-2\gamma-2N}} p^{-\gamma} \\
&= \left(1-\frac{1}{p}\right) |a|_p^{-1/2} \sum_{0\leq\gamma<\infty} e^{-|a|_p^{-1} p^{-2\gamma}} p^{-\gamma} \\
&= \left(1-\frac{1}{p}\right) |a|_p^{-1/2} S\left(|a|_p^{-1}, \frac{1}{p}\right) .
\end{aligned}$$

For $|x|_p > |a|_p^{-1/2} = p^{-N}$ we have $|2ax|_p = p^M \geq p^{N+1}$, and by virtue of the formulas (5.4), (1.7) and (3.2) of Sec. 4.3 the integral is equal to (5.1_3):

$$\chi_p(ax^2)\Bigg[\sum_{-\infty<\gamma\le -M} e^{-p^{2\gamma}}\int_{S_\gamma} dy + e^{-p^{2(1-M)}}\int_{S_{1-M}}\chi_p(-2axy)dy$$
$$+\sum_{2-M\le\gamma\le -N} e^{-p^{2\gamma}}\int_{S_\gamma}\chi_p(-2axy)dy$$
$$+\sum_{1-N\le\gamma<\infty} e^{-p^{2\gamma}}\int_{S_\gamma}\chi_p(ay^2-2axy)dy\Bigg]$$
$$=\left(1-\frac{1}{p}\right)\chi_p(ax^2)\sum_{-\infty<\gamma\le -M} e^{-p^{2\gamma}}p^{\gamma}-\chi_p(ax^2)p^{-M}e^{-p^{2(1-M)}}$$
$$+e^{-|x|_p^2}|a|_p^{-1/2}$$
$$=e^{-|x|_p^2}\lambda_p(a)|a|_p^{-1/2}$$
$$+|ax|_p^{-1}\chi_p(ax^2)\left[\left(1-\frac{1}{p}\right)\sum_{0\le\gamma<\infty}p^{-\gamma}e^{-|ax|_p^{-2}-2\gamma}-e^{-p^2|ax|_p^{-2}}\right].$$

Let $\gamma(a)$ be odd, $|a|_p = p^{2N+1}$, $N\in\mathbb{Z}$. Then $|a|_p\ge p^{2-2\gamma}$ if $\gamma\ge -N+1$ and $|a|_p = p^{1-2\gamma}$ if $\gamma = -N$. The integral is equal to

$$\chi_p(ax^2)\Bigg[\sum_{-\infty<\gamma\le -N-1} e^{-p^{2\gamma}}\int_{S_\gamma}\chi_p(-2axy)dy$$
$$+e^{-p^{-2N}}\int_{S_{-N}}\chi_p(ay^2-2axy)dy$$
$$+\sum_{1-N\le\gamma<\infty} e^{-p^{2\gamma}}\int_{S_\gamma}\chi_p(ay^2-2axy)dy\Bigg].\tag{5.5}$$

For $|x|_p\le |a|_p^{-1/2}$ i.e. $|x|_p\le p^{-N-1}$ we have $\chi_p(ax^2)=1$,

$$\chi_p(-2axy)=1,\ y\in S_\gamma,\ \gamma\le -N-1,\qquad \left|\frac{2ax}{a}\right|_p = |x|_p\ne p^{\gamma},\ \gamma\ge -N+1.$$

By virtue of the formulas (1.7) and (1.8) the integral (5.5) is expressed by the formula (5.1_2):

$$\left(1-\frac{1}{p}\right)\sum_{-\infty<\gamma\le -N-1} e^{-p^{2\gamma}}p^{\gamma}+e^{-p|a|_p^{-1}}|a|_p^{-1/2}\left[\lambda_p(a)\chi_p(-ax^2)-\frac{1}{\sqrt{p}}\right]$$

$$=\left(1-\frac{1}{p}\right)\frac{1}{\sqrt{p}}|a|_p^{-1/2}S\left(\frac{1}{p}|a|_p^{-1},\frac{1}{p}\right)+|a|_p^{-1/2}e^{-p|a|_p^{-1}}\left[\lambda_p(a)-\frac{1}{\sqrt{p}}\right].$$

For $|x|_p > p^{-N}$ we have $|2ax|_p = p^M \ge p^{N+2}$. By virtue of the formulas (1.7), (1.8) and (3.2) of Sec. 4.3 the desired integral (5.5) is expressed by the formula (5.1_3):

$$\chi_p(ax^2)\Bigg[\sum_{-\infty<\gamma\le -M} e^{-p^{2\gamma}}\int_{S_\gamma}dy+e^{-p^{2(1-M)}}\sum_{S_{1-M}}\chi_p(-2axy)dy$$

$$+\sum_{2-M\le\gamma<\infty} e^{-p^{2\gamma}}\int_{S_\gamma}\chi_p(ay^2-2axy)dy\Bigg]$$

$$=\chi_p(ax^2)\Bigg[\left(1-\frac{1}{p}\right)\sum_{\gamma=M}^{\infty} e^{-p^{-2\gamma}}p^{-\gamma}-p^{-M}e^{-p^{2(1-M)}}$$

$$+e^{-|x|_p^2}\lambda_p(a)|a|_p^{-1/2}\chi_p(-ax^2)\Bigg]$$

$$=\lambda_p(a)|a|_p^{-1/2}e^{-|x|_p^2}+\left(1-\frac{1}{p}\right)\chi_p(ax^2)|ax|_p^{-1}S\left(|ax|_p^{-2},\frac{1}{p}\right)$$

$$-\chi_p(ax^2)|ax|_p^{-1}e^{-p^2|ax|_p^{-2}}.$$

For $|x|_p = p^{-N}$ we have $|2ax|_p = p^{N+1}$, and owing to the formulas (1.7) and (1.8) the integral (5.5) is expressed again by the formula (5.1_3):

$$\chi_p(ax^2)\Bigg[\sum_{-\infty<\gamma\le -N-1} e^{-p^{2\gamma}}\left(1-\frac{1}{p}\right)p^{\gamma}+e^{-p^{-2N}}\int_{S_{-N}}\chi_p(ay^2-2axy)dy$$

$$+\sum_{-N+1\le\gamma<\infty} e^{-p^{2\gamma}}\int_{S_\gamma}\chi_p(ay^2-2axy)dy\Bigg]$$

$$=\left(1-\frac{1}{p}\right)\chi_p(ax^2)S\left(|ax|_p^{-2},\frac{1}{p}\right)$$

$$+e^{-|x|_p^2}\chi_p(ax^2)|a|_p^{-1/2}\left[\lambda_p(a)\chi_p(-ax^2)-\frac{1}{\sqrt{p}}\right]$$

as

$$\frac{1}{\sqrt{p}}|a|_p^{-1/2}e^{-|x|_p^2} = |ax|_p^{-1}e^{-p^2|ax|_p^{-2}} . \qquad \blacksquare$$

Example 14. $p = 2$ (see [239])

$$\int_{Q_2} e^{-|y|_2^2}\chi_2(a(x-y)^2)dy$$

$$= (-1)^{a_1} i|a|_2^{-1/2}e^{-4|a|_2^{-1}} + \frac{1}{2}|a|_2^{-1/2}S\left(|a|_2^{-1}, \frac{1}{2}\right), \quad |x|_2 \le |a|_2^{-1/2}, \tag{5.6$_1$}$$

$$= |a|_2^{-1/2}e^{-4|a|_2^{-1}} + \frac{i}{2}(-1)^{a_1}|a|_2^{-1/2}S\left(|a|_2^{-1}, \frac{1}{2}\right), \quad |x|_2 = 2|a|_2^{-1/2}, \tag{5.6$_2$}$$

if $\gamma(a)$ is even;

$$= \frac{1}{2\sqrt{2}}|a|_2^{-1/2}S\left(\frac{1}{2}|a|_p^{-1}, \frac{1}{2}\right) - \frac{1}{\sqrt{2}}|a|_2^{-1/2}e^{-2|a|_2^{-1}}$$
$$+ \sqrt{2}\lambda_2(a)|a|_2^{-1/2}e^{-8|a|_2^{-1}}, \quad |x|_2 \le \frac{1}{\sqrt{2}}|a|_2^{-1/2}, \tag{5.6$_3$}$$

$$= \frac{1}{\sqrt{2}}|a|_2^{-1/2}S\left(2|a|_2^{-1}, \frac{1}{2}\right) + \sqrt{2}\lambda_2(a)|a|_2^{-1/2}e^{-8|a|_2^{-1}},$$
$$|x|_2 = |a|_2^{-1/2}, \tag{5.6$_4$}$$

$$= \frac{1}{2\sqrt{2}}\lambda_2(a)|a|_2^{-1/2}S\left(2|a|_p^{-1}, \frac{1}{2}\right), \quad |x|_2 = \sqrt{2}|a|_2^{-1/2}, \tag{5.6$_5$}$$

if $\gamma(a)$ is odd;

$$= \sqrt{2}\lambda_2(a)|a|_2^{-1/2}e^{-|x|_2^2} + |ax|_2^{-1}\chi_2(ax^2)\left[S\left(4|ax|_2^{-2}, \frac{1}{2}\right) - 2e^{-16|ax|_2^{-2}}\right],$$
$$|x|_2 \ge 2|a|_2^{-1/2} . \tag{5.6$_6$}$$

The integrals (5.6) are calculated in a similar way to integrals (5.1) for $p \neq 2$.

From the formula (5.6_6) it follows the asymptotics

$$\int_{\mathbb{Q}_2} e^{-|y|_2^2}\chi_2(a(x-y)^2)dy \sim \frac{192}{7}|ax|_2^{-3}\chi_2(ax^2) + O(|x|_2^{-5}) ,$$

$$|x|_2 \to \infty . \tag{5.7}$$

6. *Analysis of the Function* $S(\alpha, q)$

The function $S(\alpha, q)$ is defined by the formula (5.2).

The function $S(\alpha,q)$ is entire on α, real for real α (and positive for $q \geq 0$) and satisfies the functional equation

$$e^{-\alpha} = S(\alpha, q) - qS(\alpha q^2, q) . \tag{6.1}$$

It has asymptotics

$$S(\alpha, q) \sim e^{-\alpha} + O(e^{-\alpha q^2}), \quad \alpha \to -\infty, \quad (|q| < 1), \tag{6.2}$$

$$S(\alpha, q) \sim \frac{1}{\sqrt{\alpha}}C(\ln\alpha, q) + O(e^{-\alpha/q^2}), \quad \alpha \to \infty (0 < q < 1) , \tag{6.3}$$

where the function

$$C(x, q) = e^{x/2} \sum_{-\infty<k<\infty} q^k e^{-e^x q^{2k}}, \quad -\infty < x < \infty \tag{6.4}$$

is real analytic positive $2|\ln q|$-periodic.

■ The listed properties of the function $S(\alpha, q)$, except the asymptotics (6.3), follow directly from the representation (5.2).

We shall prove the asymptotics (6.3). Let us represent the function S in the form

$$S(\alpha, q) = \sum_{-\infty<k<\infty} q^k e^{-\alpha q^{2k}} - \sum_{1\leq k<\infty} q^{-k} e^{-\alpha q^{-2k}} . \tag{6.5}$$

If we set $\alpha = e^x$ and use the function $C(x, q)$ (see (6.4)) then we get from (6.5)

$$S(\alpha, q) = \frac{1}{\sqrt{\alpha}} C(\ln \alpha, q) - \psi(\alpha) \tag{6.6}$$

where the remainder term

$$\psi(\alpha) = \sum_{1 \le k < \infty} q^{-k} e^{-\alpha q^{-2k}}$$

satisfies the estimates

$$\frac{1}{q} e^{-\alpha/q^2} < \psi(\alpha) < C_q e^{-\alpha/q^2}, \quad \alpha \ge 2$$

and the function $C(x, q)$ is $T = 2|\ln q|$-periodic,

$$\begin{aligned} C(x + T, q) &= e^{\frac{x+T}{2}} \sum_{-\infty < k < \infty} q^k e^{-e^{x+T} q^{2k}} \\ &= e^{x/2} \sum_{-\infty < k < \infty} q^{k-1} e^{-e^x q^{2(k-1)}} = C(x, q) . \end{aligned}$$

■

From the formulas (5.1_1), (5.1_2), (5.6_1), (5.6_3) it follows the asymptotics as $|a|_p^{-1} \to \infty$

$$\int_{\mathbb{Q}} e^{-|x-y|_p^2} \chi_p(ay^2) dy$$

$$\sim \begin{cases} \left(1 - \frac{1}{p}\right) C\left(\ln |a|_p^{-1}, \frac{1}{p}\right) + O(e^{-p^2|a|_p^{-1}}), & \gamma(a) \text{ even}, \\ \left(1 - \frac{1}{p}\right) C\left(\ln |a|_p^{-1} - \ln p, \frac{1}{p}\right) + O(|a|_p^{-1/2} e^{-p|a|_p^{-1}}), & \gamma(a) \text{ odd}. \end{cases} \tag{6.7}$$

VI. Generalized Functions (Distributions)

The theory of generalized functions on any locally compact group was presented by F. Bruhat [43] and on a locally compact disconnected field by I.M.Gelfand, M.I.Graev and I.I.Pjatetskii-Shapiro [82]. Here we expose the following [206] bases of this theory adapted to the field $\mathbb{Q}_p$ (and to the space $\mathbb{Q}_p^n$). This theory in many aspects is similar to the corresponding theory on the space $\mathbb{R}^n$ but there are some essential distinctions.

1. *Locally Constant Functions*

A complex-valued function $f(x)$ defined on $\mathbb{Q}_p$ is called *locally-constant* if for any point $x \in \mathbb{Q}_p$ there exists an integer $l(x) \in \mathbb{Z}$ such that

$$f(x + x') = f(x), \quad |x'|_p \leq p^{l(x)} .$$

The set of locally-constant functions on $\mathbb{Q}_p$ we denote by $\mathcal{E} = \mathcal{E}(\mathbb{Q}_p)$.

It is clear that any function from $\mathcal{E}$ is continuous on $\mathbb{Q}_p$. The set $\mathcal{E}$ is linear over the field $\mathbb{C}$.

Lemma 1. *Let $f \in \mathcal{E}$ and K be a compact in $\mathbb{Q}_p$. Then there exists $l \in \mathbb{Z}$ such that*

$$f(x + x') = f(x), \quad |x'|_p \leq p^l, \quad x \in K .$$

■ Let K is contained in a disc B_N. It is sufficient to prove the Lemma 1 for the compact B_N. By the Heine-Borel Lemma (see Corollary 5 from the Lemma 3 of Sec. 1.3) from the covering $\{B_{l(x)}(x), x \in B_N\}$ of the compact B_N it is possible to choose a finite disjont subcovering

$$\{B_{l(x^k)}(x^k), \quad k = 1, 2, \ldots, M\} .$$

Let us denote $l = \min_k l(x^k)$. Then for any point $x \in B_{l(x^k)}(x^k)$ and for all $x' \in B_l$ we have

$$f(x + x') = f(x^k + x - x^k + x') = f(x)$$

as

$$|x - x^k + x'|_p \leq \max(|x - x^k|_p, |x'|_p) \leq \max(p^{l(x^k)}, p^l) = p^{l(x^k)} . \quad ■$$

Examples.

1. $|x|_p \in \mathcal{E}$.
2. $|x|_p \Omega(p^\gamma |x|_p) \in \mathcal{E}$, $\gamma \in \mathbb{Z}$, where $\Omega(t) = 1$, $0 \leq t \leq 1$, $\Omega(t) = 0$, $t > 1$.

3. $\chi_p(x), \chi_p(x^2) \in \mathcal{E}$.
4. $\delta(x_0 - k) \in \mathcal{E}$, $k = 1, 2, \ldots, p-1$, where $\delta(x_0 - k) = 1$, $x_0 = k$, $\delta(x_0 - k) = 0$, $x_0 \neq k$.
5. Let K be a clopen set in $\mathbb{Q}_p$ (see Sec. 1.3) and $\theta_K(x)$ be its characteristic function: $\theta_K(x) = 1$, $x \in K$, $\theta_K(x) = 0$, $x \notin K$. Then $\theta_K \in \mathcal{E}$.

Denote by $\Delta_k(x)$ the characteristic function of the disc $B_k : \Delta_k(x) = \Omega(p^{-k}|x|_p)$, $k \in \mathbb{Z}$.

Lemma 2. *Every function f from $\mathcal{E}$ in every disc B_N is represented in the form*

$$f(x) = \sum_{1 \leq \nu \leq p^{N-l}} f(a^\nu)\Delta_l(x - a^\nu), \quad x \in B_N \tag{1.2}$$

where $l \in \mathbb{Z}$ and $a^\nu \in B_N$ such that the discs $B_l(a^\nu)$, $\nu = 1, 2, \ldots, p^{N-l}$ form the canonical covering of the disc B_N.

■ By the Lemma 1 there exists a number $l \in \mathbb{Z}$, such that

$$f(x + x') = f(x), \quad x \in B_N, \quad x' \in B_l\,.$$

By $l = N$ the statement of the Lemma is obvions (by $a^1 = 0$). Let $l < N$. Owing to Sec. 1.3 (the example 2) the disc B_N can be covered by disjont discs $B_l(a^\nu)$, $\nu = 1, 2, \ldots, p^{N-l}$ with centers a^ν of the form

$$a^1 = 0, \quad a^\nu = p^{-r}(a_0 + a_1 p + \ldots + a_{r-l-1}p^{r-l-1}), \quad r = N, N-1, \ldots, l+1$$

(the canonical covering of the disc B_N). Therefore

$$\sum_{1 \leq \nu \leq p^{N-l}} \Delta_l(x - a^\nu) = \begin{cases} 1, & x \in B_N \\ 0, & x \notin B_N. \end{cases}$$

Hence the equality (1.2) is valid:

$$f(x) = \sum_{1 \leq \nu \leq p^{N-l}} f(x)\Delta_l(x - a^\nu) = \sum_{1 \leq \nu \leq p^{N-l}} f(a^\nu)\Delta_l(x - a^\nu), \quad x \in B_N\,.$$

■

Convergence in $\mathcal{E}$ we define by the following way: $f_k \to 0$, $k \to \infty$ in $\mathcal{E}$ if for any compact $K \subset \mathbb{Q}_p$

$$f_k(x) \overset{x \in K}{\rightrightarrows} 0, \qquad k \to \infty .$$

2. *Test Functions*

$n = 1$. We call a *test function* every function from $\mathcal{E}$ with compact support.

The set of test functions is linear, we denote it by $\mathcal{D} = \mathcal{D}(\mathbb{Q}_p)$.

Let $\varphi \in \mathcal{D}$. Then by the Lemma 1 of Sec. 6.1 there exists $l \in \mathbb{Z}$, such that

$$\varphi(x + x') = \varphi(x), \quad x' \in B_l, \quad x \in \mathbb{Q}_p .$$

Such largest number l we call *the parameter of constancy* of a function φ, $l = l(\varphi)$.

Let us denote by $\mathcal{D}_N^l = \mathcal{D}_N^l(\mathbb{Q}_p)$ the set of test functions with support in the circle B_N and with parameter of constancy $\geq l$.

The following imbedding is true:

$$\mathcal{D}_N^l \subset \mathcal{D}_{N'}^{l'}, \quad N \leq N', l \leq l' .$$

Examples. 1. $\Delta_\gamma(x) \in \mathcal{D}_\gamma^\gamma$, $\gamma \in \mathbb{Z}$.

2. $\delta(|x|_p - p^\gamma) \in \mathcal{D}_\gamma^{\gamma-1}$, $\gamma \in \mathbb{Z}$, where

$$\delta(|x|_p - p^\gamma) = \begin{cases} 1, & x \in S_\gamma, \\ 0, & x \notin S_\gamma. \end{cases} \tag{2.1}$$

3. $\delta(x_0 - k)\delta(|x|_p - p^\gamma) \in \mathcal{D}_\gamma^{\gamma-1}$, $k = 1, 2, \ldots, p-1$, $\gamma \in \mathbb{Z}$, where the function $\delta(x_0 - k)$ is defined by the formula (1.1).

4. If K is a clopen set then $\theta_K \in \mathcal{D}$.

5. $\Delta_\gamma(x)\chi_p(x) \in \mathcal{D}_\gamma^l$, $l = \min(\gamma, 0)$, $\gamma \in \mathbb{Z}$.

By the Lemma 2 of Sec. 6.1 every function φ from $\mathcal{D}_N^l$ is represented in the form

$$\varphi(x) = \sum_{1 \leq \upsilon \leq p^{N-l}} \varphi(a^\upsilon)\Delta_l(x - a^\upsilon), \quad x \in \mathbb{Q}_p \tag{2.2}$$

for some $a^\upsilon \in B_N$, which do not depend on φ and such that the discs $B_l(a^\upsilon)$, $\upsilon = 1, 2, \ldots, p^{N-l}$, are disjont and cover the disc B_N.

From here it follows that *the space $\mathcal{D}_N^l$ is finite-dimensional, its dimension is p^{N-l}; the functions*

$$\Delta_l(x - a^{\upsilon}), \quad \upsilon = 1, 2, \dots, p^{N-l}$$

form an orthogonal basis in $\mathcal{D}_N^l$.

Convegence in $\mathcal{D}$ we define by the following way: $\varphi_k \to 0$, $k \to \infty$ in $\mathcal{D}$ iff

(i) $\varphi_k \in \mathcal{D}_N^l$ where N and l do not depend on k,
(ii) $\varphi_k \overset{x \in Q_p}{\longrightarrow} 0$, $k \to \infty$.

This convergence assigns the Schwartz topology in $\mathcal{D}$.

The space $\mathcal{D}$ is complete i.e. for every convergent in itself sequence $\{\varphi_k, k \to \infty\}$, $\varphi_k \in \mathcal{D}$, $\varphi_k - \varphi_l \to 0$, $k, l \to \infty$ in $\mathcal{D}$, there exists a function $\varphi \in \mathcal{D}$ such that $\varphi_k \to \varphi$, $k \to \infty$ in $\mathcal{D}$.

From the definitions it follows directly

$$\mathcal{D} = \lim_{N \to \infty} \operatorname{ind} \mathcal{D}_N, \qquad \mathcal{D}_N = \lim_{l \to -\infty} \operatorname{ind} \mathcal{D}_N^l \tag{2.3}$$

Now we shall prove: *$\mathcal{D}(\mathbb{Q}_p)$ is dense in $C(K)$.**

■ Let $f \in C(K)$ and ε be an arbitrary positive number.

There exists a number $\gamma \in \mathbb{Z}$ such that $|f(x) - f(a)| < \varepsilon$ if $x \in B_\gamma(a) \cap K$, $a \in K$. As the compact K can be covered by a finite number of disjoint discs $B_\gamma(a^{\upsilon})$ (see Corollary 3 from the Lemma 3 of Sec. 1.3) the characteristic functions $\Delta_\gamma(x - a^{\upsilon})$ of these discs obey the property

$$\sum_{\nu} \Delta_\gamma(x - a^{\upsilon}) = 1, \quad x \in K, \tag{2.4}$$

besides $\Delta_\gamma(x - a^{\upsilon}) \in \mathcal{D}$ (see Example 1). Therefore

$$f_\gamma(x) = \sum_{\nu} f(a^{\upsilon}) \Delta_\gamma(x - a^{\upsilon}) \in \mathcal{D}$$

and owing to (2.4)

$$\|f - f_\gamma\|_{C(K)} \le \max_{x \in K} \left| \sum_{\nu} [f(x) - f(a^{\upsilon})] \Delta_\gamma(x - a^{\upsilon}) \right| < \varepsilon \sum_{\nu} \Delta_\gamma(x - a^{\upsilon}) = \varepsilon.$$

■

* The definition of the space $C(K)$ see in Sec. 4.1

Let $\mathcal{O}$ be an open set in $\mathbb{Q}_p$. The space of test functions $\mathcal{D}(\mathcal{O})$ is defined as a set of test functions from $\mathcal{D}(\mathbb{Q}_p) = \mathcal{D}$ which supports are contained in $\mathcal{O}$. The space $\mathcal{D}(\mathcal{O})$ is the subspace of the space $\mathcal{D}$; its properties are similar to the properties of $\mathcal{D}$ as in the case of the field $\mathbb{R}$ (see [205]).

$\mathcal{D}(\mathcal{O})$ is dense in $L^\rho(\mathcal{O})$, $1 \le \rho < \infty$.

■ It follows from the facts that $\mathcal{D}(\mathcal{O})$ is dense in $C(K)$ and $C(K)$ is dence in $L^\rho(\mathcal{O})$ where K is an arbitrary compact contained in $\mathcal{O}$ (see Sec. 4.1). ■

In the space $\mathcal{D}(\mathcal{O})$ the **Theorem on "decomposition of unity"** is valid: *let an open set $\mathcal{O}$ be a union of no more than a countable set of disjoint circles,*

$$\mathcal{O} = \bigcup_k B_{\gamma_k}(a^k), \quad B_{\gamma_k}(a^k) \cap B_\gamma(a^i) = \phi, \quad k \neq i .$$

Then their characteristic functions $\Delta_{\gamma_k}(x - a^k)$, $k = 1, 2, \ldots$ form a decomposition of unity in $\mathcal{O}$,

$$\sum_k \Delta_{\gamma_k}(x - a^k) = 1, \quad x \in \mathcal{O} . \tag{2.5}$$

In conclusion we shall prove the following

Lemma. *In order that $\varphi_k \to 0$, $k \to \infty$ in $\mathcal{D}$, it is necessary and sufficient that the condition* (i) *and one of the conditions*

(ii$_1$) $\varphi_k(x) \to 0$, $k \to \infty$, $x \in \mathbb{Q}_p$,

(ii$_2$) $\int_{\mathbb{Q}_p} \varphi_k(x)\varphi(x)dx \to 0$, $k \to \infty$, $\varphi \in \mathcal{D}$

are satisfied.

■ The conditions (ii$_1$) and (ii$_2$) are necessary. Let us prove their sufficiency. From the condition (i) and (ii$_1$) it follows the condition (ii) by virtue of representation (2.2). It remains to prove that from the conditions (i) and (ii$_2$) it follows condition (ii$_1$). Let it be not the case. By the condition (i) $\varphi_k \in \mathcal{D}_N^l$, $k = 1, 2, \ldots$. Then there exists a subsequence $\{\varphi_{k_i}, i \to \infty\}$ such that for some point $a \in B_N$, $\varphi_{k_i}(a) \to C \neq 0$, $i \to \infty$ (a number C may be infinite). Then for the test function $\varphi = \Delta_l(x-a)$ we have a contradictory chain of equalities:

$$\lim_{k\to\infty}\int_{\mathbb{Q}_p}\varphi_k(x)\Delta_l(x-a)dx = 0 = \lim_{i\to\infty}\int_{\mathbb{Q}_p}\varphi_{k_i}(x)\Omega(p^{-l}|x-a|_p)dx$$

$$= \lim_{i\to\infty}\int_{B_l(a)}\varphi_{k_i}(x)dx = \lim_{i\to\infty}\varphi_{k_i}(x)\int_{B_l(a)}dx = Cp^l \neq 0\,. \qquad \blacksquare$$

3. *Generalized Functions (Distributions)*, $n = 1$

A generalized function (distribution) f: we call every linear functional f on $\mathcal{D}$, $f: \varphi \to (f,\varphi)$, $\varphi \in \mathcal{D}$. This set we denote by $\mathcal{D}' = \mathcal{D}'(\mathbb{Q}_p)$. $\mathcal{D}'$ is a linear set: linear combination $\lambda f + \mu g$ of generalized functions f and g from $\mathcal{D}'$ (λ and μ are any complex numbers) is defined by the equality

$$(\lambda f + \mu g, \varphi) = \lambda(f,\varphi) + \mu(g,\varphi), \quad \varphi \in \mathcal{D}\,.$$

Convergence in $\mathcal{D}'$: we define as the weak convergence of functionals: $f_k \to 0$, $k \to \infty$ in $\mathcal{D}'$ iff $(f_k,\varphi) \to 0$, $k\to\infty$, $\varphi \in \mathcal{D}$.

Lemma. *If A is a linear operator from $\mathcal{D}$ into a linear topological space M then operator A is continuous from $\mathcal{D}$ into M.*

$\blacksquare$ Let $\varphi_k \to 0$, $k \to \infty$ in $\mathcal{D}$. Then $\varphi_k \in \mathcal{D}_N^l$ for some N and l and owing to the representation (2.2) of Sec. 6.2

$$\varphi_k(x) = \sum_{1\le\nu\le p^{N-l}}\varphi_k(a^\nu)\Delta_l(x-a^\nu), \varphi_k(a^\nu) \to 0, k\to\infty\,.$$

From here by virtue of linearity of the operator A it follows that A is continuous

$$A\varphi_k(x) = \sum_{1\le\nu\le p^{N-l}}\varphi_k(a^\nu)A\Delta_l(x-a^\nu) \to 0, \quad k\to\infty \text{ in } M\,. \qquad \blacksquare$$

From the Lemma it follows that $\mathcal{D}'$ is the set of linear continuous functionals on $\mathcal{D}$, i.e. the space $\mathcal{D}'$ is a strongly conjugate space to the space $\mathcal{D}$.

Therefore by the study of the space $\mathcal{D}'$ it is possible to use general theorems of the functional analysis. In addition the theory is essentially simplified in comparison with the corresponding theory over the field $\mathbb{R}$ (cf. [205]): it is sufficient to verify the linearity of functionals, their continuity follows automatically.

The space $\mathcal{D}'$ is complete.

■ Let a sequence $\{f_k, k \to \infty\}$ of functionals $f_k \in \mathcal{D}'$ converge in itself, $f_k - f_l \to 0$, $k, l \to \infty$ in $\mathcal{D}'$, i.e.

$$(f_k, \varphi) - (f_l, \varphi) \to 0, \qquad k, l \to \infty, \quad \varphi \in \mathcal{D} .$$

Hence there exists a number $C(\varphi)$ such that

$$\lim_{k \to \infty} (f_k, \varphi) = C(\varphi) . \tag{3.1}$$

It is clear that the functional $C(\varphi)$ is linear, and thus it belongs to $\mathcal{D}'$. We put $C(\varphi) = (f, \varphi)$, $f \in \mathcal{D}'$, $\varphi \in \mathcal{D}$. The equality (3.1) shows that $f_k \to f$, $k \to \infty$ in $\mathcal{D}'$. ■

Every function $f \in L^1_{\text{loc}}$* defines a generalized function $f \in \mathcal{D}'$ by the formula

$$(f, \varphi) = \int_{\mathbb{Q}_p} f(x)\varphi(x)dx, \quad \varphi \in \mathcal{D} . \tag{3.2}$$

The correspondence (3.2) *between functions $f \in L^1_{\text{loc}}$ and generalized functions $f \in \mathcal{D}'$ is one-to-one.*

A degeneralized function f vanishes on an open set $\mathcal{O}$, $f(x) = 0$, $x \in \mathcal{O}$, if $(f, \varphi) = 0$ for all $\varphi \in \mathcal{D}(O)$. Analogously the equality of generalized functions f and g in $\mathcal{O}$, $f(x) - g(x) = 0$, $x \in \mathcal{O}$, is defined.

Since in $\mathcal{D}$ a "decomposition of unity" is valid (see Sec. 6.2) then the notion of support of a generalized function f is introduced by the standard way, like to the case of the field $\mathbb{R}$. We denote support f by supp f; $x \in$ supp f means that f does not vanish in any neighborhood of the point x.

Example 1. The Dirac δ-function

$$(\delta, \varphi) = \varphi(0), \quad \varphi \in \mathcal{D} . \tag{3.3}$$

* The definition of L^1_{loc} see in Sec. 4.1.

It is clear that supp $\delta = \{0\}$.

Conversely, *every* $f \in \mathcal{D}'$ *for which* supp $f = \{0\}$ *is represented in the form* $f = C\delta$ *where* C *is a constant.*

■ Let $\varphi \in \mathcal{D}$ and l be its parameter of constancy (see Sec. 6.2), so that $\varphi(x) = \varphi(0)$, $|x|_p \leq p^l$. Let η be another function from $\mathcal{D}$ such that $\eta(x) = 1$, $|x|_p \leq 1$. Then we have

$$(f, \varphi) = (f, \eta\varphi) = \varphi(0)(f, \eta) = C(\delta, \varphi), \quad C = (f, \eta) . \qquad ■$$

Example 2.

$$\delta_k(x) = p^k \Omega(p^k |x|_p) \to \delta(x), \quad k \to \infty \text{ in } \mathcal{D}' \tag{3.4}$$

■ Let $\varphi \in \mathcal{D}$ and l be its parameter of constancy. Then for all $k \geq -l$ we shall have

$$(\delta_k, \varphi) = \int_{\mathbb{Q}_p} p^k \Omega(p^k |x|_p) \varphi(x) dx = p^k \int_{B_{-k}} \varphi(x) dx$$

$$= p^k \varphi(0) \int_{B_{-k}} dx = \varphi(0) = (\delta, \varphi) . \qquad ■$$

Remark. The limiting relation (3.4) is equivalent to the following one

$$\int_{\mathbb{Q}_p} \delta_k(x) \varphi(x) dx \to \varphi(0), \qquad k \to \infty, \varphi \in \mathcal{D}. \tag{3.4'}$$

The limiting relation (3.4′) is valid also for all functions φ continuous at the point $x = 0$.

4. *Linear Operators in* $\mathcal{D}'$

Linear operators in $\mathcal{D}'$ are defined as conjugate ones to corresponding linear operators in $\mathcal{D}$: Let A be a linear operator from $\mathcal{D}$ into $\mathcal{D}$ then its conjugate operator $A^* : \mathcal{D}' \to \mathcal{D}'$ is defined by the formula

$$(A^* f, \varphi) = (f, A\varphi), \quad \varphi \in \mathcal{D}, \ f \in \mathcal{D}' . \tag{4.1}$$

It is clear that $A^*f \in \mathcal{D}'$ and A^* is linear and thus by the Lemma of Sec. 6.3 it is a continuous operator from $\mathcal{D}'$ into $\mathcal{D}'$.

A specific expression for the operator A^* we get by using the formula (4.1) to functions $f \in L^1_{\text{loc}}$ after performing the integral (see (3.2))

$$(f, A\varphi) = \int_{\mathbb{Q}_p} f(x)(A\varphi)(x)dx$$

to the form (A^*f, φ).

According to what has been said a *product* af, $a \in \mathcal{E}$, $f \in \mathcal{D}'$ is defined by the formula

$$(af, \varphi) = (f, a\varphi), \quad \varphi \in \mathcal{D}\,. \tag{4.2}$$

Example 1. $a(x)\delta(x) = a(0)\delta(x)$.

Example 2. If $f \in \mathcal{D}'$ and supp $f \subset B_N$ then

$$f = \Delta_N(x)f\,. \tag{4.3}$$

Example 3. If $f \in \mathcal{D}'$ and supp f is a clopen set then

$$f = \theta_{\text{supp}f}(x)f \tag{4.4}$$

where $\theta_{\text{supp}f}$ is the characteristic function of the set supp f (see Example 5 of Sec. 6.1).

Linear change of variable $y = ax + b$, $a \neq 0$ in a generalized function $f(y)$ in accordance with the formula (4.1) is defined by the equality

$$(f(ax+b), \varphi) = \left(f(y), \frac{1}{|a|_p}\varphi\left(\frac{y-b}{a}\right)\right), \qquad \varphi \in \mathcal{D}. \tag{4.5}$$

Example 4. $(\delta(x - x^0), \varphi) = \varphi(x^0)$.

We denote by $\hat{f}(x) = f(-x)$ the *reflection operation* $f(x) \to f(-x)$ in $\mathcal{D}'$ ($a = -1$, $b = 0$ in (4.5)).

Example 5. $\delta(x) = \hat{\delta}(x)$ is even.

Denote by $\mathcal{E}' = \mathcal{E}'(\mathbb{Q}_p)$ the strongly conjugate space to the space $\mathcal{E}$, i.e. the space of linear continuous functionals on $\mathcal{E}$. It clear that $\mathcal{E}' \subset \mathcal{D}'$.

Theorem. *In order that a generalized function f from $\mathcal{D}'$ belongs to $\mathcal{E}'$ it is necessary and sufficient that* supp f *is compact.*

■ **Sufficiency** immediatly follows from the formula (4.3).

Necessity. Let $f \in \mathcal{E}'$ and supp f be an unbounded set in $\mathbb{Q}_p$. Then there exists a sequence of points $\{x^k, k = 1, 2, \dots\}$, $x^k \in$ supp f, such that $|x^k|_p \to \infty$, $k \to \infty$. It means that there exist neighborhoods $B_\gamma(x^k)$, $\gamma_k \leq \gamma$ and functions $\varphi_k \in \mathcal{D}(B_\gamma(x^k))$ such that $(f, \varphi_k) = 1$, $k \to 1, 2, \dots$. On the other hand $\varphi_k \to 0$, $k \to \infty$ in $\mathcal{E}$ (see Sec. 6.1) and hence $(f, \varphi_k) \to 0$, $k \to \infty$. The contradiction shows that the supp f is bounded. ■

Let $\mathcal{O}$ be an open set in $\mathbb{Q}_p$. The space $\mathcal{D}'(\mathcal{O})$ is the set of linear (and hence continuous) functionals on $\mathcal{D}(\mathcal{O})$.

If $\mathcal{O}$ is a clopen set then every $f \in \mathcal{D}'(\mathcal{O})$ admits an extention $F \in \mathcal{D}'(\mathbb{Q}_p)$.

■ In this case $\theta_{\mathcal{O}}(x) \in \mathcal{E}$ (see Example 5 of Sec. 6.1), and a required extention F is given by the formula

$$(F, \varphi) = (f, \theta_{\mathcal{O}}\varphi), \qquad \varphi \in \mathcal{D}\,. \qquad ■$$

Example 6. The function

$$e^{-|x|_p^{-1}} \in L^1_{\text{loc}}(\mathbb{Q}_p \setminus \{0\})\,,$$

and therefore it belongs to $\mathcal{D}'(\mathbb{Q}_p \setminus \{0\})$. It admits an extention (regularisation) from $\mathcal{D}'$ by the formula

$$(f_0, \varphi) = \int_{|x|_p \leq 1} e^{-|x|_p^{-1}} [\varphi(x) - \varphi(0)] dx + \int_{|x|_p > 1} e^{-|x|_p^{-1}} \varphi(x) dx$$

This fact does not occur in the case of the field $\mathbb{R}$ (cf. [205])!

Every generalized function $f \in \mathcal{D}'(\mathcal{O})$ admits the restriction $f_{\mathcal{O}'} \in \mathcal{D}'(\mathcal{O}')$ on any open set $\mathcal{O}' \subset \mathcal{O}$ by the rule

$$(f_{\mathcal{O}'}, \varphi) = (f, \varphi), \quad \varphi \in \mathcal{D}(\mathcal{O}') .$$

Conversely, in the space $\mathcal{D}'(\mathcal{O})$ the **Theorem on "picewise sewing"** is valid: *let an open set $\mathcal{O}$ be a union of no more than a contable set of disjoint discs $B_{\gamma^k}(a^k)$, $k = 1, 2, \ldots$, and let $f_k \in \mathcal{D}'(B_{\gamma_K}(a^k))$. Then there exists a unique $f \in \mathcal{D}'(\mathcal{O})$ such that $f_{B_{\gamma_k}}(a^k) = f_k$, $k = 1, 2, \ldots$.*

■ It follows from the Theorem on "decomposition of unity" of Sec. 6.2. The desired function f is defined by the formula

$$f(x) = \sum_k \Delta_{\gamma^k}(x - a^k) f_k(x), \quad x \in \mathcal{O} . \qquad \blacksquare$$

5. *Test and Generalized Functions (Distributions)*, $n > 1$.

The theory of test and generalized functions in $\mathbb{Q}_p^n$, $n > 1$ (see Sec. 1.7) is constructed similar to one in $\mathbb{Q}^p$, all results in a suitable reformulation remain valid. The ball (see Sec. 1.7)

$$B_\gamma(a) = [x \in \mathbb{Q}_p^n : |x - a|_p \le p^\gamma], \quad a = (a_1, a_2, \ldots, a_n)$$

is the product of discs $B_\gamma(a_i)$,

$$B_\gamma(a) = B_\gamma(a_1) \times B_\gamma(a_2) \times \ldots \times B_\gamma(a_n) . \tag{5.1}$$

The parameter of constancy l (see Sec. 6.2) is a vector, $l = (l_1, l_2, \ldots, l_n)$, where l_j is a parameter of constancy with respect to the variable x_j, $j = 1, 2, \ldots, n$; $l = \min l_j$.

The formula (2.2) of Sec. 6.2 takes the form: every $f \in \mathcal{D}_N^l$, $l = (l_1, l_2, \ldots, l_n)$ is represented in the form

$$\varphi(x) = \sum_{1 \le \nu \le p^{nN - \sum_{1\le j\le n} l_j}} \varphi(a^\nu) \Delta_{l_1}(x_1 - a_1^\nu) \ldots \Delta_{l_n}(x_n - a_n^\nu), \quad x \in \mathbb{Q}_p^n \tag{5.2}$$

for some $a^\nu = (a_1^\nu, a_2^\nu, \ldots a_n^\nu) \in B_N$ which does not depend on φ.

For $l_1 = l_2 = \ldots = l_n = l$ the formula (5.2) owing to (5.1) takes the form

$$\varphi(x) = \sum_{1 \le \nu \le p^{nN-nl}} \varphi(a^\nu)\Delta_l(x - a^\nu) \tag{5.2'}$$

where $\Delta_l(x - a^\nu)$ is the characteristic function of the ball $B_l(a^\nu)$.

In the Fourier-transform theory the form $x\xi$ must be replaced on the scalar product $(x, \xi) = x_1\xi_1 + \ldots + x_n\xi_n$.

6. *The Direct Product of Generalized Functions*

Let generalized functions $f \in \mathcal{D}'(\mathbb{Q}_p^n)$ and $g \in \mathcal{D}'(\mathbb{Q}_p^m)$ be given. Their direct product is defined by the formula

$$(f(x) \times g(y), \varphi) = (f(x), (g(y), \varphi(x, y))), \quad \varphi \in \mathcal{D}(\mathbb{Q}_p^{n+m}). \tag{6.1}$$

As by virtue of the representation (5.2) of Sec. 6.5 a test function φ from $\mathcal{D}(\mathbb{Q}_p^{n+m})$ is represented in a finite sum of the form

$$\varphi(x, y) = \sum_k \varphi_k(x)\psi_k(y), \quad \varphi_k \in \mathcal{D}(\mathbb{Q}_p^n), \quad \psi_k \in \mathcal{D}(\mathbb{Q}_p^m), \tag{6.2}$$

then the operator $\varphi \to (g(y), \varphi(x, y))$ is linear from $\mathcal{D}(\mathbb{Q}_p^{n+m})$ into $\mathcal{D}(\mathbb{Q}_p^n)$, and thus the functional on the right-hand side of (6.1) is linear on $\mathcal{D}(\mathbb{Q}_p^{n+m})$. Hence $f(x) \times g(y) \in \mathcal{D}'(\mathbb{Q}_p^{n+m})$. Further owing to (6.1) and (6.2) we have

$$(f(x) \times g(y), \varphi) = \sum_k (f, \varphi_k)(g, \psi_k) = (g(y) \times f(x), \varphi), \quad \varphi \in \mathcal{D}(\mathbb{Q}_p^{n+m}) \tag{6.3}$$

so the direct product is *commutative*:

$$f(x) \times g(y) = g(y) \times f(x) \tag{6.4}$$

From (6.3) it follows also that the direct product $f(x) \times g(y)$ is continuous with respect to the joint factors f and g: if $f_k \to 0$, $k \to \infty$ in $\mathcal{D}'(\mathbb{Q}_p^n)$ and $g_k \to 0$, $k \to \infty$ in $\mathcal{D}'(\mathbb{Q}_p^m)$ then $f_k \times g_k \to 0$, $k \to \infty$ in $\mathcal{D}'(\mathbb{Q}_p^{n+m})$.

Note that for $g = 1$ the equality (6.4) is equivalent to the equality

$$\left(f(x), \int_{\mathbb{Q}_p^m} \varphi(x, y)dy\right) = \int_{\mathbb{Q}_p^m} (f(x), \varphi(x, y))dy, \quad \varphi \in \mathcal{D}(\mathbb{Q}_p^{n+m}). \tag{6.5}$$

Example. $\delta(x) = \delta(x_1) \times \delta(x_2) \times \ldots \times \delta(x_n)$.

The direct product $f(x) \times g(y)$ of generalized functions $f(x)$ from $\mathcal{D}'(\mathcal{O})$, $\mathcal{O} \subset \mathbb{Q}_p^n$ and $g(y)$ from $\mathcal{D}'(\mathcal{O}')$, $\mathcal{O}' \subset \mathbb{Q}_p^m$ is defined by the similar way.

7. *The "Kernel" Theorem*

Remind that every linear operators and functionals defined on $\mathcal{D}$ is continuous (see Sec. 6.3).

Every generalized function F from $\mathcal{D}'(\mathbb{Q}_p^{n+m})$ defines the linear operator $A\varphi : \varphi \in \mathcal{D}(\mathbb{Q}_p^n) \to \mathcal{D}'(\mathbb{Q}_p^m)$ by the formula

$$(A\varphi, \psi) = (F, \varphi(x)\psi(y)), \qquad \varphi \in \mathcal{D}(\mathbb{Q}_p^n), \quad \psi \in D(\mathbb{Q}_p^m)\,. \tag{7.1}$$

The inverse statement is also true.

Theorem. *Let $B(\varphi, \psi)$, $\varphi \in \mathcal{D}(\mathbb{Q}_p^n)$, $\psi \in \mathcal{D}(\mathbb{Q}_p^m)$ be a bilinear functional. Then there exists an unique generalized function $F \in \mathcal{D}'(\mathbb{Q}_p^{n+m})$ such that*

$$(F, \varphi(x)\psi(y)) = B(\varphi, \psi), \qquad \varphi \in \mathcal{D}(\mathbb{Q}_p^n), \quad \psi \in \mathcal{D}(\mathbb{Q}_p^m)\,. \tag{7.2}$$

■ As every $\Psi(x, y)$ from $\mathcal{D}(\mathbb{Q}_p^{n+m})$ is represented in a finite sum of the form (6.2) then the bilinear functional $B(\varphi, \psi)$ defines some linear functional

$$F : \Psi \to \sum_k B(\varphi_k, \psi_k)$$

on $\mathcal{D}(\mathbb{Q}_p^{n+m})$, and thus $F \in \mathcal{D}'(\mathbb{Q}_p^{n+m})$. The generalized function F satisfies the equality (7.2). It is clear that F is unique. ■

Corollary 1. **(Schwartz' Theorem on "Kernel")** *Let $\varphi \to A(\varphi)$ be a linear operator from $\mathcal{D}(\mathbb{Q}_p^n)$ into $\mathcal{D}'(\mathbb{Q}_p^m)$. Then there exists a unique generalized function $F \in \mathcal{D}'(\mathbb{Q}_p^{n+m})$ such that the equality (7.1) is valid.*

The generalized function $F(x, y)$ is called the *kernel* of the operator A.

Corollary 2. *Every bilinear form $B(\varphi, \psi)$, $\varphi \in \mathcal{D}(\mathbb{Q}_p^n)$, $\psi \in \mathcal{D}(\mathbb{Q}_p^m)$ is continuous.*

8. Adeles

As it is known, the Riemann zeta-function admits the representation in the form of the Euler product over all prime numbers:

$$\zeta(s) = \sum_{1\leq n<\infty} \frac{1}{n^s} = \prod_p \frac{1}{1-p^{-s}}, \qquad \mathrm{Re}\, s > 1 .$$

Relation of such type was generalized in the theory of adeles. The most simple relation containing adelic (Eulerian) product is the formula

$$\prod_p |a|_p = 1, \qquad a \in \mathbb{Q} .$$

In this section certain facts from analysis on adele group are given.

Adeles group. An adele is a sequence of the form $x = (x_\infty, x_2, \ldots, x_p, \ldots)$, where x_∞ is a real number, x_p is a p-adic number, and also beginning from some p all x_p satisfy the inequality $|x|_p \leq 1$. The set of all adeles forms the adeles ring $\mathbb{A}$, if addition and multiplication are defined componentwisely. The additive group of this ring is called the adeles group. Elements of the adeles ring $\mathbb{A}$ which have an inverse element are called ideles, that is a sequence $\lambda = (\lambda_\infty, \lambda_2, \ldots, \lambda_p, \ldots)$ is an idele if $\lambda_p \neq 0$ and $|\lambda_p|_p = 1$ for all p with the exception of a finite quantity. The set of all ideles forms a group with respect to multiplication. On the adeles group $\mathbb{A}$ one introduces a natural topology, which respect to which $\mathbb{A}$ becomes a locally-compact group. The Haar measure on $\mathbb{A}$ is denoted by da. It can be expressed in terms of measures da_p on $\mathbb{Q}_p$ by the following way:

$$dx = dx_\infty dx_2 \ldots dx_p \ldots .$$

The Haar measure on the ideles group $\mathbb{A}^*$ we denote by $d^*\lambda$, it can be expressed in terms of measures $d^*\lambda_p$ on multiplicative group $\mathbb{Q}_p^*$ by the way:

$$d^*\lambda = d^*\lambda_\infty d^*\lambda_2 \ldots d^*\lambda_p \ldots .$$

There exists a parameterization of characters of the additive group of rational numbers $\mathbb{Q}$ by means of adeles. For an adele

$$a = (a_\infty, a_2, \ldots)$$

we put

$$\chi_a(x) = \exp 2\pi i(-a_\infty x + a_2 x + \ldots a_p x + \ldots) . \tag{8.1}$$

Here the sum is considered modulo integer numbers, in this case only a finite number of summands nonvanishes. The map $a \to \chi_a(x)$ is a homomorphism of the adeles group on the group of characters of $\mathbb{Q}$. The kernel of this homomorphism consists of adeles of the form $a = (\alpha, \alpha, \dots, \alpha, \dots)$, where α is a rational number, such adeles are called principal adeles.

Let us consider the function on $\mathbb{A}$:

$$\chi_{(0)}(a) = \exp 2\pi i(-a_\infty + a_2 + \dots) . \tag{8.2}$$

It has the properties:

1) $\chi_{(0)}(a)$ is a character on $\mathbb{A}$;

2) $\chi_{(0)}(a) = 1$ if a is a principle adele, that is

$$\prod_{2 \le p \le \infty} \chi_p(\alpha) = 1, \qquad \alpha \in \mathbb{Q} , \tag{8.3}$$

where χ_p is a character on Q_p, see Sec. 3 ($\chi_\infty(\alpha) = \exp 2\pi i(-\alpha)$). The property 1) follows directly from the definition (8.2). The property 2) is proved in Sec. 3.

Any additive character on the adeles group $\mathbb{A}$ has the form $\chi_{(0)}(ax)$, where $a \in \mathbb{A}$.

The Tate formula. The Bruhat-Schwartz space $\mathcal{T}(\mathbb{A})$ consists of functions $\varphi(x)$ which can be represented in the product form

$$\varphi(x) = \varphi_\infty(x_\infty) \prod_p \varphi_p(x_p) ,$$

where $\varphi_p(x_p)$ satisfy the following conditions:

1) $\varphi_\infty(x_\infty)$ is a infinitely differentiable function on $\mathbb{R}$, decreasing if $|x_\infty| \to \infty$ more rapidly than any inverse degree of $|x_\infty|$.

2) $\varphi_p(x_p) \in \mathcal{D}(\mathbb{Q}_p)$ where the space $\mathcal{D}(\mathbb{Q}_p)$ is defined in Sec. 6.1.

3) For almost all p one has $\varphi_p(x_p) = 1$ if $|x_p|_p \le 1$ and $\varphi_p(x_p) = 0$ if $|x_p|_p > 1$.

The Fourier transform maps the space $\mathcal{T}(\mathbb{A})$ into itself. Let $\pi(\lambda)$ be a character on the ideles group $\mathbb{A}^*$ of the form $\pi(\lambda) = |\lambda|^s \theta(\lambda)$, where s is an arbitrary complex number, $|\lambda| = |\lambda_\infty|_\infty |\lambda_2|_2 \dots |\lambda_p|_p$, and $\theta(\lambda) = \theta_2(\lambda_2) \dots \theta_p(\lambda_p) \dots$ is an arbitrary character on the subgroup of ideles of the form $(1, \lambda_2, \dots)$, where $|\lambda_p|_p = 1$ for all p. Let us consider the Mellin transform of a function $\varphi \in \mathcal{T}(\mathbb{A})$,

$$\phi(\theta, s) = \int_{\mathbb{A}^*} \varphi(\lambda) |\lambda|^s \theta(\lambda) d^*\lambda . \tag{8.4}$$

The function (8.4) can be analytically continued to the whole plane of the complex variables except the points $s = 0$ and $s = 1$. The following relation (Tate formula) is valid

$$\phi(\theta, s) = \tilde{\phi}(\theta^{-1}, 1 - s), \tag{8.5}$$

where

$$\tilde{\phi}(\theta, s) = \int_{A^*} \tilde{\phi}(\lambda)|\lambda|^s \theta(\lambda) d^*\lambda \, . \tag{8.6}$$

Let us show that from (8.5) it follows the functional relation for zeta-function. We shall choose in (8.4) a function $\varphi(\lambda)$ of the following form:

$$\varphi(\lambda) = \varphi_\infty(\lambda_\infty)\Omega(|\lambda_2|_2)\Omega(|\lambda_3|_3)\dots \, , \tag{8.7}$$

where

$$\varphi_\infty(x) = \frac{1}{\sqrt{2\pi}} \exp\left(-\frac{x^2}{2}\right) ,$$

and the function $\Omega(|x|_p)$ is defined in Sec. 6.1. Note, that (8.7) is the product of vacuum vectors in real and p-adic quantum mechanics (see Sec. 11).

VII. Convolution and the Fourier Transformation

In this section we study most important linear operations over generalized functions, namely the convolution and the Fourier transform operations and connected with them the multiplication operation.

1. *Convolution of Generalized Functions*

A sequence $\{\eta_k, k \to \infty\}$ of functions $\eta_k \in \mathcal{D}$ we call *1-sequence*, if there exists $N \in \mathbb{Z}$ such that

$$\eta_k(x) = \Delta_k(x) \equiv \Omega(p^{-k}|x|_p), \qquad k \geq N \, .$$

It is clear that

$$\eta_k \to 1, k \to \infty \text{ in } \mathcal{E} \, .$$

The sequence $\{\Delta_k, k \to \infty\}$ we call *the canonical 1-sequence.*

Let f and g be generalized functions from $\mathcal{D}'$. Their *convolution* $f * g$ we call the functional defined by the equality

$$(f * g, \varphi) = \lim_{k \to \infty} (f(x) \times g(y), \Delta_k(x)\varphi(x + y)) \tag{1.1}$$

if the limit exists for all $\varphi \in \mathcal{D}$. The right-hand side of the equality (1.1) defines an linear functional on $\mathcal{D}$, and thus $f*g \in \mathcal{D}'$ (see Sec. 6.3). Note that the equality (1.1) is equivalent to the following one

$$(f*g,\varphi) = \lim_{k\to\infty}(f(x)\times g(y), \eta_k(x)\varphi(x+y)), \qquad \varphi \in \mathcal{D}$$

for any 1-sequence $\{\eta_k, k \to \infty\}$.

The convolution $g*f$ is defined by the similar way:

$$(g*f,\varphi) = \lim_{k\to\infty}(g(y)\times f(x), \Delta_k(y)\varphi(x+y)), \quad \varphi \in \mathcal{D}\,. \tag{1.1'}$$

Theorem. *If the convolution $f*g$, $f,g \in \mathcal{D}'$ exists then there exists the convolution $g*f$, and they are equal*

$$f*g = g*f\ .$$

■ Let $\varphi \in \mathcal{D}$ so that $\varphi \in \mathcal{D}_N^l$ for some $N, l \in \mathbb{Z}$. By the formula (5.2′) of Sec. 6.5 the function φ is represented in the form of finite sum

$$\varphi(x) = \sum_{1\le\upsilon\le p^{N-l}} C_\upsilon \Delta_l(x-a^\upsilon)$$

for some $a^\upsilon \in B_N$ and $C_\upsilon \in C$. therefore it is sufficient to prove the theorem for the test functions of the form

$$\varphi_\upsilon(x) = \Delta_l(x-a^\upsilon)\ .$$

By the condition of the theorem the convolution $f*g$ exists. If we apply the formula (1.2) to the 1-sequence

$$\{\Delta_k(-x+a^\upsilon), k\to\infty\}$$

we get

$$\begin{aligned}(f*g,\varphi_\upsilon) &= \lim_{k\to\infty}(f(x)\times g(y), \Delta_k(-x+a^\upsilon)\varphi_\upsilon(x+y))\\ &= \lim_{k\to\infty}(f(x)\times g(y), \Delta_k(-x+a^\upsilon)\Delta_l(x+y-a^\upsilon))\ .\end{aligned}$$

Taking into account the easily verifiable identity

$$\Omega(p^{-k}|-x+a^{\nu}|_p)\Omega(p^{-l}|x+y-a^{\nu}|_p) = \Omega(p^{-k}|y|_p)\Omega(p^{-l}|x+y-a^{\nu}|_p)$$

for $k \geq l$, we proceed with our equalities

$$\begin{aligned}(f*g, \varphi_\nu) &= \lim_{k\to\infty} (f(x) \times g(y), \Delta_k(y)\Delta_l(x+y-a^\nu)) \\ &= \lim_{k\to\infty} (f(x) \times g(y), \Delta_k(y)\varphi_\nu(x+y)) = (g*f, \varphi) .\end{aligned}$$

Here we have used commutativity of the direct product (see Sec. 6.6) and the formula (1.2). ■

The following formula is valid: if $f*g$ exists then

$$\check{f}*g = \check{f}*\check{g}$$

where $\check{f}(x) = f(-x)$ is the reflection operator (see Sec. 6.4).

If $f, g \in \mathcal{D}'$ and supp $g \subset B_N$ then the convolution $f*g$ exists and the representation

$$(f*g, \varphi) = (f(x) \times g(y), \Delta_N(y)\varphi(x+y)), \quad \varphi \in \mathcal{D} \tag{1.4}$$

is valid.

■ We use the definition (1.1) and the formula (4.3) of Sec. 6.4. Then for all $\varphi \in \mathcal{D}$ we shall have

$$\begin{aligned}&(f*g, \varphi) \\ &= \lim_{k\to\infty} (f(x) \times g(y), \Delta_k(y)\varphi(x+y)) \\ &= \lim_{k\to\infty} (f(x) \times \Delta_N(y)g(y), \Delta_k(y)\varphi(x+y)) \\ &= \lim_{k\to\infty} (f(x) \times g(y), \Delta_N(y)\Delta_k(y)\varphi(x+y)) \\ &= (f(x) \times g(y), \Delta_N(y)\varphi(x+y)) ,\end{aligned}$$

as

$$\Delta_k(y)\Delta_N(y)\varphi(x+y) \to \Delta_N(y)\varphi(x+y), \quad k \to \infty \text{ in } \mathcal{D}(\mathbb{Q}_p^2) .$$ ■

The convolution $f{*}g$ is continuous with respect to the joint factors f and g: if $f_k \to f$, $k \to \infty$ in $\mathcal{D}'$, $g_k \to g$, $k \to \infty$ in $\mathcal{D}'$, supp $g \subset B_N$ *then $f_k{*}g_k \to f{*}g$, $k \to \infty$ in $\mathcal{D}'$.*

■ It follows from the representation (1.4) and from the continuity of the direct product, see Sec. 6.6. ■

Example 1.

$$f{*}\delta = f = \delta{*}f, \qquad f \in \mathcal{D}' . \tag{1.5}$$

■ It follows from the representation (1.4) and (6.1) of Sec. 6.6

$$\begin{aligned}(f{*}\delta, \varphi) &= (f(x) \times \delta(y), \Delta_N(y)\varphi(x+y)) \\ &= (f(x), (\delta(y), D_N(y)\varphi(x+y))) = (f, \varphi) .\end{aligned}$$ ■

Example 2. If $f \in \mathcal{D}'$ then

$$f{*}\delta_k \to f, k \to \infty \text{ in } \mathcal{D}' \tag{1.6}$$

where the sequence $\{\delta_k, k \to \infty\}$ is defined in (3.4) of Sec. 6.3.

■ It follows from (1.5) and (3.4) of Sec. 6.3:

$$f{*}\delta_k \to f{*}\delta = f, \quad k \to \infty \text{ in } \mathcal{D}'$$

owing to the continuity of convolution, see Sec. 6.6. ■

If $g = \psi \in \mathcal{D}$ then $f * \psi \in \mathcal{E}$ and the formula (1.4) *takes the form*

$$(f{*}\psi)(x) = (f(y), \psi(x-y)), \quad x \in \mathbb{Q}_p ; \tag{1.7}$$

in addition the parameter of constancy of the function $f{*}g$ does not exceed the parameter of constancy of the function ψ.

■ Let $\psi \in \mathcal{D}'_N$. Then by using the formulas (6.5) of Sec. 6.6 and (1.4) we obtain the representation (1.7):

$$(f*\psi,\varphi) = (f(x),(\psi(y),\Delta_N(y)\varphi(x+y))) = \left(f(x),\int_{\mathbb{Q}_p}\psi(y)\varphi(x+y)dy\right)$$

$$= \left(f(x),\int_{\mathbb{Q}_p}\psi(\xi-x)\varphi(\xi)d\xi\right) = \int_{Q_p}(f(x),\psi(\xi-x))\varphi(\xi)d\xi, \varphi\in\mathcal{D}$$

as $\psi(\xi-x)\varphi(\xi)\in\mathcal{D}(\mathbb{Q}_p^2)$. The remaining affirmations follow from the representation (1.7) ■

Example 3. If $f\in\mathcal{D}'$ then

$$\Delta_k(f*\delta_k)\longrightarrow f,\quad k\to\infty \text{ in } \mathcal{D}'\,,$$

and also $\Delta_k(f*\delta_k)\in\mathcal{D}$.

The Example 3 shows that *every generalized function is a weak limit of test functions i.e. $\mathcal{D}$ is dense in $\mathcal{D}'$.*

The formula (1.4) allows us to write the equality

$$(f(x)\times g(y),\Delta_k(x)\varphi(x+y)) = ((\Delta_k f)*g,\varphi)$$

Therefore the definition (1.1) of the convolution $f*g$ is equivalent to the following one:

$$(\Delta_k f)*g\longrightarrow f*g,\quad k\to\infty \text{ in } \mathcal{D}'\,. \tag{1.8}$$

From the definition (1.1) of the convolution $f*g$ it follows that supp $(f*g)$ is contained in the closure of the set

$$[\xi:\xi\in\mathbb{Q}_p,\quad \xi=x+y,\quad x\in\text{supp } f,\ y\in\text{supp } g]\,.$$

In particular, if supp $f\subset B_N$ and supp $g\subset B_N$ then supp $(f*g)\subset B_N$.

Thus, the set of generalized functions with supports in B_N forms the *convolution algebra* (commutative and associative) with a unity where the δ-function plays the role of the unit element.

In conclusion we shall indicate one more criterion of the extension of the convolution. *Let functions f and g belong to L^1_{loc} and a function $q \in L^1_{\text{loc}}$ exists such that*

$$\int_{B_k} f(x-y)g(y)dy \longrightarrow q(x), \quad k \to \infty \text{ in } \mathcal{D}' .$$

Then the convolution $f*g$ exists and it is equal to q.

■ It follows from the definition of convolution (1.1)

$$(f*g,\varphi) = \lim_{k\to\infty} \int_{\mathbb{Q}_p} f(x) \int_{\mathbb{Q}_p} \Delta_k(y)g(y)\varphi(x+y)dydx$$
$$= \lim_{k\to\infty} \int_{\mathbb{Q}_p} \varphi(\xi) \int_{B_k} f(\xi-y)g(y)dyd\xi = \int_{\mathbb{Q}_p} q(\xi)\varphi(\xi)d\xi, \quad \varphi \in \mathcal{D} .$$
■

2. *The Fourier-Transform of Test Functions*

Let $\varphi \in \mathcal{D}$. Its *Fourier-transform* $F[\varphi] = \tilde{\varphi}$ is defined by the formula

$$\tilde{\varphi}(\xi) = \int_{\mathbb{Q}_p} \chi_p(\xi x)\varphi(x)dx, \quad \xi \in \mathbb{Q}_p . \tag{2.1}$$

Lemma. *If $\varphi \in \mathcal{D}^l_N$ then $\tilde{\varphi} \in \mathcal{D}^{-N}_{-l}$ i.e. the operation $\varphi \to \tilde{\varphi}$ is linear (and so continuous) from $\mathcal{D}$ in $\mathcal{D}$.*

■ Let us prove that $\tilde{\varphi}(\xi) = 0$, $|\xi|_p > p^{-l}$. By performing in (2.1) the change of variable of integration $x = x' + a$, $|a|_p = p^l$ we get

$$\tilde{\varphi}(\xi) = \int_{\mathbb{Q}_p} \chi_p(\xi(x'+a))\varphi(x'+a)dx'$$
$$= \chi_p(\xi a) \int_{\mathbb{Q}_p} \chi_p(\xi x')\varphi(x')dx' = \chi_p(\xi a)\tilde{\varphi}(\xi) . \tag{2.2}$$

As $|\xi a|_p = |\xi|_p|a|_p > 1$ then for any ξ, $|\xi|_p > p^{-l}$ there exists $a \in \mathbb{Q}_p$, $|a|_p = p^l$ such that $\chi_p(\xi a) \neq 1$ (see Sec. 3.1). Hence from (2.2) it follows that $\tilde{\varphi}(\xi) = 0$, $|\xi|_p > p^{-l}$ i.e. $\tilde{\varphi} \in \mathcal{D}_{-l}$.

Now we prove that

$$\tilde{\varphi}(\xi + \xi') = \tilde{\varphi}(\xi), \qquad \xi' \in B_{-N}, \quad \xi \in \mathbb{Q}_p\,,$$

i.e. the parameter of constancy of $\tilde{\varphi} \geq -N$. Indeed

$$\tilde{\varphi}(\xi + \xi') = \int_{B_N} \chi_p((\xi + \xi')x)\varphi(x)dx = \int_{B_N} \chi_p(\xi x)\chi_p(\xi' x)\varphi(x)dx$$

$$= \int_{\mathbb{Q}_p} \chi_p(\xi x)\varphi(x)dx = \tilde{\varphi}(\xi), \tag{2.3}$$

as $|\xi' x|_p = |\xi'|_p|x|_p \leq 1$ and $\chi_p(\xi' x) = 1$. ■

Theorem. *The Fourier-transform $\varphi \to \tilde{\varphi}$ is the linear isomorphism $\mathcal{D}$ onto $\mathcal{D}$, and also the inversion formula*

$$\varphi(x) = \int_{\mathbb{Q}_p} \chi_p(-x\xi)\tilde{\varphi}(\xi)d\xi = \tilde{\check{\varphi}}, \varphi \in \mathcal{D} \tag{2.4}$$

is valid where $\check{\varphi}(\xi) = \varphi(-\xi)$, and the Parseval-Steklov equalities are valid:

$$\int_{\mathbb{Q}_p} \varphi(x)\overline{\psi(x)}dx = \int_{\mathbb{Q}_p} \tilde{\varphi}(\xi)\overline{\tilde{\psi}(\xi)}d\xi, \quad \varphi, \psi \in \mathcal{D}, \tag{2.5}$$

$$\int_{\mathbb{Q}_p} \varphi(x)\tilde{\psi}(x)dx = \int_{\mathbb{Q}_p} \tilde{\varphi}(\xi)\psi(\xi)d\xi, \quad \varphi, \psi \in \mathcal{D}\,. \tag{2.5'}$$

■ By the Lemma the operation $\varphi \to \tilde{\varphi}$ maps $\mathcal{D}$ in $\mathcal{D}$. In order to prove that this mapping is onto and one-to-one it is sufficient to prove the inversion formula (2.4). Let $\varphi \in \mathcal{D}_N^l$, then by the Lemma $\tilde{\varphi} \in \mathcal{D}_{-l}^{-N}$.

Starting from (2.1) we have

$$\begin{aligned}\int_{\mathbb{Q}_p} \chi_p(-x\xi)\tilde{\varphi}(\xi)d\xi &= \int_{B_{-l}} \chi_p(-x\xi) \int_{B_N} \varphi(x')\chi_p(x'\xi)dx'd\xi \\ &= \int_{B_N} \varphi(x') \int_{B_{-l}} \chi_p(-x\xi)\chi_p(x'\xi)d\xi dx' \\ &= \int_{\mathbb{Q}_p} \varphi(x') \int_{B_{-l}} \chi_p(\xi(x'-x))d\xi dx' \\ &= \int_{|x'-x|\le p^l} \varphi(x') \int_{B_{-l}} \chi_p(\xi(x'-x))d\xi dx' \\ &\quad + \int_{|x'-x|\ge p^{l+1}} \varphi(x') \int_{B_{-l}} \chi_p(\xi(x'-x))d\xi dx' \ . \end{aligned} \tag{2.6}$$

In the first integral from the right in (2.6) $\varphi(x') = \varphi(x)$ and $\chi_p(\xi(x'-x)) = 1$ and it is equal to (see (2.3) of Sec. 6.2)

$$\varphi(x) \int_{B_l(x)} \int_{B_{-l}} d\xi dx' = \varphi(x)p^l p^{-l} = \varphi(x) \ .$$

The second integral from the right in (2.6) is equal to 0 owing to the formula (3.1) of Sec. 4.3.

To prove the equality (2.5) we denote $\bar{\tilde{\psi}}(\xi) = \eta(\xi) \in \mathcal{D}$, and then $\bar{\psi}(x) = \tilde{\eta}(x)$, and it takes the form (2.5′):

$$\int_{\mathbb{Q}_p} \varphi(x)\tilde{\eta}(x)dx = \int_{\mathbb{Q}_p} \tilde{\varphi}(\xi)\eta(\xi)d\xi$$

which is verified immediately:

$$\begin{aligned}\int_{\mathbb{Q}_p} \varphi(x)\tilde{\eta}(x)dx &= \int_{\mathbb{Q}_p} \varphi(x) \int_{\mathbb{Q}_p} \eta(\xi)\chi_p(\xi x)d\xi dx \\ &= \int_{\mathbb{Q}_p} \eta(\xi) \int_{\mathbb{Q}_p} \varphi(x)\chi_p(\xi x)dx d\xi = \int_{\mathbb{Q}_p} \eta(\xi)\tilde{\varphi}(\xi)d\xi \ . \end{aligned}$$

■

Example 1.

$$\delta_k = \tilde{\Delta}_k, \quad k \in \mathbb{Z}\,, \tag{2.7}$$

where the functions δ_k and Δ_k are defined in Secs. 6.3 and 7.1:

$$\delta_k(\xi) = p^k\Omega(p^k|\xi|_p), \Delta_k(x) = \Omega(p^{-k}|x|_p)\,.$$

The formula (2.7) is another way of writing of the formula (3.1) of Sec. 4.3.

Example 2. $\gamma \in \mathbb{Z}$.

$$F[\delta(|x|_p - p^\gamma)](\xi) = \left(1 - \frac{1}{p}\right)\delta_\gamma(\xi) - p^{\gamma-1}\delta(|\xi|_p - p^{1-\gamma}) \tag{2.8}$$

where the function $\delta(|x|_p - p^{1-\gamma})$ is defined in Sec. 6.2.

The formula (2.8) is another way of writing of the formula (3.2) of Sec. 4.3.

Example 3. $\gamma \in \mathbb{Z}$, $k = 1, 2, \dots, p-1$

$$F[\Delta_{1-\gamma}(x)\chi_p(kp^{-\gamma}x)](\xi) = p^{1-\gamma}\delta(|\xi|_p - p^\gamma)\delta(\xi_0 - k) \tag{2.9}$$

■ The desired quantity owing to (2.7) is equal to

$$\begin{aligned}\tilde{\Delta}_{1-\gamma}(kp^{-\gamma} + \xi) &= \delta_{1-k}(kp^{-\gamma} + \xi) = p^{1-\gamma}\Omega(p^{1-\gamma}|kp^{-\gamma} + \xi|_p)\\ &= p^{1-\gamma}\begin{cases} 1 & \text{if } |\xi|_p = p^\gamma, \xi_0 = k \\ 0 & \text{othewise}\end{cases}\\ &= p^{1-\gamma}\delta(|\xi|_p - p^\gamma)\delta(\xi_0 - k)\,.\end{aligned}$$

■

Example 4. $|4a|_p \geq p^{2-2\gamma}$

$$F[\chi_p(ax^2)\delta(|x|_p - p^\gamma)](\xi) = \lambda_p(a)|2a|_p^{-1/2}\chi_p\left(-\frac{\xi^2}{4a}\right)\delta\left(\left|\frac{\xi}{2a}\right|_p - p^\gamma\right) \tag{2.10}$$

The formula (2.10) is another way of writing formula (1.7) of Sec. 5.1.

Example 5. $p \neq 2$, $|a|_p = p^{1-2\gamma}$, $\gamma \in \mathbb{Z}$

$$F[\chi_p(ax^2)\delta(|x|_p - p^\gamma)](\xi) = |a|_p^{-1/2}\left[\lambda_p(a)\chi_p\left(-\frac{\xi^2}{4a}\right) - \frac{1}{\sqrt{p}}\right]\Delta_{1-\gamma}(\xi). \tag{2.11}$$

The formula (2.11) is another form of formula (1.8) of Sec. 5.1.

Example 6. $p \neq 2$, $|a|_p \geq p^{1-2\gamma}$, $\gamma \in \mathbb{Z}$

$$F[\chi_p(ax^2)\Delta_\gamma(x)](\xi) = \lambda_p(a)|a|_p^{-1/2}\chi_p\left(-\frac{\xi^2}{4a}\right)\Delta_\gamma\left(\frac{\xi}{a}\right). \tag{2.12}$$

The formula (2.12) is another form of the formula (2.1) of Sec. 5.2.

Example 7. $p = 2$, $\gamma \in \mathbb{Z}$

$$F[\chi_2(ax^2)\Delta_\gamma(x)](\xi)$$
$$= \lambda_2(a)|a|_2^{-1/2}\chi_2\left(-\frac{\xi^2}{4a}\right) \times \begin{cases} \delta(|\xi|_2 - 2^{1-\gamma}), & |a|_2 = 2^{1-2\gamma}, \\ \Delta_{-\gamma}(\xi), & |a|_2 = 2^{2-2\gamma}, \\ \Delta_{\gamma-1}\left(\frac{\xi}{a}\right), & |a|_2 \geq 2^{3-2\gamma}. \end{cases} \tag{2.13}$$

The formula (2.13) is another way of writing formula (2.3) of Sec. 5.2.

Example 8. $p \neq 2$, $\gamma \in \mathbb{Z}$, $\sum_{1 \leq k \leq p-1} \eta(k) = 0$

$$F[\eta(x_0)\delta(|x|_p - p^\gamma)](\xi) = p^{\gamma-1}\eta'(\xi_0)\delta(|\xi|_p - p^{1-\gamma}) \tag{2.14}$$

where

$$\eta'(\xi_0) = \sum_{1 \leq k \leq p-1} \eta(k)\exp\left(2\pi i\frac{k\xi_0}{p}\right), \quad \sum_{1 \leq k \leq p-1} \eta'(k) = 0. \tag{2.15}$$

■ The desired Fourier transform is reduced to the integral

$$I = \int_{S_\gamma} \eta(x_0)\chi_p(\xi x)dx .$$

If $|\xi|_p \le p^{-\gamma}$ then $\chi_p(\xi x) = 1$ and

$$I = \int_{S_\gamma} \eta(x_0)dx = p^{\gamma-1} \sum_{1\le k\le p-1} \eta(k) = 0 .$$

If $|\xi|_p = p^N \ge p^{2-\gamma}$ i.e. $N + \gamma - 1 \ge 1$ then

$$\begin{aligned}\{\xi x\}_p &= \{p^{-N-\gamma}(\xi_0 + \xi_1 p + \ldots)(x_0 + x_1 p + \ldots)\}_p \\ &= L(x_0, x_1, \ldots, x_{N+\gamma-2}) + \frac{\xi_0}{p} x_{N+g-1} ,\end{aligned}$$

and hence like in Example 24 of Sec. 5.1 $I = 0$. If $|\xi|_p = p^{1-\gamma}$ then $\{\xi x\}_p = \dfrac{\xi_0 x_0}{p}$, and owing to (2.15)

$$\begin{aligned}I &= \int_{S_\gamma} \eta(x_0) \exp\left(2\pi i \frac{\xi_0 x_0}{p}\right) dx \\ &= p^{\gamma-1} \sum_{1\le k\le p-1} \eta(k) \exp\left(2\pi i \frac{k\xi_0}{p}\right) = p^{\gamma-1}\eta'(\xi_0) .\end{aligned}$$ ■

Example 9. Re $\alpha > 0$

$$\begin{aligned}&F\left[|x|_p^{\alpha-1}\Omega\left(\left|\frac{x}{m}\right|_p\right)\right] \\ &= |m|_p^\alpha \Omega(|m\xi|_p)\frac{1-p^{-1}}{1-p^{-\alpha}} + |\xi|_p^{-\alpha}[1 - \Omega(|m\xi|_p)]\frac{1-p^{\alpha-1}}{1-p^{-\alpha}}\end{aligned} \quad (2.16)$$

■ By changing the variable $t = x\xi$, $dt = |\xi|_p dx$, and using the formula (3.2) of Sec. 4.3 we obtain for the desired Fourier transform the equality

$$\int\limits_{|x|_p\le|m|_p} |x|_p^{\alpha-1}\chi_p(x\xi)dx = |\xi|_p^{-\alpha} \int\limits_{|t|_p\le|m\xi|_p} |t|_p^{\alpha-1}\chi_p(t)dt$$
$$= |\xi|_p^{-\alpha} \sum_{-\infty<\gamma\le N} p^{\gamma(\alpha-1)} \int\limits_{S_\gamma} \chi_p(t)dt$$
$$= |\xi|_p^{-\alpha} \begin{cases} \left(1-\frac{1}{p}\right) \sum\limits_{-N\le\gamma<\infty} p^{-\gamma\alpha} & \text{if } N\le 0 \\ -p^{\alpha-1} + \left(1-\frac{1}{p}\right) \sum\limits_{0\le\gamma<\infty} p^{-\gamma\alpha} & \text{if } N\ge 1 \end{cases}$$
$$= |\xi|_p^{-\alpha} \begin{cases} \frac{1-p^{-1}}{1-p^{-\alpha}} p^{N\alpha}, & N\le 0 \\ \frac{1-p^{\alpha-1}}{1-p^{-\alpha}}, & N\ge 1 \end{cases}$$
$$= |\xi|_p^{-\alpha}|m\xi|_p^{\alpha}\Omega(|m\xi|_p)\frac{1-p^{-1}}{1-p^{-\alpha}} + |\xi|_p^{-\alpha}[1-\Omega(|m\xi|_p)]\frac{1-p^{\alpha-1}}{1-p^{-\alpha}}$$

where we have denoted $|m\xi|_p = p^N$. ∎

The following formula

$$F[\varphi(ax+b)](\xi) = |a|_p^{-1}\chi_p\left(-\frac{b}{a}\xi\right) F[\varphi]\left(\frac{\xi}{a}\right), \quad \varphi\in\mathcal{D}, a\ne 0 \qquad (2.17)$$

is valid, in particular

$$\tilde{\tilde{\varphi}} = \tilde{\varphi}, \qquad \varphi(x-b) = \chi_p(b\xi)\tilde{\varphi}(\xi)\ . \qquad (2.18)$$

∎ The formula (2.17) follows immediately from (2.1)

$$F[\varphi(ax+b)](\xi) = \int\limits_{\mathbb{Q}_p} \varphi(ax+b)\chi_p(\xi x)dx = |a|_p^{-1}\int\limits_{\mathbb{Q}_p} \varphi(x')\chi_p\left(\xi\frac{x'-b}{a}\right)dx'$$
$$= |a|_p^{-1}\chi_p\left(-\xi\frac{b}{a}\right)\int\limits_{\mathbb{Q}_p} \varphi(x')\chi_p\left(\frac{\xi}{a}x'\right)dx' = |a|_p^{-1}\chi_p\left(-\xi\frac{b}{a}\right)\tilde{\varphi}\left(\frac{\xi}{a}\right). \qquad ∎$$

Every function $\varphi\in\mathcal{D}_N^l$ is represented in the form

$$\varphi(x) = \sum_{1\le v\le p^{N-l}} \tilde{\varphi}(a^v)\chi_p(xa^v)\delta_{-N}(x) \qquad (2.19)$$

for some $a^{\nu} \in B_{-l}$ which do not depend on φ.

■ By the Lemma $\tilde{\varphi} \in \mathcal{D}_{-l}^{-N}$, it is represented in the form (2.2) of Sec. 6.2

$$\tilde{\varphi}(x) = \sum_{1 \leq \nu \leq p^{N-l}} \tilde{\varphi}(a^{\nu}) \Delta_{-N}(\xi - a^{\nu})$$

for some $a^{\nu} \in B_{-l}$. By applying to the last equality the inverse Fourier transform and using the formulas (2.18) and (2.7) we get the representation (2.19). ■

The set of trigonometrical polynomials

$$\left\{ \sum_{\nu-\text{finite}} C_{\nu} \chi_p(\xi^{\nu} x), \quad \xi^{\nu} \in M \right\}$$

is dense in $C(K)$, and thus in $L^2(K)$ for any compact set $K \subset \mathbb{Q}_p$; here M is any countable everywhere dense set in $\mathbb{Q}_p$.

■ It follows from the representation (2.19), because $\mathcal{D}$ is dense in $C(K)$, see Sec. 6.2. ■

3. *The Fourier-Transform of Generalized Functions*

In accordance with the formula (4.1) of Sec. 6.4 the Fourier transform $\tilde{f}$ of a generalized function $f \in \mathcal{D}'$ we define by the formula

$$(\tilde{f}, \varphi) = (f, \tilde{\varphi}), \quad \varphi \in \mathcal{D}\,. \tag{3.1}$$

As the operator $\varphi \to \tilde{\varphi}$ is linear from $\mathcal{D}$ in $\mathcal{D}$ (see Sec. 7.2) then the functional from the right in (3.1) is linear on $\mathcal{D}$, so $\tilde{f} \in \mathcal{D}'$, and the operator $f \to \tilde{f}$ is continuous from $\mathcal{D}'$ in $\mathcal{D}'$ (see Sec. 6.4).

The inversion formula

$$f = \tilde{\tilde{\tilde{f}}}, \quad f \in \mathcal{D}' \tag{3.2}$$

is valid, so the Fourier transform $f \to \tilde{f}$ is the linear isomorphism $\mathcal{D}'$ onto $\mathcal{D}'$.

■ The formula (3.2) follows from the formula (2.4) and from the definition (3.1):

$$(\tilde{\tilde{\tilde{f}}},\varphi) = (\tilde{\tilde{f}},\tilde{\varphi}) = (\tilde{f},\tilde{\tilde{\varphi}}) = (f,\tilde{\tilde{\tilde{\varphi}}}) = (f,\varphi), \quad \varphi \in \mathcal{D}\,. \qquad ■$$

In the case $f = \psi \in \mathcal{D}$ the formula (3.1) is reduced to the equality (2.5′), and therefore the Fourier transform introduced by (3.1) is in fact an extention of the classical Fourier transform (2.1).

The formulas (2.16) and (2.17) are carried over generalized functions

$$F[f(ax+b)](\xi) = |a|_p^{-1}\chi_p\left(-\frac{b}{a}\xi\right)F[f]\left(\frac{\xi}{a}\right), \quad f \in \mathcal{D}', \; a \neq 0. \tag{3.3}$$

$$\tilde{\tilde{f}} = \check{f}, \; f(x-b) = \chi_p(b\xi)\tilde{f}(\xi) \tag{3.4}$$

where the generalized function $f(ax+b)$ is defined in Sec. 6.4.

If $f \in L^1$ then the formula (3.1) is equivalent to the formula (2.1)

$$\tilde{f}(\xi) = \int_{\mathbb{Q}_p} \chi_p(\xi x)f(x)dx, \quad \xi \in \mathbb{Q}_p\,. \tag{3.5}$$

If $f \in L^1_{\text{loc}}$ and there exists $q \in L^1_{\text{loc}}$ such that

$$\int_{B_k} \chi_p(\xi x)f(x)dx \longrightarrow q(x), \quad k \to \infty \text{ in } \mathcal{D}'$$

then the Fourier-transform $\tilde{f}$ exists and is equal to q.

■ It follows from the continuity of the Fourier transform in $\mathcal{D}'$ and from the formula (3.5). In fact,

$$\Omega(p^{-k}|x|_p)f(x) \longrightarrow f(x), \quad k \to \infty \text{ in } \mathcal{D}'$$

then

$$F[\Omega(p^{-k}|x|_p)f] = \int_{B_k} \chi_p(\xi x)f(x)dx \longrightarrow \tilde{f}, \quad k \to \infty \text{ in } \mathcal{D}'\,,$$

and hence $\tilde{f} = q$ owing to uniqueness of the limit. ■

An analogy of the Riemann-Lebesque Theorem

If $f \in L^1$ then $\tilde{f}$ is continuous on $\mathbb{Q}_p$ and $\tilde{f}(\xi) \to 0$, $|\xi|_p \to \infty$.

■ The continuity of $\tilde{f}(\xi)$ follows from the representation (3.5) according to an analogy of the Lebesque Theorem on limiting passage under the sign of an integral (see Sec. 6.4) owing to majorization

$$|f(x)\chi_p(\xi x)| = |f(x)|, \quad x \in \mathbb{Q}_p \ .$$

Now we prove that $\tilde{f}(\xi) \to 0$, $|\xi|_p \to \infty$. As $\mathcal{D}$ is dense in L^1 (see Sec. 6.2) then for any $\varepsilon > 0$ there exists $\varphi \in \mathcal{D}$ such that

$$\int_{\mathbb{Q}_p} |f(x) - \varphi(x)|dx < \infty \ .$$

By the Lemma of Sec. 7.2 $\tilde{\varphi}(\xi) = 0$, $|\xi|_p > p^N$ for some $N \in \mathbb{Z}$. Therefore if $|\xi|_p > p^N$ then

$$\begin{aligned} |\tilde{f}(\xi)| &= \left| \int_{\mathbb{Q}_p} [f(x) - \varphi(x)]\chi_p(\xi x)dx + \int_{\mathbb{Q}_p} \varphi(x)\chi_p(\xi x)dx \right| \\ &\leq \int_{\mathbb{Q}_p} |f(x) - \varphi(x)|dx + |\tilde{\varphi}(\xi)| < \varepsilon \ . \end{aligned}$$ ■

Example 9.

$$\tilde{\delta} = 1, \tilde{1} = \delta \qquad \text{(cf. (4.4) of Sec. 4.4).} \tag{3.6}$$

■ It follows from (3.1) and (3.2):

$$(\tilde{\delta}, \varphi) = (\delta, \tilde{\varphi}) = \tilde{\varphi}(0) = \int_{\mathbb{Q}_p} \varphi(x)dx = (1, \varphi), \quad \varphi \in \mathcal{D} \ .$$ ■

Example 10. The formula (3.7) of Sec. 4.3 takes the form

$$F\left[\frac{1}{|x|_p^2+m^2}\right](\xi)=\left(1-\frac{1}{p}\right)\frac{|\xi|_p}{p^2+m^2|\xi|_p^2}\sum_{0\le\gamma<\infty}p^{-\gamma}\frac{p^2-p^{-2\gamma}}{p^{-2\gamma}+m^2|\xi|_p^2}\,. \tag{3.7}$$

Example 11. The formula (2.11) of Sec. 5.2 takes the form

$$F[\chi_p(ax^2)](\xi)=\lambda_p(a)|2a|_p^{-1/2}\chi_p\left(-\frac{\xi^2}{4a}\right),\quad a\neq 0. \tag{3.8}$$

Example 12.

$$\lambda_p(a)|2a|_p^{-1/2}\chi_p\left(-\frac{\xi^2}{4a}\right)\longrightarrow\delta(\xi),\quad |a|_p\longrightarrow 0 \text{ in } \mathcal{D}'. \tag{3.9}$$

■ It follows from (3.8), owing to (3.6) we have

$$F[\chi_p(ax^2)](\xi)\longrightarrow\tilde{1}=\delta,\quad a\to 0 \text{ in } \mathcal{D}'\,. \qquad ■$$

Theorem. *In order $f\in\mathcal{D}'$ and supp $f\subset B_N$, it is necessary and sufficient that $\tilde{f}\in\mathcal{E}$ and the parameter of constancy of $\tilde{f}$ is $\ge -N$, in addition*

$$\tilde{f}(\xi)=(f(x),\Delta_N(x)\chi_p(\xi x))\,. \tag{3.10}$$

■ **Necessity.** By using the formulas (4.3) of Sec. 6.4 and (6.4) of Sec. 6.6 from (3.1) we derive the representation (3.10):

$$(\tilde{f},\varphi)=(\Delta_N(x)f,\tilde{\varphi})=\left(f(x),\Delta_N(x)\int_{\mathbb{Q}_p}\varphi(\xi)\chi_p(x\xi)d\xi\right)$$

$$=\int_{\mathbb{Q}_p}(f(x),\Delta_N(x)\chi_p(\xi x))\varphi(\xi)d\xi,\quad \varphi\in\mathcal{D}.$$

Further if $\xi' \in B_{-N}$ then $|\xi' x|_p \leq 1$ and $\chi_p(\xi' x) = 1$ for $x \in B_N$, and

$$\tilde{f}(\xi + \xi') = (f(x), \Delta_N(x)\chi_p((\xi + \xi')x)) = (f(x), \Delta_N(x)\chi_p(\xi x)) = \tilde{f}(\xi),$$

so that the parameter of constancy of $\tilde{f}$ is $\geq -N$. ■

Sufficiency. Let $\tilde{f} \in \mathcal{E}$ and its parameter of constancy is $\geq -N$. By applying the inverse Fourier-transform to the equality

$$f(x + x') = \tilde{f}(x), \qquad x \in \mathbb{Q}_p, |\xi'|_p \leq p^{-N}$$

and using the formula (3.4) we get the equality

$$\chi_p(xx')f(x) = f(x), \quad x \in \mathbb{Q}_p \,. \tag{3.11}$$

But for any x, $|x|_p > p^N$ there exists some $x' \in S_{-N}$ such that $\chi_p(xx') \neq 1$. From (3.11) it follows that $f(x)$, $|x|_p > p^N$, i.e. supp $f \subset B_N$. ■

4. *The Space* L^2

The space L^2 is defined in Sec. 4.1. We introduce in L^2 the scalar product

$$(f, g) = \int_{\mathbb{Q}_p} f(x)\overline{g(x)}dx, \quad f, g \in L^2\,,$$

so that $||f||^2 = (f, f)$* The Caushy-Bunjakovsky inequality is valid:

$$|(f, g)| \leq ||f||\ ||g||, \qquad f, g \in L^2$$

In terms of scalar product the Parseval-Steklov equality (2.5) takes the form

$$(\varphi, \psi) = (\tilde{\varphi}, \tilde{\psi}), \qquad \varphi, \psi \in \mathcal{D} \tag{4.1}$$

Theorem. *The Fourier-transform maps L^2 onto L^2 one-to-one and mutually continuous. In addition*

$$\tilde{f}(\xi) = \lim_{\gamma \to \infty} \int_{B_\gamma} f(x)\chi_p(\xi x)dx \ \text{in}\ L^2\,; \tag{4.2}$$

---

* In the norm $||\ ||_2$ we shall omit the index 2, $||f||_2 = ||f||$.

the inversion formula

$$f(x) = \lim_{\gamma\to\infty} \int_{B_\gamma} \tilde{f}(\xi)\chi_p(-x\xi)d\xi \ \text{in}\ L^2 \tag{4.3}$$

and the Parseval-Steklov equality

$$(f,g) = (\tilde{f},\tilde{g}), \|f\| = \|\tilde{f}\|, \qquad f,g \in L^2 \tag{4.4}$$

are valid.

■ Let $f \in L^2$. Then $f_\gamma = \Omega(p^{-\gamma}|x|_p)f \in L^1$ for all $\gamma \in \mathbb{Z}$ and $f_\gamma \to f$, $\gamma \to \infty$ in L^2:

$$\int_{\mathbb{Q}_p} |f_\gamma(x)|dx = \int_{B_\gamma} |f(x)|dx \le \left[\int_{B_\gamma} dx \int_{B_\gamma} |f(x)|^2 dx\right]^{1/2} \le p^{\gamma/2}\|f\|,$$

$$\|f - f_\gamma\|^2 = \int_{|x|>p^\gamma} |f(x)|^2 dx \longrightarrow 0, \gamma \to \infty\ .$$

Therefore $\|f_\gamma\| \longrightarrow \|f\|$, $\gamma \to \infty$,

$$\tilde{f}_\gamma(\xi) = \int_{B_\gamma} f(x)\chi_p(\xi x)dx \in L^2 \cap \mathcal{E} \ \text{(see (3.5))},$$

and the Parseval-Steclov equality (4.1) is valid (as $\mathcal{D}$ is dense in L^2, see Sec. 6.2):

$$\|\tilde{f}_\gamma\| = \|f_\gamma\| \longrightarrow \|f\|, \quad \gamma \to \infty\ , \tag{4.5}$$

$$\|\tilde{f}_\gamma - \tilde{f}_{\gamma'}\| = \|f_\gamma - f_{\gamma'}\| \longrightarrow 0, \quad \gamma,\gamma' \to \infty. \tag{4.6}$$

The limit relation (4.6) shows that the sequence $\{\tilde{f}_\gamma, \gamma \to \infty\}$ converges in itself in L^2. By the Riesz-Fisher Theorem there exists $F \in L^2$ such that $\tilde{f}_\gamma \to F$, $\gamma \to \infty$ in L^2. On the other hand $\tilde{f}_\gamma \to \tilde{f}$, $\gamma \to \infty$ in $\mathcal{D}'$. Hence $\tilde{f} = F \in L^2$, $\tilde{f}_\gamma \to \tilde{f}$, $\gamma \to \infty$ in L^2, and the representation (4.2) is proved. From $\|\tilde{f}_\gamma\| \to \|\tilde{f}\|$, $\gamma \to \infty$ and from (4.5) the equalities (4.4) follow. The inversion formula $f = \tilde{\tilde{f}}$ owing to (4.2) takes the form (4.3). Thus the operator $f \to \tilde{f}$ maps L^2 onto L^2 one-to-one continuously. ■

Corollary. *The Fourier-transform $f \to \tilde{f}$ is a unitary operator in L^2.*

A special orthonormal basis in L^2 will be constructed in Sec. 9.4.

Lemma. *For any $f \in L^2$ the equality is valid:*

$$\lim_{\gamma \to \infty} p^{-\gamma/2} \int_{B_\gamma} |f(x)| dx = 0 \,. \tag{4.7}$$

■ Let $f \in L^2$ and $\varepsilon > 0$. There exists $N \in \mathbb{Z}$ such that

$$\int_{\mathbb{Q}_p \setminus B_N} |f(x)|^2 dx < \frac{\varepsilon^2}{4}.$$

Supposing $\gamma > N$ and applying the Cauchy-Bunjakovski inequality we get

$$p^{-\gamma/2} \int_{B_\gamma} |f(x)| dx = p^{-\gamma/2} \int_{B_\gamma \setminus B_N} |f(x)| dx + p^{-\gamma/2} \int_{B_N} |f(x)| dx$$

$$\leq p^{-\gamma/2} \left[\int_{B_\gamma \setminus B_N} dx \int_{B_\gamma \setminus B_N} |f(x)|^2 dx \right]^{1/2} + p^{-\gamma/2} \left[\int_{B_N} dx \int_{B_N} |f(x)|^2 dx \right]^{1/2}$$

$$\leq \frac{\varepsilon}{2} + p^{\frac{N-\gamma}{2}} \|f\| < \varepsilon \qquad \text{if } p^{\frac{n-\gamma}{2}} \|f\| < \frac{\varepsilon}{2}.$$ ■

5. *Multiplication of Generalized Functions*

A sequence $\{\omega_k, k \to \infty\}$, $\omega_k \in \mathcal{D}$ is called the *δ-sequence*, if there exists $N \in \mathbb{Z}$ such that

$$\omega_k(x) = \delta_k(x) = p^k \Omega(p^k |x|_p), \quad k \geq N.$$

The sequence $\{\delta_k, k \to \infty\}$ is called *the canonical δ-sequence.*

As it was established in Sec. 6.3

$$\omega_k \longrightarrow \delta, k \to \infty \text{ in } \mathcal{D}' .$$

If $\{\eta_k, k \to \infty\}$ is a 1-sequence then $\{\tilde{\eta}_k, k \to \infty\}$ is a δ-sequence, and vice versa (see (2.7) of Sec. 7.2).

Let generalized functions f and g from $\mathcal{D}'$ be given. Their *product* $f \cdot g$ we call the functional defined by the equality

$$(f \cdot g, \varphi) = \lim_{k\to\infty} (g, (f * \delta_k)\varphi) \tag{5.1}$$

if the limit exists for any $\varphi \in \mathcal{D}$. The right-hand side of (5.1) defines a linear functional on $\mathcal{D}$, so that $f \cdot g \in \mathcal{D}'$. The equality (5.1) is equivalent to the equality

$$(f \cdot g, \varphi) = \lim_{k\to\infty} (g, (f * \omega_k)\varphi), \quad \varphi \in \mathcal{D}$$

for any δ-sequence $\{\omega_k, k \to \infty\}$.

The definition (5.1) can be reformulated in the form

$$f \cdot g = \lim_{k\to\infty} (f * \delta_k) g$$

if the limit exists in $\mathcal{D}'$.

The product $g \cdot f$ is defined by the similar way:

$$g \cdot f = \lim_{k\to\infty} f(g * \delta_k) \tag{5.1'}$$

if the limit exists in $\mathcal{D}'$.

Example 1. Let f and g be continuous on $\mathbb{Q}_p$. Then their product $f \cdot g$ exists and coincides with the pointwise product $f(x)g(x)$.

■ Owing to (5.1), (1.7) and (3.4′) of Sec. 6.3 for any $\varphi \in \mathcal{D}$ we have

$$(f \cdot g, \varphi) = \lim_{k\to\infty} (g, (f * \delta_k)\varphi) = \lim_{k\to\infty} \int_{\mathbb{Q}_p} g(x)\varphi(x) \int_{\mathbb{Q}_p} f(y)\delta_k(x-y)\,dy\,dx$$

$$= \lim_{k\to\infty} \int_{\mathbb{Q}_p} g(x)\varphi(x) \int_{\mathbb{Q}_p} \delta_k(\xi) f(x-\xi)\,d\xi\,dx$$

$$= \lim_{k\to\infty} \int_{\mathbb{Q}_p} \delta_k(\xi) \int_{\mathbb{Q}_p} g(x)\varphi(x) f(x-\xi)\,dx\,d\xi = \int_{\mathbb{Q}_p} f(x)g(x)\varphi(x)\,dx,$$

from where we derive $(f \cdot g)(x) = f(x)g(x)$. ■

Remark. The statement in Example 1 remains valid for those f and g from $L^1_{\rm loc}$ for which the function

$$\int_{\mathbb{Q}_p} g(x)\varphi(x)f(x-\xi)dx, \quad \varphi \in \mathcal{D}$$

is continuous in 0, e.g. if $f \in L^p_{\rm loc}$ and $g \in L^q_{\rm loc}$, $\frac{1}{p} + \frac{1}{q} = 1$, $p \geq 1$.

Example 2. If $f \in \mathcal{D}'$ and $a \in \mathcal{E}$ then the product $a \cdot f$ exists and it coincides with the product af introduced in Sec. 6.4.

■ Owing to (5.1), (1.7) and (3.4′) of Sec. 6.3 for any $\varphi \in \mathcal{D}$ we have

$$\begin{aligned}(a \cdot f, \varphi) &= \lim_{k\to\infty} (f, (a^* \delta_k)\varphi) = \lim_{k\to\infty} \left(f(x), \varphi(x) \int_{\mathbb{Q}_p} a(y)\delta_k(x-y)dy \right) \\ &= \lim_{k\to\infty} \left(f(x), \varphi(x) \int_{\mathbb{Q}_p} \delta_k(\xi)a(x-\xi)d\xi \right) \\ &= \lim_{k\to\infty} \int_{\mathbb{Q}_p} \delta_k(\xi)(f(x), \varphi(x)a(x-\xi))d\xi = (f, \varphi a) = (af, \varphi)\end{aligned}$$

from where we derive that $a \cdot f = af$. Here we used the equality (6.4) of Sec. 6.6. ■

If $f, g \in \mathcal{D}'$ and supp $g \subset B_N$ *then the following equality is valid*

$$F[f * g] = \tilde{f} \cdot \tilde{g} \,. \tag{5.3}$$

■ By using the formulas (3.1), (1.4) and (3.10) for all $\varphi \in \mathcal{D}$ we have

$$\begin{aligned}
(F[f*g], \varphi) &= (f*g, \tilde{\varphi}) = (f(x) \times g(y), \Delta_N(y)\tilde{\varphi}(x+y)) \\
&= \left(f(x), \left(g(y), \Delta_N(y) \int_{\mathbb{Q}_p} \varphi(\xi)\chi_p(\xi(x+y))d\xi \right)\right) \\
&= \left(f(x), \int_{\mathbb{Q}_p} (g(y), \Delta_N(y)\chi_p(\xi y))\varphi(\xi)\chi_p(\xi x)d\xi \right) \\
&= \left(f(x), \int_{\mathbb{Q}_p} \tilde{g}(\xi)\varphi(\xi)\chi_p(\xi x)d\xi) \right) \\
&= (f, F[\tilde{g}\varphi]) = (\tilde{f}, \tilde{g}\varphi) = (\tilde{g}\tilde{f}, \varphi)
\end{aligned}$$

from where we derive $F[f*g] = \tilde{g}\tilde{f} = \tilde{g} \cdot \tilde{f}$ as $\tilde{g} \in \mathcal{E}$ (see example 2). Here we used the equality (6.5) of Sec. 6.6. ■

Theorem. *Let $f, g \in \mathcal{D}'$. In order that the product $f \cdot g$ exists it is necessary and sufficient that the convolution $\tilde{f}*\tilde{g}$ exists. In addition the following equality is valid*

$$F[f \cdot g] = \tilde{f}*\tilde{g}, \qquad F[f*g] = \tilde{f} \cdot \tilde{g} \tag{5.4}$$

■ It follows from the definition of the convolution, the product, the Fourier-transform and from the equalities

$$F[(f*\delta_k) \cdot g] = (\Delta_k \tilde{f})*\tilde{g}, F[(\Delta_k f)*g] = (\tilde{f}*\delta_k) \cdot \tilde{g}, \qquad k \in \mathbb{Z}\,. \quad ■$$

Corollary. *If the product $f \cdot g$ exists then the product $g \cdot f$ exists and they are equal:*

$$f \cdot g = g \cdot f\,. \tag{5.5}$$

■ It follows from the commutativity of convolution, see Sec. 7.1. ■

VIII. Homogeneous Generalized Functions

1. *Homogeneous Generalized Functions*

Let $\pi(x)$ be a multiplicative character of the field $\mathbb{Q}_p$, $\pi(xy) = \pi(x)\pi(y)$ (see Sec. 3.2). A generalized function f from $\mathcal{D}'$ is called *homogeneous of degree* $\pi(x)$ if for any $\varphi \in \mathcal{D}$ and $t \in \mathbb{Q}_p^*$ the equality

$$\left(f(x), \varphi\left(\frac{x}{t}\right)\right) = \pi(t)|t|_p(f, \varphi) \tag{1.1}$$

is fulfilled, i.e. owing to (4.5) of Sec. 6.4

$$f(tx) = \pi(t)f(x), \qquad t \in \mathbb{Q}_p^*. \tag{1.1'}$$

Remark. The equality (1.1′) allows to speak about a value of a generalized function f at the point $x = 1$:

$$f(1) = \frac{f(t)}{\pi(t)}, \qquad t \in \mathbb{Q}_p^*.$$

Now we shall describe all homogeneous generalized functions.

According to Sec. 3.2 all multiplicative characters $\pi(x)$ can be represented in the form

$$\pi(x) \equiv \pi_\alpha(x) = |x|_p^{\alpha-1}\pi_1(x) \tag{1.2}$$

where $\pi(p) = p^{1-\alpha}$ and $\pi_1(x)$ is a normed multiplicative character of the field $\mathbb{Q}_p$ such that

$$\pi_1(x) = \pi_1(|x|_p x), \qquad \pi_1(p) = \pi_1(1) = 1, \qquad |\pi_1(x)| = 1; \tag{1.3}$$

$\pi_1(x)$ is a character of the group S_0.

If $\pi(x) \not\equiv 1$ is a multiplicative character of the field $\mathbb{Q}_p$ then

$$\int_{S_\gamma} \pi(x)dx = 0, \qquad \gamma \in \mathbb{Z}. \tag{1.4}$$

■ As $\pi(x) \not\equiv 1$ then there exists $a \in \mathbb{Q}_p$ such that $|a|_p = 1$, $\pi(a) \neq 1$. Changing the variable $x = ax'$ of integration in the integral (1.4) we obtain the equality (1.4):

$$\int_{S_\gamma} \pi(x)dx = \int_{S_\gamma} \pi(ax')d(ax') = \pi(a)\int_{S_\gamma} \pi(x')dx' . \qquad ■$$

For every character $\pi_\alpha(x)$ we define the generalized function π_α from $\mathcal{D}'$ by the formula

$$(\pi_\alpha, \varphi) = \int_{\mathbb{Q}_p} |x|_p^{\alpha-1} \pi_1(x)\varphi(x)dx, \qquad \varphi \in \mathcal{D}. \tag{1.5}$$

For $\operatorname{Re}\alpha > 0$ *the integral* (1.5) *converges absolutely and defines a holomorphic function; for the others α we define it by means of analytic continuation.*

Later on we shall use the following definition: a generalized function $f_\alpha \in \mathcal{D}'$ which depends on a complex parameter α is called holomorphic in a domain $\mathbb{O} \subset \mathbb{C}$ if for all $\varphi \in \mathcal{D}$ the function $\alpha \to (f_\alpha, \varphi)$ is holomorphic in $\mathbb{O}$.

The function π_α is holomorphic in the domain $\Re\alpha > 0$. Its analytical continuation is given by the formula

$$(\pi_\alpha, \varphi) = \int_{B_0} |x|_p^{\alpha-1} \pi_1(x)[\varphi(x) - \varphi(0)]dx + \int_{\mathbb{Q}_p \setminus B_0} |x|_p^{\alpha-1} \pi_1(x)\varphi(x)dx$$
$$+ \varphi(0) \int_{B_0} |x|_p^{\alpha-1} \pi_1(x)dx, \qquad \varphi \in \mathcal{D}. \tag{1.6}$$

The last integral to the right in (1.6) *for $\pi_1(x) \not\equiv 1$by virtue of the formulas* (1.3) *of Sec.* 4.1 *and* (1.4) *is equal to 0:*

$$\int_{B_0} |x|_p^{\alpha-1} \pi_1(x)dx = \sum_{-\infty<\gamma\leq 0} p^{\gamma(\alpha-1)} \int_{S_\gamma} \pi_1(x)dx = 0;$$

for $\pi_1(x) \equiv 1$ this integral by virtue of the formula (2.5) *of Sec.* 4.2 *is equal to*

$$\int_{B_0} |x|_p^{\alpha-1} dx = \frac{1-p^{-1}}{1-p^{-\alpha}}. \tag{1.7}$$

Therefore the representation (1.6) *can be rewritten in the form*

$$(\pi_\alpha, \varphi) = \int_{B_0} |x|_p^{\alpha-1} \pi_1(x)[\varphi(x) - \varphi(0)]dx + \int_{\mathbb{Q}_p \setminus B_0} |x|_p^{\alpha-1} \pi_1(x)\varphi(x)dx$$
$$+ \begin{cases} \frac{1-p^{-1}}{1-p^{-\alpha}}\varphi(0) & \text{if } \pi_1(x) \equiv 1 \\ 0 & \text{if } \pi_1(x) \not\equiv 1. \end{cases} \tag{1.8}$$

The generalized function π_α is entire if $\pi_1(x) \not\equiv 1$; for $\pi_1(x) \equiv 1$ it is holomorphic everywhere except the points

$$\alpha_k = \frac{2\pi i k}{\ln p}, \qquad k \in \mathbb{Z} \tag{1.9}$$

where it has simple poles with the residue $\frac{p-1}{p \ln p}\delta(x)$; π_α is the homogeneous generalized function of degree $\pi_\alpha(x)$.

■ It follows from the representation (1.8). The first integral in (1.8) is an entire function on α as $\varphi(x) - \varphi(0)$ is equal to 0 in a vicinity of 0. The second integral in (1.8) is also an entire function on α as $\varphi(x)$ has a compact support. Therefore for $\pi_1(x) \equiv 1$ the function (π_α, φ) has the form

$$(\pi_\alpha, \varphi) = \varphi_\alpha + \frac{1-p^{-1}}{1-p^{-\alpha}}\varphi(0)$$

where φ_α is an entire function. Therefore (π_α, φ) is a holomorphic function on α everywhere except the points α_k, $k = 0, \pm 1, \ldots$, defined by the equality (1.9) where it has simple poles with residue

$$\operatorname*{res}_{\alpha=\alpha_k} (\pi_\alpha, \varphi) = \operatorname*{res}_{\alpha=\alpha_k} \frac{1-p^{-1}}{1-p^{-\alpha}}\varphi(0) = \frac{p-1}{p \ln p}(\delta, \varphi).$$

Homogeneity of the generalized function π_α follows from (1.1) and (1.5): for $\Re\alpha > 0$

$$\begin{aligned}\left(\pi_\alpha(x), \varphi\left(\frac{x}{t}\right)\right) &= \int_{\mathbb{Q}_p} |x|_p^{\alpha-1}\pi_1(x)\varphi\left(\frac{x}{t}\right) dx \\ &= \int_{\mathbb{Q}_p} |tx'|_p^{\alpha-1}\pi_1(tx')\varphi(x')d(tx') \\ &= |t|_p^{\alpha-1}|t|_p\pi_1(t) \int_{\mathbb{Q}_p} |x'|_p^{\alpha-1}\pi_1(x')\varphi(x')dx' \\ &= \pi_\alpha(t)|t|_p(\pi_\alpha, \varphi), \qquad \varphi \in \mathcal{D}, t \in \mathbb{Q}_p^*,\end{aligned}$$

for other α- by the principle of analytical continuation. ■

δ-function is homogeneous of degree $|x|_p^{-1}$.

■ In fact,

$$\left(\delta, \varphi\left(\frac{x}{t}\right)\right) = \varphi(0) = |t|_p |t|_p^{-1}(\delta, \varphi), \qquad \varphi \in \mathcal{D}, t \in \mathbb{Q}_p^* . \qquad ■$$

Let us introduce the generalized function $\mathcal{P}\frac{1}{|x|_p}$ by the formula

$$\left(\mathcal{P}\frac{1}{|x|_p}, \varphi\right) = \int_{B_0} \frac{\varphi(x) - (\varphi)(0)}{|x|_p} dx + \int_{\mathbb{Q}_p \setminus B_0} \frac{\varphi(x)}{|x|_p} dx, \qquad \varphi \in \mathcal{D}. \quad (1.10)$$

The generalized function $\mathcal{P}\frac{1}{|x|_p}$ is not homogeneous of degree $|x|_p^{-1}$.

$$■ \left(\mathcal{P}\frac{1}{|x|_p}, \varphi\left(\frac{x}{t}\right)\right) = \int_{B_0} \frac{\varphi(\frac{x}{t}) - \varphi(0)}{|x|_p} dx + \int_{\mathbb{Q}_p \setminus B_0} \frac{\varphi(\frac{x}{t})}{|x|_p} dx$$

$$= \int_{|x'|_p \le |t|_p^{-1}} \frac{\varphi(x') - \varphi(0)}{|x'|_p} dx' + \int_{|x'|_p > |t|_p^{-1}} \frac{\varphi(x')}{|x'|_p} dx'$$

$$= \left(\mathcal{P}\frac{1}{|x|_p}, \varphi\right) - \varphi(0) \int_{p \le |x'|_p < |t|_p^{-1}} \frac{dx'}{|x'|_p} \ne \left(\mathcal{P}\frac{1}{|x|_p}, \varphi\right)$$

if $0 < |t|_p < 1$, $\varphi \in \mathcal{D}$, $\varphi(0) \ne 0$. ■

Lemma. *If f_i, $i = 1, 2, \ldots, n$, are homogeneous generalized functions of degree $\pi_{\alpha_i}(x)$ respectively where all α_i are different, and they satisfy the equation $\sum_{1 \le i \le n} f_i = 0$ then $f_1 = f_2 = \ldots = f_n = 0$.*

■ For all $\varphi \in \mathcal{D}$ owing to (1.1) we have

$$\sum_{1 \le i \le n} \left(f_i, \varphi\left(\frac{x}{t}\right)\right) = \sum_{1 \le i \le n} \pi_{\alpha_i}(t)|t|_p (f_i, \varphi) = 0, \qquad t \in \mathbb{Q}_p^*.$$

From here putting $t = p^N$, $N \in \mathbb{Z}$ and using (1.3) we get

$$\sum_{1 \le i \le n} p^{\alpha_i N}(f_i, \varphi) = 0, \qquad N \in \mathbb{Z},$$

and hence $(f_1,\varphi) = (f_2,\varphi) = \ldots = (f_n,\varphi) = 0$, i.e. $f_1 = f_2 = \ldots = f_n = 0$.
■

The following Theorem gives a description of all homogeneous generalized functions.

Theorem. *Every homogeneous generalized functions of degree* $\pi_\alpha(x) = |x|_p^{\alpha-1}\pi_1(x)$ *is* $C\pi_\alpha$ *if* $\pi_1(x) \not\equiv 1$ *or if* $\pi_1(x) \equiv 1$ *then* $\alpha \neq 0$; *every homogeneous generalized function of degree* $|x|_p^{-1}$ *is* $C\delta$, *where* C *is an arbitrary constant.*

■ Let $f \not\equiv 0$ be a homogeneous generalized function of degree π_α and also $\pi_1(x) \not\equiv 1$ or if $\pi_1(x) = 1$ then $\alpha \neq 0$.

At first we shall prove: there exists a number $C \neq 0$ such that for any $\varphi \in \mathcal{D}$, $\varphi(0) = 0$ the equality is valid

$$(f,\varphi) = C(\pi_\alpha,\varphi). \tag{1.11}$$

The support of f contains points different from 0. (Otherwise we would have $f = C\delta$ (see Sec. 6.3) and f would be a homogeneous generalized function of degree $|x|_p^{-1}$ which is excluded.) Therefore there exists a function $\omega \in \mathcal{D}$, $\omega(0) = 0$ such that $(f,\omega) = 1$ and $(\pi_\alpha,\omega) \neq 0$. By virtue of (1.1) we have

$$\left(f(x),\omega\left(\frac{x}{t}\right)\right) = \pi_\alpha(t)|t|_p, \qquad t \in \mathbb{Q}_p^*,$$

and hence

$$\int_{\mathbb{Q}_p} \left(f(x),\omega\left(\frac{x}{t}\right)\right)\frac{\varphi(t)}{|t|_p}dt = (\pi_\alpha,\varphi), \qquad \varphi \in \mathcal{D}, \quad \varphi(0) = 0. \tag{1.12}$$

As $\omega(x)$ and $\varphi(x)$ vanish in a vicinity of 0 so

$$\omega\left(\frac{x}{t}\right)\varphi(t)|t|_p^{-1} \in \mathcal{D}(\mathbb{Q}_p^2)\,.$$

Then owing to the formula (6.5) of Sec. 6.6 from the equality (1.12) we derive

$$(\pi_\alpha,\varphi) = \left(f(x), \int_{Q_p} \omega\left(\frac{x}{t}\right)\frac{\varphi(t)}{|t|_p}dt\right). \tag{1.13}$$

In the inner integral in (1.13) we change the variable of integration (for any fixed $x \neq 0$)

$$t = \frac{x}{x'},\, dt = |x|_p |x'|_p^{-2} dx', \qquad \text{(see Sec. 4.2)}.$$

As a result we obtain

$$(\pi_\alpha, \varphi) = \left(f(x), \int_{\mathbb{Q}_p} \omega(x')\varphi\left(\frac{x}{x'}\right) \frac{dx'}{|x'|_p} \right)$$
$$= \int_{\mathbb{Q}} \frac{\omega(x')}{|x'|_p} \left(f(x), \varphi\left(\frac{x}{x'}\right)\right) dx'. \tag{1.14}$$

Here we again used the formula (6.5) of Sec. 6.6 as

$$\omega(x')\varphi\left(\frac{x}{x'}\right)|x'|_p^{-1} \in \mathcal{D}(\mathbb{Q}_p^2).$$

By using the property (1.1) ($x' \neq 0$!) on the right side of the equality (1.14) we get the equality (1.11)

$$(\pi_\alpha, \varphi) = \int_{\mathbb{Q}_p} \omega(x')\pi_\alpha(x')(f, \varphi)dx' = (\pi_\alpha, \varphi)(f, \varphi)$$

in which $C = 1/(\pi_\alpha, \omega)$.

From the equality (1.11) it follows: either $f = C\pi_\alpha$ or $\operatorname{supp}(f - C\pi_\alpha) = \{0\}$. But the last is impossible otherwise we would have $f - C\pi_\alpha = C_1\delta$ for some constant C_1 which by the lemma is possible only for $C_1 = 0$ and thus $f - C\pi_\alpha = 0$.

Let now $f \not\equiv 0$ be a homogeneous generalized function of degree $|x|_p^{-1}$. Let us suppose that supp f contains points different from 0. Then repeating literally the previous arguments we are convinced that the formula (1.11) is true also for $\pi_\alpha(x) = |x|_p^{-1}$:

$$(f, \varphi) = C \int_{\mathbb{Q}_p} \varphi(x) \frac{dx}{|x|_p}, \qquad \varphi(\mathcal{D}), \quad \varphi(0) = 0 \quad (C \neq 0). \tag{1.15}$$

Using the definition (1.10) of the generalized function $\mathcal{P}\frac{1}{|x|_p}$ we rewrite the equality (1.15) in the form

$$(f, \varphi) = \left(C\mathcal{P}\frac{1}{|x|_p}, \varphi, \right), \qquad \varphi \in \mathcal{D}, \quad \varphi(0) = 0.$$

Therefore

$$f - C\mathcal{P}\frac{1}{|x|_p} = C_1\delta$$

where C_1 is some constant. The generalized function $f - C_1\delta$ is homogeneous of degree $|x|_p^{-1}$ but $\mathcal{P}\frac{1}{|x|_p}$ is not. This contradiction proves that supp f does not contain points different from 0, and thus $f = C\delta$. ■

Let us calculate the following integral (see Sec. 4):

$$\int_{|x|_p>1} |x|_p^{\alpha-1}dx = \sum_{1\leq\gamma<\infty} p^{\gamma(\alpha-1)}\int_{S_\gamma} dx$$

$$= \left(1-\frac{1}{p}\right)\sum_{1\leq\gamma<\infty} p^{\gamma\alpha} = -\frac{1-p^{-1}}{1-p^{-\alpha}} \tag{1.16}$$

if $\Re\alpha < 0$; for $\Re\alpha \geq 0$, $\alpha \neq \alpha_k = \frac{2k\pi i}{\ln p}$, $k \in \mathbb{Z}$ the integral (1.16) we define by means of analytical continuation. By virtue of the formulas (1.7) and (1.16) the following equality is valid

$$\int_{\mathbb{Q}_p} |x|_p^{\alpha-1}dx = \int_{|x|_p\leq 1} |x|_p^{\alpha-1}dx + \int_{|x|_p>1} |x|_p^{\alpha-1}dx = 0,$$

$$\alpha \neq \alpha_k, k \in \mathbb{Z}. \tag{1.17}$$

Therefore the formula (1.8) for $\pi_1(x) \equiv 1$ takes the form

$$(|x|_p^{\alpha-1}, \varphi) = \int_{\mathbb{Q}_p} |x|_p^{\alpha-1}[\varphi(x) - \varphi(0)]dx,$$

$$\varphi \in \mathcal{D}, \alpha \neq \alpha_k, k \in \mathbb{Z}. \tag{1.18}$$

We shall call the generalized function $|x|_p^{\alpha-1}$ *homogeneous of degree* $\alpha - 1$,

$$|tx|_p^{\alpha-1} = |t|_p^{\alpha-1}|x|_p^{\alpha-1}, \qquad t \in \mathbb{Q}_p^*, \quad \alpha \neq \alpha_k, \quad k \in \mathbb{Z}.$$

2. *The Fourier-Transform of Homogeneous Generalized Functions and Γ-Function*

By the Theorem of Sec. 8.1 all homogeneous generalized functions of degree $\pi_\alpha(x) = |x|_p^{\alpha-1}\pi_1(x)$ are $C\pi_\alpha$ if either $\pi_1(x) \not\equiv 1$ or $\alpha \neq 0$; in case $\pi_1(x) \equiv 1$ and $\alpha = 1$ they are $C\delta$. Here C is an arbitrary constant.

The Fourier-transform of a homogeneous generalized function π_α is a homogeneous generalized function $\tilde{\pi}_\alpha$ of degree $\pi_\alpha^{-1}(x)|x|_p^{-1} = |x|_p^{-\alpha}\pi_1^{-1}(x)$.

■ It follows from the formulas (3.3) of Sec. 7.3 and (1.1′) that

$$\tilde{\pi}_\alpha(t\xi) = |t|_p^{-1} F\left[\pi_\alpha\left(\frac{x}{t}\right)\right] = |t|_p^{-1}\pi_\alpha\left(\frac{1}{t}\right)\tilde{\pi}_\alpha(\xi)$$
$$= |t|_p^{-\alpha}\pi_1^{-1}(t)\tilde{\pi}_\alpha(\xi), \qquad t \neq 0 . \qquad ■$$

Therefore $\pi_\alpha(\xi)$ is proportional to the homogeneous generalized function $|\xi|_p^{-\alpha}\pi_1^{-1}(\xi)$, i.e.

$$\tilde{\pi}_\alpha(\xi) = \Gamma_p(\pi_\alpha)|\xi|_p^{-\alpha}\pi_1^{-1}(\xi). \tag{2.1}$$

A factor of proportionality $\Gamma_p(\pi_\alpha)$ is called the Γ-*function* of the character $\pi_\alpha(x)$.

Putting in the formula (2.1) $\xi = 1$ and using (1.3) we obtain the equality

$$\Gamma_p(\pi_\alpha) = \tilde{\pi}_\alpha(1) = \int_{\mathbb{Q}_p} |x|_p^{\alpha-1}\pi_1(x)\chi_p(x)dx. \tag{2.2}$$

The integral on the right-hand side of the equality (2.2) is understood as the sum of the analytic continuation on the parameter α of the integrals on the right-hand side of the equality (2.3):

$$\int_{\mathbb{Q}_p} |x|_p^{\alpha-1}\pi_1(x)\chi_p(x)dx$$
$$= \int_{B_0} |x|_p^{\alpha-1}\pi_1(x)\chi_p(x)dx + \int_{\mathbb{Q}_p\setminus B_0} |x|_p^{\alpha-1}\pi_1(x)\chi_p(x)dx. \tag{2.3}$$

The following functional relation for the Γ-function is valid

$$\Gamma_p(\pi_\alpha)\Gamma_p(\pi_\alpha^{-1}|x|_p^{-1}) = \pi_1(-1). \tag{2.4}$$

■ If we apply the inverse Fourier-transform to the equality (2.1) then we obtain (see Sec. 7.3)

$$\tilde{\tilde{\pi}}_\alpha = \pi_\alpha = \Gamma_p(\pi_\alpha)F[|\xi|_p^{-\alpha}\pi_1^{-1}(-\xi)] = \frac{\Gamma_p(\pi_\alpha)}{\pi_1(-1)}F[\pi_\alpha^{-1}|\xi|_p^{-1}]. \tag{2.5}$$

As $\pi_\alpha^{-1}|\xi|_p^{-1}$ is a homogeneous generalized function of degree $\pi_\alpha^{-1}(\xi)|\xi|_p^{-1}$ (see Sec. 8.1) then its Fourier-transform is a homogeneous generalized function of degree

$$[\pi_\alpha^{-1}(x)|x|_p^{-1}]^{-1}|x|_p^{-1} = \pi_\alpha(x).$$

Hence using again the formula (2.1) we have

$$F[\pi_\alpha^{-1}|\xi|_p^{-1}] = \Gamma_p(\pi_\alpha^{-1}|\xi|_p^{-1})\pi_\alpha,$$

from here and (2.5) the formula (2.1) follows:

$$\pi_\alpha = \frac{\Gamma_p(\pi_\alpha)}{\pi_1(-1)}\Gamma_p(\pi_\alpha^{-1}|\xi|_p^{-1})\pi_\alpha\,. \qquad \blacksquare$$

Note that the formula (2.4) reminds the relation

$$\Gamma(t)\Gamma(1-t) = \frac{\pi}{\sin \pi t}$$

for the classical Γ-function.

Let the range k of a character $\pi_1(x) \not\equiv 1$ be positive. Then

$$\Gamma_p(\pi_\alpha) = p^{\alpha k} a_{p,k}(\pi_1) \tag{2.6}$$

where the numbers

$$a_{p,k}(\pi_1) = \int_{S_0} \pi_1(t)\chi_p(p^{-k}t)dt \tag{2.7}$$

satisfy the relations

$$a_{p,k}(\pi_1)a_{p,k}(\pi_1^{-1}) = p^{-k}\pi_1(-1) \tag{2.8}$$

$$|a_{p,k}(\pi_1)| = p^{-k/2}. \tag{2.9}$$

$\blacksquare$ Taking into account the equality (1.4) from (2.2) and (2.3) we get

$$\Gamma_p(\pi_\alpha) = \sum_{-\infty<\gamma\leq 0} p^{\gamma(\alpha-1)}\int_{S_\gamma}\pi_1(x)dx + \sum_{1\leq\gamma<\infty} p^{\gamma(\alpha-1)}\int_{S_\gamma}\pi_1(x)\chi_p(x)dx$$

$$= \sum_{1\leq\gamma<\infty} p^{\gamma(\alpha-1)}\int_{S_\gamma}\pi_1(x)\chi_p(x)dx.$$

Performing in the last integrals change of variable of integration $x = p^{-\gamma}t$, $dx = p^{\gamma}dt$ we get

$$\begin{aligned}\Gamma_p(\pi_\alpha) &= \sum_{1\le\gamma<\infty} p^{\gamma\alpha} \int_{S_0} \pi_1(p^{-\gamma}t)\chi_p(p^{-\gamma}t)dt \\ &= \sum_{1\le\gamma<\infty} p^{\gamma\alpha}\pi_1(p^{-\gamma}) \int_{S_0} \pi_1(t)\chi_p(p^{-\gamma}t)dt \\ &= \sum_{1\le\gamma<\infty} p^{\gamma\alpha} \int_{S_0} \pi_1(t)\chi_p(p^{-\gamma}t)dt \qquad (2.10)\end{aligned}$$

as owing to (1.3) $\pi_1(p^{-\gamma}) = [\pi_1(p)]^{-\gamma} = 1$.

Now we prove that

$$\int_{S_0} \pi_1(t)\chi_p(p^{-\gamma}t)dt = 0, \qquad \gamma \ne k, \quad \gamma \in \mathbb{Z}_+. \qquad (2.11)$$

Let $\gamma \ge 1$. We have

$$\chi_p(p^{-\gamma}t) = \exp[2\pi i(p^{-\gamma}t_0 + p^{-\gamma+1}t_1 + \ldots + p^{-1}t_{\gamma-1})], \qquad t \in S_0.$$

As the range of the character $\pi_1(t)$ is equal to $k \ge 1$ then $\pi_1(t)$ depend only on $t_0, t_1, \ldots, t_{k-1}$. Further for $k \ge 2$ the equality is valid (see (2.3) of Sec. 3.2)

$$\sum_{0\le t_{k-1}\le p-1} \pi_1(t_0 + t_1p + \ldots + t_{k-1}p^{k-1}) = 0.$$

Therefore for $1 \le \gamma < k$ (2.11) follows from

$$\begin{aligned}\int_{S_0} \pi_1(t)\chi_p(p^{-\gamma}t)dt = C &\sum_{1\le t_0\le p-1} \sum_{0\le t_1\le p-1} \\ \cdots &\sum_{0\le t_{k-2}\le p-1} \exp[2\pi i(p^{-\gamma}t_0 + p^{-\gamma+1}t_1 + \ldots + p^{-1}t_{\gamma-1})] \\ \cdot &\sum_{0\le t_{k-1}\le p-1} \sum_{0\le t_{k-1}\le p-1} \pi_1(t_0 + t_1p + \ldots + t_{k-1}p^{k-1}) = 0;\end{aligned}$$

for $1 \le k < \gamma$ (2.11) follows from

$$\int_{S_0} \pi_1(t)\chi_p(p^{-\gamma}t)dt = C \sum_{1\le t_0\le p-1} \sum_{0\le t_1\le p-1} \cdots \sum_{0\le t_{\gamma-2}\le p-1} \pi_1(t_0 + t_1 p + \ldots + t_{k-1}p^{k-1}) \cdot \sum_{0\le t_{\gamma-1}\le p-1} \exp[2\pi i(p^{-\gamma}t_0 + p^{-\gamma+1}t_1 + \ldots + p^{-1}t_{\gamma-1})] = 0.$$

Now the equality (2.6) follows from (2.10) and (2.11).

To obtain the equalities (2.8) and (2.9) we calculate the Fourier-transform of $\delta(|x|_p - p^k)\pi_1(x)$:

$$\begin{aligned} &F[\delta(|x|_p - p^k)\pi_1] \\ &= \int_{S_k} \pi_1(x)\chi_p(\xi x)dx = p^k \int_{S_0} \pi_1(p^{-k}t)\chi_p(p^{-k}\xi t)dt \\ &= p^k \int_{S_0} \pi_1(t)\chi_p(p^{-k-N}\xi' t)dt = p^k \int_{S_0} \pi_1\left(\frac{t}{\xi'}\right)\chi_p(p^{-k-N}t)dt \\ &= p^k \pi_1\left(\frac{1}{\xi'}\right) \int_{S_0} \pi_1(t)\chi_p(p^{-k-N}t)dt \\ &= \begin{cases} p^k \pi_1^{-1}(\xi')a_{p,k}(\pi_1), & N = 0, \\ 0, & N \ne 0, \end{cases} \end{aligned}$$

i.e.

$$F[\delta(|x|_p - p^k)\pi_1] = p^k \delta(|\xi|_p - 1)\pi_1^{-1}(\xi)a_{p,k}(\pi_1). \tag{2.12}$$

In the calculation of equality (2.12) we have denoted $|\xi|_p = p^N$ and $\xi' = |\xi|_p \xi$ and have used the formulas (2.7) and (2.11).

From (2.12) by virtue of the Parseval-Steklov equality (see Sec. 7.4) the equality (2.9) follows:

$$\begin{aligned} &\int_{S_k} |\pi_1(x)|^2 dx = p^k \left(1 - \frac{1}{p}\right) \\ &= p^{2k}|a_{p,k}(\pi_1)|^2 \int_{S_0} |\pi_1^{-1}(\xi')|^2 d\xi' = p^{2k}\left(1 - \frac{1}{p}\right)|a_{p,k}(\pi_1)|^2. \end{aligned}$$

To prove the equality (2.8) we apply to the equality (2.12) the inverse Fourier-transform

$$\delta(|x|_p - p^k)\pi_1(x) = p^k F[\delta(|\xi|_p - 1)\pi_1^{-1}(-\xi)]a_{p,k}(\pi_1)$$
$$= p^k a_{p,k}(\pi_1)\pi_1^{-1}(-1)F[\delta(|\xi|_p - 1)\pi_1^{-1}].$$

In order to calculate the Fourier-transform of the function $\delta(|\xi|_p - 1)$ $\pi_1^{-1}(\xi)$ we denote $|x|_p = p^N$, $x' = |x|_p x$, and from (2.7) and (2.11) as above we get

$$F[\delta(|\xi|_p - 1)\pi_1^{-1}] = \int_{S_0} \pi_1^{-1}(\xi)\chi_p(\xi x)d\xi$$
$$= \int_{S_0} \pi_1^{-1}(\xi)\chi_p(p^{-N}x'\xi)d\xi = \int_{S_0} \pi_1^{-1}\left(\frac{t}{x'}\right)\chi_p(p^{-N}t)dt$$
$$= \pi_1^{-1}\left(\frac{1}{x'}\right)\int_{S_0} \pi_1^{-1}(t)\chi_p(p^{-N}t)dt = \pi_1(x)a_{p,k}(\pi_1^{-1})\delta(|\xi|_p - p^k)$$

because the rank of the character π_1^{-1} is equal to k. From here and (2.13) the equality (2.8) follows

$$\pi_1(x) = p^k a_{p,k}(\pi_1)\pi_1^{-1}(-1)\pi_1(x)a_{p,k}(\pi_1^{-1}) \ . \qquad \blacksquare$$

From the formulas (2.6) and (2.8) we derive

$$\Gamma_p(\pi_\alpha)\Gamma_p(\pi_\alpha^{-1}) = p^k\pi_1(-1). \tag{2.14}$$

Comparing the formulas (2.14) and (2.4) we get

$$\Gamma_p(\pi_\alpha^{-1}) = p^k\Gamma_p(\pi_\alpha^{-1}|x|_p^{-1})$$

or by changing $\pi_\alpha^{-1}|x|_p^{-1}$ by π_α

$$\Gamma_p(\pi_\alpha|x|_p) = p^k\Gamma_p(\pi_\alpha). \tag{2.15}$$

The formula (2.15) can be interpreted as an analogy of equality $\Gamma(t+1) = t\Gamma(t)$ for the classical Γ-function.

Now we consider the special case $\pi_1(x) \equiv 1$, i.e. the rank of the character $\pi_1(x)$ is equal to 0. In this case we denote

$$\Gamma_p(\pi_\alpha) = \Gamma_p(|x|_p^{\alpha-1}) \equiv \Gamma_p(\alpha).$$

The formula (2.4) takes the form

$$\Gamma_p(\alpha)\Gamma_p(1-\alpha) = 1. \tag{2.16}$$

Now we prove the formula

$$\Gamma_p(\alpha) = \frac{1-p^{\alpha-1}}{1-p^{-\alpha}}. \tag{2.17}$$

■ To this end we use the representation (2.2)–(2.3) and the formulas (2.5) of Sec. 4.2 and (3.2) of Sec. 4.3

$$\begin{aligned}\Gamma_p(\alpha) &= \int_{B_0} |x|_p^{\alpha-1}\chi_p(x)dx + \int_{|x|_p>1} |x|_p^{\alpha-1}\chi_p(x)dx \\ &= \int_{B_0} |x|_p^{\alpha-1}dx + \sum_{1\le\gamma<\infty} p^{\gamma(\alpha-1)} \int_{S_\gamma} \chi_p(x)dx \\ &= \frac{1-p^{-1}}{1-p^{-\alpha}} - p^{\alpha-1} = \frac{1-p^{\alpha-1}}{1-p^{-\alpha}}.\end{aligned}$$ ■

From (2.17) it follows: $\Gamma_p(\alpha)$ is holomorphic in the complex plane α except at the points

$$\alpha_k = \frac{2k\pi i}{\ln p}, \qquad k \in \mathbb{Z} \tag{2.18}$$

where it has simple poles with residue $\frac{p-1}{p\ln p}$; $\Gamma_p(\alpha)$ has simple zeros at the points $1+\alpha_k$, $k \in \mathbb{Z}$.

From the formula (2.1) we derive

$$F[|x|_p^{\alpha-1}] = \Gamma_p(\alpha)|\xi|_p^{-\alpha}, \qquad \alpha \neq \alpha_k, \quad k \in \mathbb{Z}. \tag{2.19}$$

The generalized function

$$f_\alpha = \frac{|x|_p^{\alpha-1}}{\Gamma_p(\alpha)}$$

is holomorphic on α *except the points* $1+\alpha_k$, $k \in \mathbb{Z}$ *where it has simple poles with residue* $-\frac{p-1}{p\ln p}|x|_p^{\alpha_k}$*; in addition*

$$f_{\alpha_k} = \delta, \qquad k \in \mathbb{Z}. \tag{2.20}$$

■ It follows from the results of Secs. (8.1)–(8.2). We have

$$\begin{aligned}\operatorname*{res}_{\alpha=1+\alpha_k} \frac{|x|_p^{\alpha-1}}{\Gamma_p(\alpha)} &= \lim_{\alpha\to 1+\alpha_k}(\alpha-1-\alpha_k)\frac{|x|_p^{\alpha-1}}{\Gamma_p(\alpha)} \\ &= \lim_{\varepsilon\to+0} \frac{\varepsilon(1-p^{-\varepsilon-1-\alpha_k})}{1-p^{\varepsilon+\alpha_k}}|x|_p^{\varepsilon+\alpha_k} \\ &= -\frac{p-1}{p\ln p}|x|_p^{\alpha_k}.\end{aligned}$$

Further from the formula (1.8) of Sec. 8.1 for $\varphi \in \mathcal{D}$ we have

$$(f_\alpha,\varphi) = \frac{\varphi_\alpha}{\Gamma_p(\alpha)} + \frac{1-p^{-1}}{(1-p^{-\alpha})\Gamma_p(\alpha)}\varphi(0) = \frac{1-p^{-\alpha}}{1-p^{\alpha-1}}\varphi_\alpha + \frac{1-p^{-1}}{1-p^{\alpha-1}}\varphi(0),$$

where φ_α is an entire function. From here it follows the formula (2.20)

$$(f_\alpha,\varphi) \longrightarrow \varphi(0) = (\delta,\varphi), \qquad \alpha \to \alpha_k,$$

i.e. $f_\alpha \to \delta$, $\alpha \to \alpha_k$ in $\mathcal{D}'$. ■

Example. Let (see Sec. 3.2)

$$\pi_\alpha(x) = |x|_p^{\alpha-1}\operatorname{sgn}_\varepsilon x, \qquad p \neq 2$$

where $\varepsilon \notin \mathbb{Q}_p^{*2}$. Then

$$\Gamma_p(\alpha) = \begin{cases} \pm\sqrt{\operatorname{sgn}_\varepsilon(-1)}\,p^{\alpha-1/2} & \text{if } \varepsilon = p, p\eta, \\ \frac{1+p^{\alpha-1}}{1+p^{-\alpha}} & \text{if } \varepsilon = \eta. \end{cases} \tag{2.21}$$

■ For either $\varepsilon = p$ or $\varepsilon = p\eta$, $\pi_1^2(x) = \pi_1(x^2) \equiv 1$ and the rank of the character $\pi_1(x)$ is equal to 1 (As every $x \in \mathbb{Q}_p$ of the form $x = 1 + pt$,

$|t|_p \leq 1$ is the square of a p-adic number, see Sec. 1.4). Now using the equality (2.8) for $k = 1$

$$a_{p,1}(\pi_1)a_{p,1}(\pi_1) = p^{-1}\pi_\alpha(-1)$$

and from (2.6) we get (2.21).

For $\varepsilon = \eta$, $\mathrm{sgn}_\eta x = 1$ if $\gamma(x)$ is even and $\mathrm{sgn}_\eta x = -1$ if $\gamma(x)$ is odd. See the Lemma below. Thus $\pi_1(x) \equiv 1$ and

$$\pi_\alpha(x) = |x|_p^{\alpha - 1 + \frac{\pi i}{\ln p}}$$

and therefore

$$\Gamma_p(\pi_\alpha) = \Gamma_p\left(\alpha + \frac{\pi i}{\ln p}\right) = \frac{1 - p^{\alpha + \frac{\pi i}{\ln p} - 1}}{1 - p^{-\alpha - \frac{\pi i}{\ln p}}} = \frac{1 + p^{\alpha - 1}}{1 + p^{-\alpha}}. \qquad \blacksquare$$

Lemma. *Let $p \neq 2$ and ε be a unity, $\varepsilon \notin \mathbb{Q}_p^{*2}$. A number $x \in \mathbb{Q}_p^*$ belongs to $\mathbb{Q}_{p,\varepsilon}^*$ if and only if $\gamma(x)$ is even.*

■ Let $x \in \mathbb{Q}_p^*$ belongs to $\mathbb{Q}_{p,\varepsilon}^*$, i.e. it is represented in the form

$$x = a^2 - \varepsilon b^2, \qquad a, b \in \mathbb{Q}_p, \quad (a, b) \neq 0. \tag{2.22}$$

If $\gamma(x)$ would be even then the congruence

$$a_0^2 - \varepsilon_0 b_0^2 \equiv 0 \pmod{p} \tag{2.23}$$

would be valid and thus ε_0 is a quadratic residue modulo p that contradicts to the fact that $\varepsilon \notin \mathbb{Q}_p^{*2}$ (see Sec. 1.4).

Conversely, let $x \in \mathbb{Q}_p^*$ and $\gamma(x)$ be even. By the Chevalle Theorem (see [37] ~ 1) the congruence

$$a_0^2 - \varepsilon_0 b_0^2 - x_0 c_0^2 \equiv 0 \pmod{p}$$

has nontrivial solution (a_0, b_0, c_0), $c_0 \neq 0$ (otherwise if $c_0 = 0$ the congruence (2.23) would have a nontrivial solution that is excluded). Hence the equation

$$a_0^2 - \varepsilon b_0^2 - x c^2 = 0$$

is solvable by $c \neq 0$. From here it follows the representation (2.22)

$$x = \left(\frac{a_0}{c}\right)^2 - \varepsilon\left(\frac{b_0}{c}\right)^2 . \qquad \blacksquare$$

3. *Convolution of Homogeneous Generalized Functions and B-Function*

Let

$$\pi_\alpha(x) = |x|_p^{\alpha-1}\pi_1(x) \text{ and } \pi'_\beta(x) = |x|_p^{\beta-1}\pi'_1(x)$$

be multiplicative character of the field $\mathbb{Q}_p$, and π_α and π_β are the corresponding generalized functions, and α and $\beta \neq \alpha_k$ (see Sec. 8.1).

The convolution $\pi_\alpha * \pi'_\beta$ *exists, and it is holomorphic in the tube domain* $\operatorname{Re}\alpha > 0$, $\operatorname{Re}\beta > 0$, $\operatorname{Re}(\alpha+\beta) < 1$, *and is homogeneous generalized function of degree*

$$\pi_\alpha(x)\pi'_\beta(x)|x|_p = |x|_p^{\alpha+\beta-1}(\pi_1\pi'_1)(x),$$

and represented by the formula

$$(\pi_\alpha * \pi'_\beta)(x) = \int_{\mathbb{Q}_p} \pi_\alpha(y)\pi'_\beta(x-y)dy$$
$$= B_p(\pi_\alpha, \pi'_\beta)|x|_p^{\alpha+\beta-1}(\pi_1\pi'_1)(x) \tag{3.1}$$

where

$$B_p(\pi_\alpha, \pi'_\beta) = \int_{\mathbb{Q}_p} |t|_p^{\alpha-1}|1-t|_p^{\beta-1}\pi_1(t)\pi'_1(1-t)dt. \tag{3.2}$$

$B_p(\pi_\alpha, \pi'_\beta)$ is called the B-function of the character $\pi_\alpha(x)$ and $\pi'_\beta(x)$.

■ By virtue of the results of Sec. 7.1 the convolution $\pi_\alpha * \pi'_\beta$ of the locally-integrable functions $\pi_\alpha(x)$ and $\pi'_\beta(x)$ exists, it is a locally-integrable function, and it is expressed by the integral (3.1):

$$(\pi_\alpha * \pi'_\beta)(x) = \int_{\mathbb{Q}_p} |y|_p^{\alpha-1}\pi_1(y)|x-y|_p^{\beta-1}\pi'_1(x-y)dy.$$

The last integral is absolutely convergent and defines a holomorphic function of the parameters (α, β) in the tube domain $\mathrm{Re}\,\alpha > 0$, $\mathrm{Re}\,\beta > 0$, $\mathrm{Re}(\alpha+\beta) < 1$. In fact,

$$\int_{B_k} \pi_\alpha(y)\pi'_\beta(x-y)dy = \int_{B_k} |y|_p^{\alpha-1}\pi_1(y)|x-y|_p^{\beta-1}\pi'_1(x-y)dy$$
$$= \int_{B_k} |tx|_p^{\alpha-1}\pi_1(tx)|x-tx|_p^{\beta-1}\pi'_1(x-tx)d(tx)$$
$$= |x|_p^{\alpha+\beta-1}\pi_1(x)\pi'_1(x) \int_{|t|_p\le p^k|x|_p^{-1}} |t|_p^{\alpha-1}|1-t|_p^{\beta-1}\pi_1(t)\pi'_1(1-t)dt$$
$$\overset{x\in K}{\Longrightarrow} |x|_p^{\alpha+\beta-1}\pi_1(x)\pi'_1(x)\int_{\mathbb{Q}_p} |t|_p^{\alpha-1}|1-t|_p^{\beta-1}\pi_1(t)\pi'_1(1-t)dt$$

where K is an arbitrary compact of $\mathbb{Q}_p$.

From the representation (3.1) it follows that the convolution $\pi_\alpha * \pi'_\beta$ is a homogeneous generalized function of degree $|x|_p^{\alpha+\beta-1}\pi_1(x)\pi'_1(x)$. ■

For the parameters (α, β) lying outside the domain $\mathrm{Re}\,\alpha > 0$, $\mathrm{Re}\,\beta > 0$, $\mathrm{Re}(\alpha+\beta) < 1$ the convolution $\pi_\alpha * \pi'_\beta$ is defined by means of analytical continuation of the right-hand side of the equality (3.1) on (α, β).

Our nearest goal is to express the B-function by the Γ-functions and to define a maximal domain of analyticity of the convolution $\pi_\alpha * \pi'_\beta$.

As $\pi_\alpha * \pi'_\beta$ is a homogeneous function of the degree

$$\pi_\alpha(x)\pi'_\beta(x) = |x|_p^{\alpha+\beta-1}(\pi_1\pi'_1)(x)$$

then applying the formula (2.1) to the equality (3.1) we get

$$F[\pi_\alpha * \pi'_\beta] = \Gamma_p(\pi_\alpha\pi'_\beta|x|_p)B_p(\pi_\alpha, \pi'_\beta)|\xi|_p^{-\alpha-\beta}(\pi_1\pi'_1)^{-1}(\xi). \qquad (3.3)$$

On the other hand by applying the Fourier-transform (see Sec. 7.5) to the convolution $\pi_\alpha * \pi'_\beta$ and by using again the formula (2.1) we get

$$F[\pi_\alpha * \pi'_\beta] = \tilde{\pi}_\alpha \cdot \tilde{\pi}'_\beta = \Gamma_p(\pi_\alpha)\Gamma_p(\pi'_\beta)|\xi|_p^{-\alpha}\pi_1^{-1}(\xi)|\xi|_p^{-\beta}\pi_1'^{-1}(\xi)$$
$$= \Gamma_p(\pi_\alpha)\Gamma_p(\pi'_\beta)|\xi|_p^{-\alpha-\beta}(\pi_1\pi'_1)^{-1}(\xi). \qquad (3.4)$$

By comparing the equalities (3.3) and (3.4) we obtain the formula

$$\mathcal{B}_p(\pi_\alpha, \pi'_\beta) = \frac{\Gamma_p(\pi_\alpha)\Gamma_p(\pi'_\beta)}{\Gamma_p(\pi_\alpha \pi'_\beta |x|_p)}. \tag{3.5}$$

In addition the equality (3.1) takes the form

$$(\pi_\alpha * \pi'_\beta)(x) = \frac{\Gamma_p(\pi_\alpha)\Gamma_p(\pi'_\beta)}{\Gamma_p(\pi_\alpha \pi'_\beta |x|_p)} |x|_p^{\alpha+\beta-1}(\pi_1\pi'_1)(x) \,. \tag{3.6}$$

For $(\pi_1\pi'_1)(x) \not\equiv 1$ the generalized function $|x|_p^{\alpha+\beta-1}(\pi_1\pi'_1)(x)$ is entire (see Sec. 8.1) and $\Gamma_p(\pi_\alpha\pi'_\beta|x|_p)$ is entire non-vanishing function (see Sec. 8.2), and therefore the domain of analyticity of the function (3.6) is defined by singular points of the function $\Gamma_p(\pi_\alpha)\Gamma_p(\pi'_\beta)$: if $\pi_1(x) \not\equiv 1$ and $\pi'_1(x) \not\equiv 1$ it is an entire function; if $\pi_1(x) \equiv 1$ and $\pi'_1(x) \not\equiv 1$ the singular points are the straight lines $\alpha = \alpha_k$, $k = 0, \pm 1, \ldots$; if $\pi_1(x) \not\equiv 1$ and $\pi'_1(x) \equiv 1$ the singular points are the straight lines $\beta = \alpha_k$, $k = 0, \pm 1, \ldots$.

For $\pi_1(x) \equiv \pi'_1(x) \equiv 1$ singular points are the straight lines $\alpha = \alpha_k$, $k = 0, \pm 1, \ldots$, $\beta = \beta_k$, $k = 0, \pm 1, \ldots$, $\alpha + \beta = 1 + \alpha_k$, $k = 0, \pm 1, \ldots$.

For $\pi_1(x) \not\equiv 1$ and $\pi'_1(x) = \pi_1^{-1}(x)$ singular points are the straight lines $\alpha + \beta = 1 + \alpha_k$, $k = 0, \pm 1, \ldots$. In this case

$$\Gamma_p(\pi_\alpha)\Gamma_p(\pi'_\beta) = p^{(\alpha+\beta-1)k}\pi_1(-1) \tag{3.7}$$

where k is the range of the character π_α.

■ The formula (3.7) follows from the formulas (2.6) and (2.8)

$$\Gamma_p(\pi_\alpha)\Gamma_p(\pi'_\beta) = p^{\alpha k} a_{p,k}(\pi_1) p^{\beta k} a_{p,k}(\pi_1^{-1}) = p^{(\alpha+\beta-1)k}\pi_1(-1) \,. \qquad ■$$

Let $\pi_1(x) \equiv \pi'_1(x) \equiv 1$. Like in Sec. 8.2 we denote

$$\mathcal{B}_p(\pi_\alpha, \pi'_\beta) \equiv \mathcal{B}_p(|x|_p^{\alpha-1}, |x|_p^{\beta-1}) = \mathcal{B}_p(\alpha, \beta).$$

By this the formulas (3.1) and (3.5) take the classical form

$$|x|_p^{\alpha-1} * |x|_p^{\beta-1} = \mathcal{B}_p(\alpha,\beta)|x|_p^{\alpha+\beta-1}, \tag{3.8}$$

$$\mathcal{B}_p(\alpha,\beta) = \frac{\Gamma_p(\alpha)\Gamma_p(\beta)}{\Gamma_p(\alpha+\beta)}. \tag{3.9}$$

If we take in account the formula (2.16) we get for $B_p(\alpha,\beta)$ the symmetrical form

$$B_p(\alpha,\beta) = \Gamma_p(\alpha)\Gamma_p(\beta)\Gamma_p(\gamma) \tag{3.10}$$

where $\alpha+\beta+\gamma = 1$.

Now we shall prove for $\operatorname{Re}\alpha, \operatorname{Re}\beta, \operatorname{Re}(\alpha+\beta) \neq -1+\alpha_k$, $k \in \mathbb{Z}$ the equalities

$$|x|_p^\alpha \cdot |x|_p^\beta = |x|_p^{\alpha+\beta}, \tag{3.11}$$

$$F[|x|_p^\alpha \cdot |x|_p^\beta] = F[|x|_p^\alpha] * F[|x|_p^\beta] = F[|x|_p^{\alpha+\beta}]. \tag{3.12}$$

■ For $\operatorname{Re}\alpha > -1$, $\operatorname{Re}\beta > -1$, $\operatorname{Re}(\alpha+\beta) > -1$ the product $|x|_p^\alpha \cdot |x|_p^\beta$ exists in $\mathcal{D}'$ (in the sense of Sec. 7.5), and it is equal to $|x|_p^{\alpha+\beta}$ so the equalities (3.11) and (3.12) are valid. In fact, the functions $|x|_p^\alpha$ and $|x|_p^\beta$ belong to L^1_{loc}, and the function (see (5.2) of Sec. 7.5)

$$\int_{\mathbb{Q}_p} |x|_p^\alpha \varphi(x)|x-\xi|_p^\beta dx$$

is continuous at the point $\xi = 0$ for every $\varphi \in \mathcal{D}$ owing to the majorization

$$|x|_p^\alpha |\varphi(x)||x-\xi|_p^\beta \leq C\max(|x|_p^\alpha, |x|_p^\beta), \qquad x \in \operatorname{supp}\varphi, \quad |\xi|_p \leq 1$$

and the Lebesgue Theorem on limiting passage under the sign of an integral (see Sec. 4.4).

For the other α and β the equalities (3.11) and (3.12) are obtained by analytical continuation of the right-hand sides on α and β. ■

Analogously, if $f \in \mathcal{E}$ and $\operatorname{Re}\alpha, \operatorname{Re}\beta, \operatorname{Re}(\alpha+\beta) \neq -1+\alpha_k$, $k \in \mathbb{Z}$ then

$$|x|_p^\alpha \cdot [|x|_p^\beta f(x)] = |x|_p^{\alpha+\beta} f(x) = |x|_p^\beta \cdot [|x|_p^\alpha f(x)], \tag{3.13}$$

$$\begin{aligned} F[|x|_p^\alpha \cdot (|x|_p^\beta f)] &= F[|x|_p^\alpha] * F[|x|_p^\beta] * \tilde{f} \\ &= (F[|x|_p^\alpha] * F[|x|_p^\beta]) * \tilde{f} = F[|x|_p^\beta \cdot (|x|_p^\alpha f)]. \end{aligned} \tag{3.14}$$

4. *Homogeneous Generalized Functions of Several Variables*

The results of Sec. 8.1–8.3 without essential changes are carried over the homogeneous function $|x|_p^{\alpha-n}$ of degree $\alpha - n$ depending on n variables $x = (x_1, x_2, \ldots, x_n)$,

$$|x|_p = \max(|x_1|_p, |x_2|_p, \ldots, |x_n|_p) \qquad \text{(see Sec. 1.7)}.$$

At first we shall calculate the integral

$$\int_{B_0} |x|_p^{\alpha-n} dx = \frac{1-p^{-n}}{1-p^{-\alpha}}, \qquad \operatorname{Re}\alpha > 0. \tag{4.1}$$

The formula (4.1) for $n = 1$ coincides with (1.7).

■ By induction on n. For $n = 1$ the formula (4.1) is true. By supposing it true for n we shall prove it for $n + 1$. Denote $x = (\tilde{x}, x_{n+1})$, $\tilde{x} = (x_1, x_2, \dots, x_n)$. By using the Fubini Theorem (see Sec. 4.4) we are convinced that the formula (4.1) is valid for $n + 1$:

$$\begin{aligned}
\int_{B_0} |x|_p^{\alpha-n-1} dx &= \int_{B_0} |(\tilde{x}, x_{n+1})|_p^{\alpha-n-1} d\tilde{x}\, dx_{n+1} \\
&= \int_{B_0} \left[\int_{|\tilde{x}|_p \le |x_{n+1}|_p} |x_{n+1}|_p^{\alpha-n-1} d\tilde{x} + \int_{|\tilde{x}|_p > |x_{n+1}|_p} |\tilde{x}|_p^{\alpha-n-1} d\tilde{x} \right] dx_{n+1} \\
&= \int_{B_0} |x_{n+1}|_p^{\alpha-1} dx_{n+1} \\
&\quad + \int_{B_0}\int_{B_0} |\tilde{x}|_p^{\alpha-n-1} d\tilde{x}\, dx_{n+1} - \int_{B_0} \int_{|\tilde{x}|_p \le |x_{n+1}|_p} |\tilde{x}|_p^{\alpha-n-1} d\tilde{x}\, dx_{n+1} \\
&= \frac{1-p^{-1}}{1-p^{-\alpha}} + \int_{B_0} \frac{1-p^{-n}}{1-p^{-\alpha+1}} dx_{n+1} - \int_{B_0} |x_{n+1}|_p^{\alpha-1} \int_{B_0} |\tilde{t}|_p^{\alpha-n-1} d\tilde{t}\, dx_{n+1} \\
&= \frac{1-p^{-1}}{1-p^{-\alpha}} + \frac{1-p^{-n}}{1-p^{-\alpha+1}} - \frac{1-p^{-1}}{1-p^{-\alpha}} \cdot \frac{1-p^{-n}}{1-p^{-\alpha+1}} = \frac{1-p^{-n-1}}{1-p^{-\alpha}}. \qquad ■
\end{aligned}$$

Let us introduce the generalized function $|x|_p^{\alpha-n}$ from $\mathcal{D}'(\mathbb{Q}_p^n)$ by the formula (cf. (1.6))

$$\begin{aligned}
(|x|_p^{\alpha-n}, \varphi) &= \int_{B_0} |x|_p^{\alpha-n} [\varphi(x) - \varphi(0)] dx \\
&\quad + \int_{Q_p^n \setminus B_0} |x|_p^{\alpha-n} \varphi(x) dx + \varphi(0) \frac{1-p^{-n}}{1-p^{-\alpha}}, \qquad \varphi \in \mathcal{D}(\mathbb{Q}_p^n).
\end{aligned} \tag{4.2}$$

The generalized function $|x|_p^{\alpha-n}$ is holomorphic everywhere except the points (1.9)

$$\alpha_k = \frac{2k\pi i}{\ln p}, \quad k \in \mathbb{Z},$$

where it has simple poles with the residue

$$\frac{p^n - 1}{p^n \ln p}\delta(x).$$

The Fourier-transform formula is valid:

$$F[|x|_p^{\alpha-n}] = \Gamma_p^n(\alpha)|\xi|_p^{-\alpha}, \qquad \alpha \neq \alpha_k, \quad k \in \mathbb{Z} \tag{4.3}$$

where Γ_p^n is the n-dimensional Γ-function ($\Gamma_p^1 = \Gamma_p$)

$$\Gamma_p^n(\alpha) = \int_{\mathbb{Q}_p^n} |x|_p^{\alpha-n}\chi_p(x_1)dx = \frac{1-p^{\alpha-n}}{1-p^{-\alpha}}. \tag{4.4}$$

For $n = 1$ the formula (4.3) coincides with the formula (2.19).

The formula (4.4) follows from the more general formula ($m \neq 0$)

$$F[|(x,m)|_p^{\alpha-n}] = \Gamma_p^n(\alpha)\Omega(|m\xi|_p)[|\xi|_p^{-\alpha} - |pm|_p^{\alpha}] \tag{4.5}$$

where x, $\xi \in \mathbb{Q}_p^n$, $|(x,m)|_p = \max(|x|_p, |m|_p)$.

In fact,for $\operatorname{Re}\alpha > n$ the following limit relations

$$|(x,m)|_p^{\alpha-n} \to |x|_p^{\alpha-n}, \quad \Omega(|m\xi|_p) \to 1, \quad |m|_p^{\alpha}\Omega(|m\xi|_p) \to 0, \quad m \to \infty$$

in $\mathcal{D}'(\mathbb{Q}_p^n)$ are valid, and the equality (4.3) follows from the continuity of the Fourier-transform operation. For other α one uses an analytical continuation on α.

■ To prove the formula (4.5) we use the induction on n. At first we shall prove it for $n = 1$,

$$\int_{\mathbb{Q}_p} |(x,m)|_p^{\alpha-1}\chi_p(x\xi)dx = \Gamma_p(\alpha)\Omega(|m\xi|_p)[|\xi|_p^{-\alpha} - |pm|_p^{\alpha}]. \tag{4.6}$$

It follows from the formulas (3.1) of Sec. 4.3, (2.16) of Sec. 7.2 and (2.19):

$$\int_{\mathbb{Q}_p} [\max(|x|_p, |m|_p)]^{\alpha-1} \chi_p(x\xi) dx$$

$$= \int_{|x|_p \le |m|_p} |m|_p^{\alpha-1} \chi_p(x\xi) dx + \int_{|x|_p > |m|_p} |x|_p^{\alpha-1} \chi_p(x\xi) dx$$

$$= |m|_p^{\alpha-1} |m|_p \Omega(|m\xi|_p) + F[|x|_p^{\alpha-1}] - \int_{|x|_p \le |m|_p} |x|_p^{\alpha-1} \chi_p(x\xi) dx$$

$$= |m|_p^{\alpha} \Omega(|m\xi|_p) + \Gamma_p(\alpha) |\xi|_p^{-\alpha} - |m|_p^{\alpha} \Omega(|m\xi|_p) \frac{1-p^{-1}}{1-p^{-\alpha}}$$

$$- \Gamma_p(\alpha) |\xi|_p^{-\alpha} [1 - \Omega(|m\xi|_p)]$$

$$= \Gamma_p(\alpha) |\xi|_p^{-\alpha} \Omega(|m\xi|_p) + |m|_p^{\alpha} \Omega(|m\xi|_p) \left(1 - \frac{1-p^{-1}}{1-p^{-\alpha}}\right).$$

Let the formula (4.5) be true for n, and we prove it for $n+1$ i.e. for $x, \xi \in \mathbb{Q}_p^{n+1}$

$$F[|(x,m)|_p^{\alpha-n-1}] = \Gamma_p^{n+1}(\alpha) \Omega(|m\xi|_p) [|\xi|_p^{-\alpha} - |pm|_p^{\alpha}] . \tag{4.7}$$

Denoting

$$J_n^m(\xi) = F[|(x,m)|_p^{\alpha-n}],\ \tilde{m} = \max(|x_{n+1}|_p, |m|_p),$$
$$x = (\tilde{x}, x_{n+1}), \qquad \xi = (\tilde{\xi}, \xi_{n+1}), \qquad \tilde{x}, \tilde{\xi} \in \mathbb{Q}_p^n$$

and using the Fubini Theorem we get

$$J_{n+1}^m(\xi) = \int\limits_{\mathbb{Q}_p^{n+1}} |(x,m)|_p^{\alpha-n-1}\chi_p((x,\xi))dx = \int\limits_{\mathbb{Q}_p} \chi_p(x_{n+1}\xi_{n+1})\cdot$$

$$\int\limits_{\mathbb{Q}_p^n} \{\max[|\tilde{x}|_p, \max(|x_{n+1}|_p, |m|_p)]\}^{\alpha-n-1}\chi_p((\tilde{x},\tilde{\xi}))d\tilde{x}dx_{n+1}$$

$$= \int\limits_{\mathbb{Q}_p} J_n^m(\tilde{\xi})\chi_p(t\xi_{n+1})dt$$

$$= \Gamma_p^n(\alpha-1)\int\limits_{\mathbb{Q}_p} \Omega(|\tilde{m}\tilde{\xi}|_p)[|\tilde{\xi}|_p^{-\alpha+1} - |p\tilde{m}|_p^{\alpha-1}]\chi_p(t\xi_{n+1})dt$$

$$= \Gamma_p^n(\alpha-1)\int\limits_{|t|_p\le|m|_p} \Omega(|m\tilde{\xi}|_p)[|\tilde{\xi}|_p^{-\alpha+1} - |pm|_p^{\alpha-1}]\chi_p(t\xi_{n+1})dt$$

$$+ \Gamma_p^n(\alpha-1)\int\limits_{|t|_p>|m|_p} \Omega(|t\tilde{\xi}|_p)[|\tilde{\xi}|_p^{-\alpha+1} - |pt|_p^{\alpha-1}]\chi_p(t\xi_{n+1})dt$$

$$= \Gamma_p^n(\alpha-1)\Omega(|m\tilde{\xi}|_p)[|\tilde{\xi}|_p^{-\alpha+1} - |pm|_p^{\alpha-1}]|m|_p\Omega(|\xi_{n+1}m|_p)$$

$$+ \Gamma_p^n(\alpha-1)|\tilde{\xi}|_p^{-\alpha+1}I_1 - \Gamma_p^n(\alpha-1)p^{-\alpha+1}I_2 \tag{4.8}$$

where

$$I_2 = \int\limits_{|t|_p>|m|_p} \Omega(|t\tilde{\xi}|_p)\chi_p(t\xi_{n+1})dt$$

$$= \Omega(|m\tilde{\xi}|_p)\int\limits_{|m|_p<|t|_p\le|\tilde{\xi}|_p^{-1}} \chi_p(t\xi_{n+1})dt$$

$$= \Omega(|m\tilde{\xi}|_p)\left[\int\limits_{|t|_p\le|\tilde{\xi}|_p^{-1}} \chi_p(t\xi_{n+1})dt - \int\limits_{|t|_p\le|m|_p} \chi_p(t\xi_{n+1})dt\right]$$

$$= \Omega(|m\tilde{\xi}|_p)\left[|\tilde{\xi}|_p^{-1}\Omega(|\tilde{\xi}|_p^{-1}|\xi_{n+1}|_p - |m|_p + \Omega(|m\xi_{n+1}|_p)\right]$$

Here we have used the formula (3.1) of Sec. 4.3.

By using the formula (2.16) Sec. 7.2 we calculate the integral I_2.

$$
\begin{aligned}
I_2 &= \int\limits_{|t|_p > |m|_p} |t|_p^{\alpha-1} \chi_p(t\xi_{n+1}) dt \\
&= \Omega(|m\tilde{\xi}|_p) \int\limits_{|m|_p < |t|_p \le |\tilde{\xi}|_p^{-1}} |t|_p^{\alpha-1} \chi_p(t\xi_{n+1}) dt \\
&= \Omega(|m\tilde{\xi}|_p) \left[\int\limits_{|t|_p \le |\tilde{\xi}|_p^{-1}} |t|_p^{\alpha-1} \chi_p(t\xi_{n+1}) dt - \int\limits_{|t|_p \le |m|_p} |t|_p^{\alpha-1} \chi_p(t\xi_{n+1}) dt \right] \\
&= \Omega(|m\tilde{\xi}|_p) \left\{ |\tilde{\xi}|_p^{-\alpha} \Omega(|\tilde{\xi}|_p^{-1} |\xi_{n+1}|_p) \frac{1-p^{-1}}{1-p^{-\alpha}} + \Gamma_p(\alpha) |\xi_{n+1}|_p^{-\alpha} \right. \\
&\quad \cdot [1 - \Omega(|\tilde{\xi}|_p^{-1} |\xi_{n+1}|_p)] - |m|_p^{\alpha} \Omega(|m\xi_{n+1}|_p) \frac{1-p^{-1}}{1-p^{-\alpha}} \\
&\quad \left. - \Gamma_p(\alpha) |\xi_{n+1}|_p^{-\alpha} [1 - \Omega(|m\xi_{n+1}|_p)] \right\} \\
&= \Omega(|m\tilde{\xi}|_p) \left(|\xi_{n+1}|_p^{-\alpha} \frac{1-p^{\alpha-1}}{1-p^{-\alpha}} - |m|_p^{\alpha} \frac{1-p^{-1}}{1-p^{-\alpha}} \right) \\
&\quad + \Omega(|m\tilde{\xi}|_p) \Omega(|\tilde{\xi}|_p^{-1} |\xi_{n+1}|_p) \left(|\tilde{\xi}|_p^{-\alpha} \frac{1-p^{-1}}{1-p^{-\alpha}} - |\xi_{n+1}|_p^{-\alpha} \frac{1-p^{\alpha-1}}{1-p^{-\alpha}} \right).
\end{aligned}
$$

By substituting values of the integrals I_1 and I_2 in (4.8) we obtain the desired expression (4.7):

$$\begin{aligned}
J^m_{n+1}(\xi) =&\Gamma^n_p(\alpha-1)\Omega(|m\xi|_p)(|m|_p|\tilde{\xi}|_p^{-\alpha+1} - |m|_p|pm|_p^{\alpha-1}) \\
&+\Gamma^n_p(\alpha-1)\Omega(|m\tilde{\xi}|_p)\Omega(|\tilde{\xi}|_p^{-1}|\xi_{n+1}|_p)|\tilde{\xi}|_p^{-\alpha} \\
&-\Gamma^n_p(\alpha-1)\Omega(|m\xi|_p)|m|_p|\tilde{\xi}|_p^{-\alpha+1} - \Gamma^n_p(\alpha-1)\Omega(|m\xi|_p) \\
&\cdot\left(|\xi_{n+1}|_p^{-\alpha}\frac{p^{-\alpha+1}-1}{1-p^{-\alpha}} - |pm|_p^{\alpha}\frac{p-1}{1-p^{-\alpha}}\right) \\
&-\Gamma^n_p(\alpha-1)\Omega(|m\tilde{\xi}|_p)\Omega(|\tilde{\xi}|_p^{-1}|\xi_{n+1}|_p) \\
&\left(|\tilde{\xi}|_p^{-\alpha}\frac{p^{-\alpha+1}-p^{-\alpha}}{1-p^{-\alpha}} - |\xi_{n+1}|_p^{-\alpha}\frac{p^{-\alpha+1}-1}{1-p^{-\alpha}}\right) \\
=&-\Gamma^{n+1}_p(\alpha)\Omega(|m\xi|_p)|pm|_p^{\alpha} \\
&+\Gamma^{n+1}_p(\alpha)\Omega(|m\tilde{\xi}|_p)\Omega(|\tilde{\xi}|_p^{-1}|\xi_{n+1}|_p)(|\xi|_p^{-\alpha} - |\xi_{n+1}|_p^{-\alpha}) \\
&+\Gamma^{n+1}_p(\alpha)\Omega(|m\xi|_p)|\xi_{n+1}|_p^{-\alpha} \\
=&\Gamma^{n+1}_p(\alpha)\Omega(|m\xi|_p)(|\xi|_p^{-\alpha} - |pm|_p^{\alpha})\,. \qquad \blacksquare
\end{aligned}$$

If we act like in Sec. 8.2 and use the formula (4.5) then we obtain the following generalization of the formula (3.8)

$$\begin{gathered}
|x|_p^{\alpha-n} * |x|_p^{\beta-n} = B^n_p(\alpha,\beta)|x|_p^{\alpha+\beta-n}, \\
\alpha \neq \alpha_k, \qquad \beta \neq \alpha_k, \qquad \alpha+\beta \neq \alpha_k
\end{gathered} \tag{4.9}$$

where

$$B^n_p(\alpha,\beta) = \Gamma^n_p(\alpha)\Gamma^n_p(\beta)\Gamma^n_p(n-\alpha-\beta). \tag{4.10}$$

If we take into account the property (see (4.4), cf. (2.16))

$$\Gamma^n_p(\alpha)\Gamma^n_p(n-\alpha) = 1 \tag{4.11}$$

then the equality (4.10) takes the classical form

$$B^n_p(\alpha,\beta) = \frac{\Gamma^n_p(\alpha)\Gamma^n_p(\beta)}{\Gamma^n_p(\alpha+\beta)} \tag{4.12}$$

Note the symmetrical expression for

$$B^n_p(\alpha,\beta) = \Gamma^n_p(\alpha)\Gamma^n_p(\beta)\Gamma^n_p(\gamma)$$

where $\alpha+\beta+\gamma = n$.

Remark. The formulas of Sec. 8.4 to within notations are obtained in the paper [22] by V.A.Smirnov. They are used in the perturbation theory in the p-adic Euclidean quantum field theory with the propagator of the form

$$|(x, m)|_p^\alpha .$$

Chapter 2

PSEUDO-DIFFERENTIAL OPERATORS ON THE FIELD OF p-ADIC NUMBERS

Pseudo-differential operator (on the field of p-adic numbers) in an open set $\mathcal{O} \subset \mathbb{Q}_p^n$ we call the operator A of the form

$$(A\psi)(x) = \int_{\mathbb{Q}_p^n} a(\xi, x)\tilde{\psi}(\xi)\chi_p(-(\xi, x))d\xi, \quad x \in \mathcal{O}$$

which acts on complex-valued functions $\psi(x)$ of p-adic arguments $x \in \mathcal{O}$. Here we suppose that functions $\psi(x)$ are extended by zero from the set $\mathcal{O}$ on whole space $\mathbb{Q}_p^n$, and $\tilde{\psi}(\xi)$ are their Fourier transforms,

$$\tilde{\psi}(\xi) = \int_{\mathcal{O}} \psi(x)\chi_p((\xi, x))dx.$$

The function $a(\xi, x)$, $\xi \in \mathbb{Q}_p^n$, $x \in \mathcal{O}$ is called *symbol* of the operator A.

We use here some notions and results on the spectral theory of operators from the books by Reed and Simon [175], Dunford and Schwartz [60] and Yosida [237].

IX. The Operator D^α

The operator $D^\alpha : \psi \to D^\alpha\psi$ is defined as convolution of generalized functions $f_{-\alpha}$ and ψ (see Sec. 8.3):

$$D^\alpha\psi = f_{-\alpha} * \psi, \quad \alpha \neq -1.$$

D^α is a pseudo-differential operator with the symbol $|\xi|_p^\alpha$ owing to the formula of the Fourier transform of convolution (see Sec. 7.5)

$$F[D^\alpha\psi] = |\xi|_p^\alpha \cdot \tilde{\psi}.$$

The operator D^α has been introduced and studied by V.S.Vladimirov in paper [206].

1. *The Operator* D^α, $\alpha \neq -1$

By virtue of the results of Sec. 8 the generalized function

$$f_\alpha(x) = \frac{|x|_p^{\alpha-1}}{\Gamma_p(\alpha)}$$

is holomorphic on α everywhere on the real line except the simple pole at $\alpha = 1$ with residue $-\frac{p-1}{p \ln p}$; in addition $f_0(x) = \delta(x)$ and (see (3.8) and (3.9) of Sec. 8.3)

$$f_\alpha * f_\beta = f_{\alpha+\beta}, \quad \alpha, \beta, \alpha+\beta \neq 1, \quad \alpha, \beta \in \mathbb{R}.$$

Let a generalized function ψ from $\mathcal{D}'(\mathbb{Q}_p)$ be such that the convolution $f_{-\alpha} * \psi$ exists ($\alpha \neq -1$). The operator $D^\alpha\psi = f_{-\alpha} * \psi$ we call for $\alpha > 0$ the operator of (fractional) *differentiation of order* α, and for $\alpha < 0$ — the operator of (fractional) *integration of order* $-\alpha$; *for* $\alpha = 0$, $D^0\psi = \delta * \psi$, ψ *is the identity operator.*

Example 1. *The analogy of the first derivative* ($\alpha = 1$, $D^1 = D$) :

$$D\psi = f_{-1} * \psi = -\frac{p^2}{p+1}|x|_p^{-2} * \psi.$$

If $\psi \in \mathcal{D}$ then this formula owing to (1.18) of Sec. 8.1 takes the form

$$(D\psi)(x) = -\frac{p^2}{p+1}(|y|_p^{-2}, \psi(x-y)) = \frac{p^2}{p+1}\int_{\mathbb{Q}_p} \frac{\psi(x) - \psi(y)}{|x-y|_p^2}dy$$

or owing to (2.19) of Sec. 8.2

$$(D\psi)(x) = \int_{\mathbb{Q}_p} |\xi|_p \tilde{\psi}(\xi)\chi_p(-\xi x)d\xi.$$

Example 2. The derivative $D^\alpha\psi$, $\alpha > 0$, $\psi \in \mathcal{D}$ is given by the expression

$$(D^\alpha\psi)(x) = \frac{p^\alpha - 1}{1 - p^{-\alpha-1}} \int\limits_{\mathbb{Q}_p} \frac{\psi(x) - \psi(y)}{|x-y|_p^{\alpha+1}} dy \tag{1.1}$$

or equivalently

$$(D^\alpha\psi)(x) = \int\limits_{\mathbb{Q}_p} |\xi|_p^\alpha \tilde{\psi}(\xi)\chi_p(-\xi x)d\xi. \tag{1.2}$$

Example 3. The primitive $D^\alpha\psi$, $\alpha < 0$, $\alpha \neq -1$ where $\psi \in \mathcal{D}$ is given by the expression (see (2.19) of Sec. 8.2)

$$(D^\alpha\psi)(x) = \frac{1 - p^\alpha}{1 - p^{-\alpha-1}} \int\limits_{\mathbb{Q}} |x-y|_p^{-\alpha}\psi(y)dy \tag{1.3}$$

or equivalently

$$(D^\alpha\psi)(x) = \begin{cases} \int\limits_{\mathbb{Q}_p} |\xi|_p^\alpha \tilde{\psi}(\xi)\chi_p(-\xi x)d\xi, & -1 < \alpha < 0, \\ \int\limits_{\mathbb{Q}_p} |\xi|_p^\alpha [\tilde{\psi}(\xi)\chi_p(-\xi x) - \tilde{\psi}(0)]d\xi, & \alpha < -1. \end{cases} \tag{1.4}$$

If a generalized function ψ belongs to $\mathcal{E}'$ and $\alpha, \beta, \alpha+\beta \neq -1$ then the equalities are valid

$$D^\alpha D^\beta \psi = D^{\alpha+\beta}\psi = D^\beta D^\alpha \psi. \tag{1.5}$$

■ It follows from the formulas (3.13)–(3.14) of Sec. 8.3 and (2.19) of Sec. 8.2:

$$\begin{aligned} F[D^{\alpha+\beta}\psi] &= F[f_{-\alpha-\beta} * \psi] = \tilde{f}_{-\alpha-\beta} \cdot \tilde{\psi} = |\xi|_p^{\alpha+\beta}\tilde{\psi} \\ &= |\xi|_p^\alpha \cdot (|\xi|_p^\beta \tilde{\psi}) = |\xi|_p^\alpha \cdot F[f_{-\beta} * \psi] = F[f_{-\alpha} * D^\beta\psi] \\ &= F[D^\alpha(D^\beta\psi)] = F[D^\beta(D^\alpha\psi)] \end{aligned}$$

because $\tilde{\psi} \in \mathcal{E}$ (see Sec. 7.3). ■

From the formulas (1.5) for $\beta = -\alpha$ it follows the equalities

$$D^{\alpha}D^{-\alpha}\psi = \psi = D^{-\alpha}D^{\alpha}\psi, \qquad \psi \in \mathcal{E}', \quad \alpha \neq \pm 1. \tag{1.6}$$

Remark. The equalities (1.5) are valid also for those generalized functions $\psi \in \mathcal{D}'$ for which the convolutions $f_{-\alpha}*(f_{-\beta}*\psi)$, $f_{-\beta}*(f_{-\alpha}*\psi)$ and $f_{-\alpha-\beta}*\psi$ exists in $\mathcal{D}'$. The existence of these convolutions is essential for the validity of the equalities (1.5) as the following example shows: if $\alpha > 0$ and $\psi = 1$ then the convolution $f_{-\alpha}*1 = F[|\xi|_p^{\alpha}\delta(\xi)] = 0$, the convolution $f_{\alpha}*1$ does not exist, and the equalities (1.6) are not valid:

$$D^{-\alpha}(D^{\alpha}1) = D^{-\alpha}0 = 0, \qquad D^{\alpha}(D^{-\alpha}1) \quad \text{do not exist.}$$

Example 4. $\alpha \in \mathbb{R}$, $a \in \mathbb{Q}_p{}^*$

$$D^{\alpha}\chi_p(ax) = |a|_p^{\alpha}\chi_p(ax). \tag{1.7}$$

■ It follows from the equalities (2.19) of Sec. 8.2, (2.8) of Sec. 9.2 (for $\alpha = -1$) and

$$\begin{aligned} F[\chi_p(ax)] &= \delta(\xi + a) \qquad \text{(see Sec. 7.3)}, \\ F[D^{\alpha}\chi_p(ax)] &= F[f_{-\alpha}*\chi_p(ax)] = F[f_{-\alpha}]\cdot F[\chi_p(ax)] \\ &= \begin{cases} |\xi|_p^{\alpha}\cdot\delta(\xi+a), & \alpha \neq -1, \\ \left[\mathcal{P}\frac{1}{|\xi|_p} + \frac{1}{p}\delta(\xi)\right]\cdot\delta(\xi+a), & \alpha = -1, \end{cases} \\ &= |a|_p^{\alpha}\delta(\xi + a), \end{aligned} \tag{1.8}$$

and from the existence of the products

$$|\xi|_p^{\alpha}\cdot\delta(\xi+\alpha) = |a|_p^{\alpha}\cdot\delta(\xi+\alpha), \qquad \mathcal{P}\frac{1}{|\xi|}\cdot\delta(\xi+a) = |\xi|_p^{-1}\delta(\xi+\alpha),$$

$$\delta(\xi)\delta(\xi+a) = 0.$$

By using the inverse Fourier-transform to the equality (1.8) we obtain the formula (1.7). ■

Example 5. $\alpha \in \mathbb{R}$, $\gamma \in \mathbb{Z}$. Let

$$\Phi(x) = F[\delta(|\xi|_p - p^\gamma)f(\xi)], \quad f \in \mathcal{D}'.$$

Then

$$D^\alpha \Phi(x) = p^{\gamma\alpha}\Phi(x). \tag{1.9}$$

■ Like in Example 4 we have

$$\begin{aligned}
\tilde{\Phi}(\xi) &= \delta(|\xi|_p - p^\gamma)f(-\xi),\\
F[D^\alpha \Phi] &= \tilde{f}_{-\alpha}\cdot\tilde{\Phi}\\
&= \begin{cases} |\xi|_p^\alpha \cdot \delta(|\xi|_p - p^\gamma)f(-\xi), & \alpha \neq -1,\\ \left[\mathcal{P}\frac{1}{|\xi|_p} + \frac{1}{p}\delta(\xi)\right]\cdot \delta(|\xi|_p - p^\gamma)f(-\xi), & \alpha = -1, \end{cases}\\
&= p^{\gamma\alpha}\delta(|\xi|_p - p^\gamma)f(-\xi) = p^{\gamma\alpha}\tilde{\Phi}(\xi)\ .
\end{aligned}$$ ■

Example 6. $\alpha \in \mathbb{R}$, $\gamma \in \mathbb{Z}$, $|2a|_p \geq p^{2-2\gamma}$

$$D^\alpha[\delta(|x|_p - p^\gamma)\chi_p(ax^2)] = p^{\gamma\alpha}|2a|_p^\alpha \delta(|x|_p - p^\gamma)\chi_p(ax^2). \tag{1.10}$$

■ It follows from the formula (1.9) for $f(\xi) = \chi_p(a\xi^2)$ owing to the formula (2.10) of Sec. 7.2. ■

Example 7. $\alpha \in \mathbb{R}$, $p \neq 2$, $\gamma \in \mathbb{Z}$, $\sum\limits_{1\leq k\leq p-1} \eta(k) = 0$

$$D^\alpha[\eta(x_0)\delta(|x|_p - p^\gamma)] = p^{\alpha(1-\gamma)}\eta(x_0)\delta(|x|_p - p^\gamma). \tag{1.11}$$

■ As in Example 6 it follows from the formula (2.14) of Sec. 7.2. ■

2. ***Operator*** D^{-1}

Our goal is to extend the operator D^α on $\alpha = -1$, so that on the class of generalized functions $\psi \in \mathcal{E}'$, $(\psi, 1) = 0$ it would be continuous on α at the point $\alpha = -1$, and the equalities (1.5) would be valid for all real α and β (see [206]).

At first we shall prove the Lemma.

Lemma 1. *In order that a generalized function $f \in \mathcal{E}'$, supp $f \subset B_N$ satisfies the condition*

$$(f, 1) == 0, \tag{2.1}$$

it is necessary and sufficient that

$$\tilde{f}(\xi) = 0, \quad \xi \in B_{-N}. \tag{2.2}$$

■ Necessity of the condition (2.2) follows from results of Sec. 7.3. In fact, let the condition (2.1) be fulfilled. As parameter of constancy of the function $\tilde{f}(\xi)$ is no less than $-N$ then $\tilde{f}(\xi) = \tilde{f}(0) = (f, \Delta_N) = (f, 1) = 0$ for all $\xi \in B_{-N}$ owing to the representatiom (3.10) of Sec. 7.3.

Conversely, if the condition (2.2) is fulfilled then by choosing

$$\varphi \in \mathcal{D}, \quad \text{supp}\, \varphi \subset B_{-N}, \quad \int_{\mathbb{Q}_p} \varphi(\xi) d\xi = 1$$

we obtain the condition (2.1)

$$0 = (\tilde{f}, \varphi) = (f, \tilde{\varphi}) = \left(f(x), \Delta_N(x) \int_{\mathbb{Q}_p} \varphi(x) \chi_p(x\xi) d\xi \right)$$

$$= \left(f(x), \Delta_N(x) \int_{\mathbb{Q}_p} \varphi(\xi) d\xi \right) = (f, \Delta_N) = (f, 1) = 0 \,. \qquad ■$$

On test functions $\varphi \in \mathcal{D}$ such that $\int_{\mathbb{Q}_p} \varphi dx = 0$ the following limit relation is valid

$$(f_\alpha, \varphi) \to -\frac{p-1}{p \ln p} \int_{\mathbb{Q}_p} \ln |x|_p \varphi(x) dx, \quad \alpha \to 1. \tag{2.3}$$

■ The existence of a limit and the formula (2.3) follows from the limit relation

$$(f_\alpha, \varphi) = \frac{1}{\Gamma_p(\alpha)} \int_{\mathbb{Q}_p} |x|_p^{\alpha-1} \varphi(x) dx$$
$$= \int_{\mathbb{Q}_p} \frac{\exp[(\alpha-1)\ln|x|_p] - 1}{1 - \exp[-(\alpha-1)\ln p]} (1 - p^{-\alpha}) \varphi(x) dx$$
$$\to -\frac{p-1}{p \ln p} \int_{\mathbb{Q}_p} \ln|x|_p \varphi(x) dx, \qquad \alpha \to 1.$$

The legitimacy of limiting passage under the sign of the integral is guaranteed by the Lebesgue Theorem (see Sec. 4.4) owing to the majorization

$$\left| \frac{\exp[(\alpha-1)\ln|x|_p] - 1}{1 - \exp[-(\alpha-1)\ln p]} \right| \le C |\ln|x|_p|, \qquad x \in \text{supp } \varphi, \quad |\alpha - 1| \le 1$$

for some $C = C(p, \text{supp } \varphi)$. ■

Let us denote by f_1 the regular generalized function $-\frac{p-1}{p \ln p} \ln|x|_p$,

$$(f_1, \varphi) = -\frac{p-1}{p \ln p} \int_{\mathbb{Q}_p} \ln|x|_p \varphi(x) dx, \quad \varphi \in \mathcal{D}. \tag{2.4}$$

Put

$$D^{-1}\psi = f_1 * \psi, \quad \psi \in \mathcal{E}'. \tag{2.5}$$

The limit relation

$$D^{-\alpha}\psi \to D^{-1}\psi, \quad \alpha \to 1 \text{ in } \mathcal{D}' \tag{2.6}$$

is valid if $\psi \in \mathcal{E}'$ *satisfies the condition (2.1).*

■ Let supp $\psi \subset B_N$ and $\varphi \in \mathcal{D}$. By using the convolution theory (see Sec. 7.1) for all $\alpha > 0$, $\alpha \ne 1$ we have

$$(D^{-\alpha}\psi, \varphi) = (f_\alpha * \psi, \varphi) = (f_\alpha(x), (\psi(y), \varphi(x+y)))$$
$$= \int_{\mathbb{Q}_p} f_\alpha(x) g(x) dx \tag{2.7}$$

as

$$g(x) = (\psi(y), \varphi(x+y)) = (\psi(y), \Delta_N(y)\varphi(x+y)) \in \mathcal{D}.$$

Further owing to the formula (6.5) of Sec. 6.6 (as $\Delta_N(y)\varphi(x+y) \in \mathcal{D}(\mathbb{Q}_p^2)$) and the condition (2.1) we have

$$\int_{\mathbb{Q}_p} g(x)dx = \int_{\mathbb{Q}_p} (\psi(y), \Delta_N(y)\varphi(x+y))dx = \left(\psi(y), \Delta_N(y) \int_{\mathbb{Q}_p} \varphi(x+y)dx\right)$$
$$= \left(\psi(y), \Delta_N(y) \int_{\mathbb{Q}_p} \varphi(x)dx\right) = (\psi, 1) \int_{\mathbb{Q}_p} \varphi(x)dx = 0.$$

So the function g satisfies the condition (2.1). Therefore by virtue of (2.3) from (2.7) and (2.5) we derive (2.6):

$$\begin{aligned}(D^{-\alpha}\psi, \varphi) &= \int_{\mathbb{Q}_p} f_\alpha(x)g(x)dx \to \int_{\mathbb{Q}_p} f_1(x)g(x)dx \\ &= \int_{\mathbb{Q}_p} f_1(x)(\psi(y), \Delta_N(y)\varphi(x+y))dx \\ &= (f_1 * \psi, \varphi) = (D^{-1}\psi, \varphi), \qquad \alpha \to 1\,.\end{aligned}$$ ■

Now we shall prove *the formula*

$$\tilde{f}_1(\xi) = \mathcal{P}\frac{1}{|\xi|_p} + \frac{1}{p}\delta(\xi) \tag{2.8}$$

where the generalized function $\mathcal{P}\frac{1}{|\xi|_p}$ *is defined in* (1.10) *of* Sec. 8.1.

■ From the formula (3.6) of Sec. 4.3 it follows that the Fourier-transform of the function $f_1(x)$ coincides with $|\xi|_p^{-1}$ for $\xi \neq 0$. Therefore (see Sec. 6.3)

$$\tilde{f}_1(\xi) - \mathcal{P}\frac{1}{|\xi|_p} = C\delta(\xi). \tag{2.9}$$

For definition of the constant C we apply both sides of the equality (2.9) to the test function $\Delta_0 = \tilde{\Delta}_0$ (the function Δ_k are defined in Sec. 7.1). By using the integral (2.6) of Sec. 4.2 we get

$$\begin{aligned} C &= (\tilde{f}_1, \Delta_0) - \left(\mathcal{P}\frac{1}{|\xi|_p}, \Delta_0)\right) \\ &= (f_1, \tilde{\Delta}_0) - \int\limits_{|\xi|_p \le 1} \frac{\Delta_0(\xi) - \Delta_0(0)}{|\xi|_p} d\xi - \int\limits_{|\xi|_p > 1} \frac{\Delta_0(\xi)}{|\xi|_p} d\xi \\ &= -\frac{p-1}{p \ln p}(\ln |x|_p, \Delta_0) = -\frac{p-1}{p\ln p}\int\limits_{B_0} \ln |x|_p dx = \frac{1}{p}\,. \end{aligned}$$ ■

If $\psi \in \mathcal{E}'$, $(\psi, 1) = 0$ then by Lemma 1 $\tilde{\psi}(\xi)$ vanishes in a vicinity of the point $\xi = 0$. Therefore by virtue of (2.8)

$$\tilde{f}_1(\xi)\tilde{\psi}(\xi) = \left(\mathcal{P}\frac{1}{|\xi|_p} + \frac{1}{p}\delta(\xi)\right) \cdot \tilde{\psi}(\xi) = |\xi|_p^{-1}\tilde{\psi}(\xi).$$

From here as above it follows that *the formulas* (1.5) *are valid for all real* α *and* β *if* $\psi \in \mathcal{E}'$, $(\psi, 1) = 0$.

Lemma 2. *If* $\varphi \in \mathcal{D}$, *supp* $\varphi \subset B_N$ *then for* $|x|_p > p^N$

$$(f_\alpha(x'), \varphi(x - x')) = \begin{cases} \frac{1}{\Gamma_p(\alpha)}|x|_p^{\alpha-1} \int\limits_{\mathbb{Q}_p} \varphi dx', & \alpha \ne 1, \\ -\frac{p-1}{p\ln p} \ln |x|_p \int\limits_{\mathbb{Q}_p} \varphi dx', & \alpha = 1. \end{cases} \tag{2.10}$$

■ Let $\alpha \ne 1$. For $|x|_p > p^N$, $\varphi(x) = 0$ and $|x - y|_p = |x|_p$, $y \in \text{supp}\,\varphi$, and from the formulas (1.1) and (1.3) the equality (2.10) follows

$$\begin{aligned} &(f_\alpha(x'), \varphi(x - x')) \\ &= \frac{1}{\Gamma_p(\alpha)} \int\limits_{|x'| \le 1} |x'|_p^{\alpha-1} \varphi(x - x') dx' + \frac{1}{\Gamma_p(\alpha)} \int\limits_{|x'| > 1} |x'|_p^{\alpha-1}\varphi(x - x')dx' \\ &= \frac{1}{\Gamma_p(\alpha)} \int\limits_{\mathbb{Q}_p} |x - y|_p^{\alpha-1}\varphi(y)dy = \frac{1}{\Gamma_p(\alpha)}|x|_p^{\alpha-1} \int\limits_{\mathbb{Q}_p} \varphi(y)dy. \end{aligned}$$

The case $\alpha = 1$ is considered analogously. ■

The following equalities are valid

$$f_1 * f_{-\alpha} = f_{1-\alpha} = f_{-\alpha} * f_1, \qquad \alpha \geq 0. \tag{2.11}$$

■ For $\alpha = 0$, $f_0 = \delta$, and the equalities (2.11) are valid. Let $\alpha > 0$. Then by the definition of convolution (see Sec. 7.1) for all $\varphi \in \mathcal{D}$ and for any 1-sequence $\{\eta_k, k \to \infty\}$ we have

$$\begin{aligned}(f_1 * f_{-\alpha}, \varphi) &= \lim_{k\to\infty} (f_1(x) \times f_{-\alpha}(y), \eta_k(x)\varphi(x+y)) \\ &= \lim_{k\to\infty} \int_{\mathbb{Q}_p} \eta_k(x) f_1(x)(f_{-\alpha}(y), \varphi(x+y))dx \\ &= \int_{\mathbb{Q}_p} f_1(x)(f_{-\alpha}(y), \varphi(x+y))dx.\end{aligned}$$

The legitimacy of limiting passage under the sign of the integral and existence of the last integral follow from the Lebesgue Theorem (see Sec. 4.4) and the Lemma 2 owing to the majorization

$$\begin{aligned}|\Delta_k(x) f_1(x)(f_{-\alpha}(y), \varphi(x+y))| &\leq C_\varphi |x|_p^{-\alpha-1} |\ln |x|_p|, \\ &\qquad |x|_p > R_\varphi, \quad k \geq N.\end{aligned}$$

Thus the convolution $f_1 * f_{-\alpha}$ exists, and hence the convolution $f_{-\alpha} * f_1$ exists, and they are equal. To calculate it we argue by the following way. The product $\tilde{f}_1 \cdot \tilde{f}_{-\alpha}$ exists, and the equality is valid

$$F[f_1 * f_{-\alpha}] = \tilde{f}_1 \ \tilde{f}_{-\alpha} \qquad \text{(see Sec. 7.5)}.$$

By taking into account the formulas (2.19) of Sec. 8.2 and (2.8) we rewrite the last equality in the form

$$F[f_1 * f_{-\alpha}] = \left(\mathcal{P}\frac{1}{|\xi|_p} + \frac{1}{p}\delta(\xi)\right) \cdot |\xi|_p^\alpha .$$

It is easy to see that for $\alpha > 0$

$$\mathcal{P}\frac{1}{|\xi|_p} \cdot |\xi|_p^\alpha = |\xi|_p^{\alpha-1}, \quad \delta(\xi) \cdot |\xi|_p^\alpha = 0.$$

Therefore

$$F[f_1 * f_{-\alpha}] = |\xi|_p^{\alpha-1}$$

from where owing to (2.19) of Sec. 8.2 the equalities (2.11) follow

$$f_1 * f_{-\alpha} = F[|\xi|_p^{\alpha-1}] = f_{1-\alpha}\,.$$

■

From (2.11) for $\alpha = -1$ the relations follow

$$f_1 * f_{-1} = \delta = f_{-1} * f_1. \tag{2.12}$$

Now we shall prove the statement: if $\psi \in \mathcal{E}'$ then for $\alpha \geq 0$ there exist $D^{-1}D^{\alpha}\psi$ and $D^{\alpha}D^{-1}\psi$ and they are equal to $D^{\alpha-1}\psi$:

$$D^{-1}D^{\alpha}\psi = D^{\alpha-1}\psi = D^{\alpha}D^{-1}\psi \tag{2.13}$$

or equivalently

$$f_1 * (f_{-\alpha} * \psi) = f_{1-\alpha} * \psi = f_{-\alpha} * (f_1 * \psi). \tag{2.14}$$

■ We use the convolution theory of Sec. 7.1. Let $\varphi \in \mathcal{D}$. Then for all $x \in \mathbb{Q}_p$ we have

$$\begin{aligned}((f_{-\alpha} * \psi)(y'), \varphi(x+y')) &= (f_{-\alpha}(y), (\psi(y'), \varphi(x+y+y'))) \\ &= (f_{-\alpha}(y), g(x+y))\end{aligned}$$

where $g(x) = (\psi(y'), \varphi(x+y')) \in \mathcal{D}$. Therefore the convolution $f_1 * (f_{-\alpha} * \psi)$ exists and is equal to $f_{1-\alpha} * \psi$:

$$\begin{aligned}(f_1 * (f_{-\alpha} * \psi), \varphi) &= \int_{\mathbb{Q}_p} f_1(x)(f_{-\alpha}(y), g(x+y))dx \\ &= (f_1 * f_{-\alpha}, g) = (f_{1-\alpha}(x), (\psi(y), \varphi(x+y))) = (f_{1-\alpha} * \psi, \varphi)\end{aligned}$$

owing to the equality (2.11). The similar consideration is valid also for the convolution

$$f_{-\alpha} * (f_1 * \psi) = f_{1-\alpha} * \psi = (f_{-\alpha} * f_1) * \psi\,.$$

■

From the equalities (2.13) for $\alpha = 1$ the equalities follow

$$D^{-1}D\psi = \psi = DD^{-1}\psi, \quad \psi \in \mathcal{E}'. \tag{2.15}$$

We summarize our results in the following Theorem.

Theorem. *The formulas* (1.5),

$$D^{\alpha}D^{\beta}\psi = D^{\alpha+\beta}\psi = D^{\beta}D^{\alpha}\psi, \quad \psi \in \mathcal{E}', \tag{1.5}$$

are valid if $\alpha, \beta, \alpha+\beta \neq -1$ *or* $\alpha \geq 0, \beta = -1$. *If* $\psi \in \mathcal{E}'$ *satisfies the condition* (2.1) $(\psi, 1) = 0$ *then the formulas* (1.5) *are valid for all real* α *and* β; *in addition* $D^{\alpha}\psi$ *is continuous on* α *in* $\mathcal{D}'$

Remark. The operator D^{α}, $\alpha \neq -1$ is defined by the homogeneous generalized function $f_{-\alpha}$ of degree $-\alpha - 1$; for $\alpha = -1$ this property is lost: the operator D^{-1} is defined by the function $-\frac{p-1}{p \ln p} \ln |x|_p$ which does not possess property of homogeneity.

3. *Equation* $D^{\alpha}\psi = g$

Let us consider the equation

$$D^{\alpha}\psi = g, \qquad g \in \mathcal{E}', \quad \alpha \in \mathbb{R} \tag{3.1}$$

with respect to an unknown generalized function $\psi \in \mathcal{D}'$. A solution of the equation (3.1) is seeked in the class of those $\psi \in \mathcal{D}'$, for which convolution $f_{-\alpha} * \psi$ exists in $\mathcal{D}'$ (and it is equal to g) (see Sec. 7).

Theorem 1. *For* $\alpha > 0$ *any solution of the equation (3.1) is expressed by the formula*

$$\psi = D^{-\alpha}g + C \tag{3.2}$$

where C *is an arbitrary constant; for* $\alpha \leq 0$ *a solution of the equation* (3.1) *is unique and it is expressed by the formula* (3.2) *for* $C = 0$.

■ The fact that $D^{-\alpha}g$ is a solution of the equation (3.1) follows from the formulas (1.6) and (2.15). It remains to investigate solutions of the homogeneous equation

$$D^{\alpha}\psi = 0. \tag{3.3}$$

By applying to the equation (3.3) the Fourier-transform (see Sec. 7.5) we get $\tilde{f}_{-\alpha}(\xi) \cdot \tilde{\psi}(\xi) = 0$. As $\tilde{f}_{-\alpha}(\xi) \neq 0$ for $\xi \neq 0$ (see (2.19) of Sec. 8.2 for

$\alpha \neq -1$ and (2.8) for $\alpha = -1$) then $\tilde{\psi}(\xi) = 0$, $\xi \neq 0$, and thus $\tilde{\psi} = C\delta(\xi)$, $\psi = C$.

Now we shall prove that for $\alpha \leq 0$, $C = 0$. For $\alpha = 0$ it is the case. Let $\alpha < 0$ and $C \neq 0$, i.e. $f_{-\alpha} * 1 = 0$. As

$$f_{-\alpha} \in L^1(B_0), \qquad \int\limits_{1<|x|_p} f_{-\alpha}(\xi)d\xi = -\infty$$

then if we choose a function $\varphi \in \mathcal{D}$, $\int\limits_{\mathbb{Q}_p} \varphi(x)dx = 1$ we obtain the following contradiction

$$0 = (f_{-\alpha}*1, \varphi) = \lim_{k\to\infty} \int\limits_{\mathbb{Q}_p} f_{-\alpha}(x)\Delta_k(x) \int\limits_{\mathbb{Q}_p} \varphi(x+y)dydx$$
$$= \int\limits_{B_0} f_{-\alpha}(x)dx + \lim_{k\to\infty} \int\limits_{1<|x|\leq p^k} f_{-\alpha}(x)dx = -\infty . \qquad \blacksquare$$

Theorem 1′. *In order that there exists a solution ψ of the equation (3.1) such that $\tilde{\psi}(\xi)$ vanishes in some vicinity of $\xi = 0$, it is necessary and sufficient that g satisfies the condition (2.1). In addition the solution $\psi = D^{-\alpha}g$ is unique, continuous with respect to $\alpha \in \mathbb{R}$ in $\mathcal{D}'$ and the formulas are valid*

$$D^{\beta}\psi = D^{\beta-\alpha}g = D^{-\alpha}D^{\beta}g. \tag{3.4}$$

■ **Necessity.** Let ψ be a solution of the equation (3.1) and $\tilde{\psi} \equiv 0$ in a vicinity of $\xi = 0$. Then

$$F[D^{\alpha}\psi] = F[f_{-\alpha}*\psi] = \tilde{f}_{-\alpha}(\xi) \cdot \tilde{\psi}(\xi) = \tilde{g}(\xi) .$$

From where it follows that $\tilde{g}(\xi) = 0$ in a vicinity of $\xi = 0$. By the Lemma 1 of Sec. 9.2 $g \in \mathcal{E}'$ and satisfies the condition (2.1).

Sufficiency. Let $g \in \mathcal{E}'$ satisfy the condition (2.1). By the Lemma 1 of Sec. 9.2 $\tilde{g}(\xi) \equiv 0$ on a vicinity of $\xi = 0$. As constant $C \neq 0$ does not satisfy the condition (2.1) then by the Theorem 1 there exists a unique solution $\psi = D^{-\alpha}g$ of the equation (3.1) which satisfies the condition (2.1):

$$\tilde{\psi}(\xi) = F[D^{-\alpha}g] = F[f_\alpha * g] = \tilde{f}_\alpha(\xi) \cdot \tilde{g}(\xi).$$

Last statements of the Theorem follow at once from the Theorem of Sec. 9.2. ■

4. *Spectrum of the Operator D^α in $\mathbb{Q}_p$, $\alpha > 0$*

The operator D^α, $\alpha > 0$ is pseudo-differential with the symbol $|\xi|_p^\alpha$ (see Sec. 9.1):

$$D^\alpha \psi = \int_{\mathbb{Q}_p} |\xi|_p^\alpha \chi_p(-\xi x)\tilde{\psi}(\xi)d\xi.$$

It is defined on those functions ψ from the Hilbert space $L^2(\mathbb{Q}_p) = L^2$ (see Sec. 7.4) for which $|\xi|_p^\alpha \tilde{\psi} \in L^2$. This set, we denote it by $\mathcal{D}(D^\alpha)$, is called the *domain of definition* of the operator D^α in $\mathbb{Q}_p$. In addition

$$(D^\alpha\psi, \varphi) = (D^{\alpha/2}\psi, D^{\alpha/2}\varphi) \int_{\mathbb{Q}_p} |\xi|_p^\alpha \tilde{\psi}(\xi)\bar{\tilde{\varphi}}(\xi)d\xi, \quad \psi, \varphi \in \mathcal{D}(D^\alpha), \tag{4.1}$$

$$\|D^\alpha\psi\|^2 = (D^\alpha\psi, D^\alpha\psi) = \int_{\mathbb{Q}_p} |\xi|_p^{2\alpha}|\tilde{\psi}(\xi)|^2 d\xi, \quad \psi \in \mathcal{D}(D^\alpha). \tag{4.2}$$

The equalities (4.1) and (4.2) follow directly from the Parseval-Steklov equality:

$$(D^\alpha\psi, \varphi) = (F[D^\alpha\psi], F[\varphi]) = (|\xi|_p^\alpha \tilde{\psi}, \tilde{\varphi});$$

here $(\cdot, \cdot)$ is the scalar product in the Hilbert space L^2 (see Sec. 7.4).

The defined operator D^α is self-adjoint and positive:

$$(D^\alpha\psi, \psi) = \|D^{\alpha/2}\psi\|^2 > 0, \quad 0 \neq \psi \in \mathcal{D}(D^\alpha)$$

so the spectrum of the operator D^α is situated on the semi-axis $\lambda \geq 0$.

Let us consider the eigen-value problem in $\mathbb{Q}_p$

$$D^\alpha\psi = \lambda\psi, \quad \psi \in L^2(Q_p). \tag{4.3}$$

Let $\lambda = 0$. It was shown in Sec. 9.3 that for $\lambda = 0$ the equation (4.3) have the unique linearly independent solution in $\mathcal{D}'\psi \equiv 1$. Thus $\psi \equiv 1$ is a generalized eigen-function of the operator D^α which corresponds to the eigen value $\lambda = 0$. However $\lambda = 0$ is not an eigen-value of the operator D^α.

Further, ran D^α(range of values of D^α) *is dense in* L^2*.

■ It follows from the fact that the equation

$$D^\alpha \psi = \varphi, \quad \varphi \in \mathcal{D}, \quad \int_{\mathbb{Q}_p} \varphi dx = 0$$

has a solution $\psi = D^{-\alpha}\varphi$ from the domain $\mathcal{D}(D^\alpha)$ as $\tilde{\varphi} \in \mathcal{D}$ and

$$\tilde{\psi} = F[D^{-\alpha}\varphi] = F[f_\alpha * \varphi] = \tilde{f}_\alpha \cdot \tilde{\varphi} = |\xi|_p^{-\alpha}\tilde{\varphi} \in \mathcal{D}, \quad \psi \in \mathcal{D}$$

(by the Lemma 1 of Sec. 9.3 $\tilde{\varphi}(\xi) = 0$ in a vicinity of $\xi = 0$), and the set of functions $\left\{\varphi \in \mathcal{D} : \int_{\mathbb{Q}_p} \varphi dx = 0\right\}$ is dense in L^2 (see the Lemma below). ■

Lemma. *The set of test functions φ from $\mathcal{D}$ satisfying the condition $\int_{\mathbb{Q}_p} \varphi dx = 0$ is dense in L^2.*

■ By virtue of the Parseval-Steklov equality $\|\varphi\| = \|\tilde{\varphi}\|$ (see Sec. 7.4) and the Lemma 1 of Sec. 9.3 it is sufficient to prove that the set

$$\mathcal{D}(\mathbb{Q}_p \backslash \{0\}) = \{\tilde{\varphi} \in \mathcal{D} : \tilde{\varphi}(\xi) \equiv 0, \quad \xi \in B_N, \exists N \in \mathbb{Z}\}$$

is dense in L^2. But in Sec. 6.2 it was proved that $\mathcal{D}(\mathbb{Q}_p \backslash \{0\})$ is dense in $L^2(\mathbb{Q}_p \backslash \{0\}) \simeq L^2(\mathbb{Q}_p) = L^2$. ■

Let $\lambda > 0$. If we apply to the equation (4.3) the Fourier-transform we get the equivalent equation

$$(|\xi|_p^\alpha - \lambda)\tilde{\psi}(\xi) = 0.$$

From here we conclude that desired points of spectrum λ (in this case – eigen-values) have the form

$$\lambda_N = p^{N\alpha}, N \in \mathbb{Z}, \tag{4.4}$$

* However this statement follows from the facts that self-adjoint operator D^α has no remainder spectrum and $\lambda = 0$ is not an eigen-value of D^α.

and the corresponding (normed) eigen-functions have the representation

$$\tilde{\psi}(\xi) = \delta(|\xi|_p - p^N)\rho(\xi), \int_{S_N} |\rho(\xi)|^2 d\xi = 1, \quad N \in \mathbb{Z}. \tag{4.5}$$

5. *Orthonormal Basis of Eigen-Functions of the Operator* D^α [207]

In (4.5) we choose as functions ρ the following system of locally constant functions on S_N: for $p \neq 2$

$$\rho^l_{N,k,\varepsilon_l}(\xi) = p^{-N/2}\sqrt{\frac{p}{p-1}}\lambda_p(\varepsilon_l p^l) \cdot \chi_p\left[-\frac{p^{2N-l}}{4\varepsilon_l}(\xi + kp^{l-N-1})^2\right], \tag{5.1'}$$

$$l = 2, 3, \ldots, \quad k = 1, 2, \ldots, p-1,$$

$$\varepsilon_l = \varepsilon_0 + \varepsilon_1 p + \ldots + \varepsilon_{l-2}p^{l-2}, \quad \varepsilon_j = 0, 1, \ldots, p-1, \quad \varepsilon_0 \neq 0,$$

$$\rho^1_{N,k,0}(\xi) = p^{\frac{1-N}{2}}\delta(\xi_0 - k), \quad l = 1, \quad k = 1, 2, \ldots, p-1, \quad \varepsilon_l = 0; \tag{5.1''}$$

for $p = 2$

$$\rho^l_{N,k,\varepsilon_l}(\xi) = 2^{\frac{1-N}{2}}\lambda_2(\varepsilon_l 2^l)\chi_2\left[-\frac{2^{2N-l-2}}{\varepsilon^l}(\xi + 2^{l-N-k})^2\right],$$

$$l = 2, 3, \ldots, \quad k = 0, 1,$$

$$\varepsilon_l = 1 + \varepsilon_1 2 + \ldots + \varepsilon_{l-2}2^{l-2}, \quad \varepsilon_j = 0, 1, \tag{5.2'}$$

$$\rho^l_{N,k,0}(\xi) = 2^{\frac{1-N1}{2}}\chi_2(k2^{N-2}\xi), \quad l = 1, \quad k = 0, 1, \quad \varepsilon'_l = 0.$$

As a result we obtain the following system of normed eigen-functions which the Fourier-transforms owing to (4.5) and (5.1) are equal: for $p \neq 2$

$$\tilde{\psi}^l_{N,k,\varepsilon_l}(\xi) = p^{-N/2}\sqrt{\frac{p}{p-1}}\lambda_p(\varepsilon_l p^l)\delta(|\xi|_p - p^N) \cdot \chi_p\left[-\frac{p^{2N-l}}{4\varepsilon_l}(\xi + kp^{l-N-1})^2\right], \tag{5.3'}$$

$$\tilde{\psi}^1_{N,k,0}(\xi) = p^{\frac{1-N}{2}}\delta(|\xi|_p - p^N)\delta(\xi_0 - k); \tag{5.3''}$$

for $p = 2$

$$\tilde{\psi}^l_{N,k,\varepsilon_l}(\xi) = 2^{\frac{1-N}{2}}\lambda_2(\varepsilon_l 2^l)\delta(|\xi|_2 - 2^N) \cdot \chi_2\left[-\frac{2^{2N-l-2}}{\varepsilon_1}(\xi + 2^{l-N-k})^2\right], \tag{5.4'}$$

$$\tilde{\psi}^1_{N,k,0}(\xi) = 2^{\frac{1-N}{2}}\delta(|\xi|_2 - 2^N)\chi_2(k2^{N-2}\xi). \tag{5.4''}$$

By applying the inverse Fourier-transform to the functions (5.3) and (5.4) and using the formulas (2.7)–(2.9) of Sec. 7.2, we get the following eigen-functions of the operator D^α corresponding to the eigen-value $\lambda_N = p^{\alpha N}$, $N \in \mathbb{Z}$: for $p \neq 2$ I kind ($l = 2, 3, \ldots,\ \ k = 1, 2, \ldots, p-1,\ \ \varepsilon_l = \varepsilon_0 + \varepsilon_1 p + \ldots + \varepsilon_{l-2} p^{l-2}$)

$$\psi^l_{N,k,\varepsilon_l}(x) = p^{\frac{N-l}{2}} \sqrt{\frac{p}{p-1}} \delta(|x|_p - p^{l-N}) \chi_p(\varepsilon_l p^{l-2N} x^2 + k p^{l-N-1} x), \quad (5.5')$$

II kind ($l = 1,\ k = 1, 2, \ldots, p-1,\ \varepsilon_l = 0$)

$$\psi^1_{N,k,0}(x) = p^{\frac{N-1}{2}} \Omega(p^{N-1}|x|_p) \chi_p(k p^{-N} x); \quad (5.5'')$$

for $p = 2$ I kind ($l = 2, 3, \ldots,\ \ k = 0, 1,\ \ \varepsilon_l = 1 + \varepsilon_1 2 + \ldots + \varepsilon_{l-2} 2^{l-2}$)

$$\psi^l_{N,k,\varepsilon_l}(x) = 2^{\frac{N-l}{2}} \delta(|x|_2 - 2^{l+1-N}) \chi_2(\varepsilon_l 2^{l-2N} x^2 + 2^{l-N-k} x), \quad (5.6')$$

II kind ($l = 1,\ k = 0, 1,\ \varepsilon_l = 0$)

$$\psi^1_{N,k,0}(x) = 2^{\frac{N-1}{2}} [\Omega(2^N |x - k 2^{N-2}|_2) - \delta(|x - k 2^{N-2}|_2 - 2^{1-N})]. \quad (5.6'')$$

The eigen-functions (5.5)–(5.6) belong to the space $\mathcal{D}$ and satisfy the condition

$$\int_{\mathbb{Q}_p} \psi^l_{N,k,\varepsilon_l}(x) dx = 0 \quad (5.7)$$

because owing to (5.3)–(5.4) $\tilde{\psi}^l_{N,k,\varepsilon_l}(0) = 0$ (see Lemma 1 of Sec. 9.2). The I kind eigen-functions satisfy also the condition

$$\int_{\mathbb{Q}_p} \tilde{\psi}^l_{N,k,\varepsilon_l}(\xi) d\xi = 0. \quad (5.8)$$

Orthogonality of the eigen-functions for distinct N follows from the formulas (5.3)–(5.4), for fixed N and distinct l it follows from the formulas (5.5)–(5.6). Orthogonality of the I kind functions for fixed N and l and distinct ε follows from the formula (2.10) of Sec. 7.2:

$$\begin{aligned}
&(\psi^l_{N,k,\varepsilon_l}, \psi^l_{N,k',\varepsilon'_l}) \\
&\sim \int_{\mathbb{Q}_p} \delta^2(|x|_p - p^{l-N})\chi_p(\varepsilon_l p^{l-2N}x^2 + kp^{l-N-1}x) \\
&\quad \cdot \bar{\chi}_p(\varepsilon'_l p^{l-2N}x^2 + k'p^{l-N-1}x)dx \\
&= \int_{S_{l-N}} \chi_p[(\varepsilon_l - \varepsilon'_l)p^{l-2N}x^2 + (k-k')p^{l-N-1}x]dx \\
&\sim \delta\left(\left|\frac{(k-k')p^{l-N-1}}{2(\varepsilon_l - \varepsilon'_l)p^{l-2N}}\right|_p - p^{l-N}\right) = 0
\end{aligned}$$

as $\varepsilon_l \neq \varepsilon'_l$, so

$$|\varepsilon_l - \varepsilon'_l| \geq p^{-l+2}, \qquad |4(\varepsilon_l - \varepsilon'_l)p^{l-2N}| \geq p^{2-2(l-N)}$$

$$\left|\frac{(k-k')p^{l-N-1}}{2(\varepsilon_l - \varepsilon'_l)p^{l-2N}}\right|_p = \frac{p^{N+1-l}|k-k'|_p}{p^{2N-l}|\varepsilon_l - \varepsilon'_l|_p} \leq p^{1-N}p^{l-2}|k-k'|_p < p^{l-N}.$$

For $p = 2$ one considers analogously. For fixed N, l and ε_l and distinct k the functions (5.6) are orthogonal by virtue of the formula (2.8) of Sec. 7.2:

$$\begin{aligned}
(\psi^l_{N,0,\varepsilon_l}, \psi^l_{N,1,\varepsilon_l}) &\sim \int_{S_{l+1-N}} \chi_2(2^{l-N}x)\chi_2(-2^{l-N-1}x)dx \\
&= \int_{S_{l+1-N}} \chi_2(2^{l-N-1}x)dx = 0, \quad l = 2, 3, \ldots
\end{aligned}$$

$$(\psi^1_{N,0,0}, \psi^1_{N,1,0}) = (\tilde{\psi}^1_{N,0,0}, \tilde{\psi}^1_{N,1,0}) \sim \int_{S_N} \chi_2(2^{N-2}\xi)d\xi = 0, \quad l = 1.$$

Hence, *for* $p = 2$ *the eigen-functions* (5.6) *form an orthonormal system in* $L^2(\mathbb{Q}_p)$.

For $p \neq 2$ the II kind eigen-functions (5.5″) are orthonormal for distinct k (and fixed N) by virtue of the formulas (5.3″). The I kind eigen-functions (5.5′) for distinct k and fixed N, l and ε are not orthogonal.

To construct an orthonormal system of eigen-functions it is sufficient to orthogonalize the system of functions

$$\{\psi^l_{N,k,\varepsilon_l}, \quad k = 1, 2, \ldots, p-1\}.$$

We shall prove now that as such orthonormal system one possibly takes the following I kind eigen-functions:

$$\varphi^{l}_{N,k,\varepsilon_l}(x) = p^{\frac{N+1-l}{2}}\delta(|x|_p - p^{l-N})\delta(x_0 - k)\cdot\chi_p(\varepsilon_l p^{l-2N}x^2),$$
$$k = 1,2,\dots,p-1. \tag{5.9}$$

At first we establish the relations

$$\psi^{l}_{N,k,\varepsilon_l}(x) = \frac{1}{\sqrt{p-1}}\sum_{1\le j\le p-1}\chi_p\left(\frac{kj}{p}\right)\varphi^{l}_{N,j,\varepsilon_l}(x), \tag{5.10}$$

$$\psi^{l}_{N,k,\varepsilon_l}(x) = \frac{\sqrt{p-1}}{p}\sum_{1\le j\le p-1}\left[\chi_p\left(-\frac{kj}{p}\right)-1\right]\psi^{l}_{N,j,\varepsilon_l}(x), \tag{5.11}$$

$$\begin{aligned}\tilde{\psi}^{l}_{N,k,\varepsilon_l}(\xi) &= p^{-\frac{N-1}{2}}\lambda_p(\varepsilon_p p^l)\delta(|\xi|_p - p^N)\delta(\xi_0 - k')\chi_p\left(-\frac{p^{2N-l}}{4\varepsilon_l}\xi^2\right)\\ &= \varphi^{l}_{l-N,k',\varepsilon'_l}(\xi)\end{aligned} \tag{5.12}$$

where $\varepsilon'_l = -\frac{1}{4\varepsilon_l}$ and k' is defined to be the congruence

$$-2\varepsilon_0 k \equiv k' \pmod p, \quad k' = 1,2,\dots,p-1. \tag{5.13}$$

■ The formula (5.10) follows from (5.3′) and (5.9) owing to the relations

$$\begin{aligned}\chi_p(kp^{l-N-1}x) &= \chi_p\left(\frac{kx_0}{p}\right)\sum_{1\le j\le p-1}\delta(x_0 - j)\\ &= \sum_{1\le j\le p-1}\chi_p\left(\frac{kj}{p}\right)\delta(x_0 - j), \qquad x \in S_{l-N}.\end{aligned}$$

Further, by using the relation

$$\begin{aligned}\sum_{1\le s\le p-1}\chi_p\left(\frac{ks}{p}\right)\chi_p\left(-\frac{js}{p}\right) &= \sum_{1\le s\le p-1}\chi_p\left(\frac{(k-j)s}{p}\right)\\ &= \begin{cases}-1, & k\ne j\\ p-1, & k=j\end{cases} = p\delta_{kj} - 1,\end{aligned} \tag{5.14}$$

from (5.10) we derive the formula (5.11)

$$\begin{aligned}
&\sum_{1\le s\le p-1} \chi_p\left(-\frac{ks}{p}\right)\psi^l_{N,s,\varepsilon}(x) \\
&= \frac{1}{\sqrt{p-1}} \sum_{1\le s\le p-1} \chi_p\left(-\frac{ks}{p}\right) \sum_{1\le j\le p-1} \chi_p\left(\frac{js}{p}\right)\varphi^l_{N,j,\varepsilon_l}(x) \\
&= \frac{1}{\sqrt{p-1}} \sum_{1\le j\le p-1} (p\delta_{kj}-1)\varphi^l_{N,j,\varepsilon_l}(x) \\
&= \frac{p}{\sqrt{p-1}}\varphi^l_{N,k,\varepsilon_l}(x) - \frac{1}{\sqrt{p-1}} \sum_{1\le j\le p-1} \varphi^l_{N,j,\varepsilon_l}(x) \\
&= \frac{p}{\sqrt{p-1}}\varphi^l_{N,k,\varepsilon_l}(x) + \sum_{1\le j\le p-1} \psi^l_{N,j,\varepsilon_l}(x)
\end{aligned}$$

owing to the equality

$$\begin{aligned}
&-\frac{1}{\sqrt{p-1}} \sum_{1\le j\le p-1} \varphi^l_{N,j,\varepsilon_l}(x) \\
&= -p^{N-l}\sqrt{\frac{p}{p-1}}\delta(|x|_p - p^{l-N})\chi_p(\varepsilon_l p^{l-2N}x^2) \sum_{1\le j\le p-1} \delta(x_0 - j) \\
&= p^{N-l}\sqrt{\frac{p}{p-1}}\delta(|x|_p - p^{l-N})\chi_p(\varepsilon_l p^{l-2N}x^2) \sum_{1\le j\le p-1} \chi_p\left(\frac{x_0 j}{p}\right) \\
&= \sum_{1\le j\le p-1} \psi^l_{N,j,\varepsilon_l}(x).
\end{aligned}$$

At last the formula (5.12) follows from (5.11) and (5.3′):

$$\begin{aligned}
\tilde{\varphi}^l_{N,k,\varepsilon_l}(x) &= \frac{\sqrt{p-1}}{p} \sum_{1\le j\le p-1} \left[\chi_p\left(-\frac{kj}{p}\right) - 1\right] \tilde{\psi}^l_{N,j,\varepsilon_l}(x) \\
&= p^{-\frac{N+1}{2}}\lambda_p(\varepsilon_l p^l)\delta(|\xi|_p - p^N)\chi_p\left(-\frac{p^{2N-l}}{4\varepsilon_l}\xi^2\right) \\
&\quad\cdot \sum_{1\le j\le p-1} \left[\chi_p\left(-\frac{kj}{p}\right) - 1\right] \chi_p\left(-\frac{p^{N-1}}{2\varepsilon_l}\xi j\right).
\end{aligned} \tag{5.15}$$

By virtue of (5.13) and (5.14) for $\xi \in S_N$ we have

$$\sum_{1\leq j\leq p-1}\left[\chi_p\left(-\frac{kj}{p}\right)-1\right]\chi_p\left(-\frac{p^{N-1}}{2\varepsilon_l}\xi j\right)$$

$$=\sum_{1\leq j\leq p-1}\left[\chi_p\left(\frac{k'}{2\varepsilon_0 p}j\right)-1\right]\chi_p\left(-\frac{\xi_0 j}{2\varepsilon_0 p}\right)$$

$$=\sum_{1\leq \alpha\leq p-1}\left[\chi_p\left(\frac{k'}{p}\alpha\right)-1\right]\chi_p\left(-\frac{\xi_0}{p}\alpha\right)=p\delta_{k'\xi_0}=p\delta(\xi_0-k').$$

From here and (5.15) the formula (5.12) follows. ■

From the representation (5.11) it follows that the functions (5.9) are eigen-functions of the operator D^α. Thus it was constructed the orthonormal system (5.9), (5.5′) of eigen-functions of the operator D^α for $p \neq 2$: $N \in \mathbb{Z}$

I kind ($l = 2,3,\ldots,\ k = 1,2,\ldots,p-1,\ \varepsilon_l = \varepsilon_0+\varepsilon_1 p+\ldots+\varepsilon_{l-2}p^{l-2}$)

$$\varphi^l_{N,k,\varepsilon_l}(x)=p^{\frac{N+1-l}{2}}\delta(|x|_p-p^{l-N})\cdot\delta(x_0-k)\chi_p(\varepsilon_l p^{l-2N}x^2), \qquad (5.16')$$

II kind ($l = 1,\ k = 1,2,\ldots,p-1,\ \varepsilon_l = 0$)

$$\varphi^1_{N,k,0}(x)=p^{\frac{N-1}{2}}\delta(|x|_p-p^{1-N})\chi_p(kp^{-N}x)=\psi^1_{N,k,0}(x). \qquad (5.16'')$$

Completeness. Let us prove the completeness of the orthonormal systems (5.6) and (5.16) in $L^2(\mathbb{Q}_p)$. It is sufficient to prove the completeness of their Fourier-transform on any circle $S_N, N \in \mathbb{Z}$ as

$$L^2(\mathbb{Q}_p)=\sum_{-\infty<\gamma<\infty}\oplus L^2(S_\gamma). \qquad (5.17)$$

To prove the last one it is sufficient to prove the Parseval-Steklov equality for any functions from the dense set

$$\{\chi_p(\xi\sigma),\ |\sigma|_p\geq p^{-N}\} \quad \text{in } L^2(S_N) \qquad \text{(see Sec. 7.2).}$$

Let $p \neq 2$. If $|\sigma|_p = p^{-N}$ then $\chi_p(\xi\sigma) = 1$ on S_N and hence owing to (5.2″)

$$1=\sum_{1\leq k\leq p-1}\delta(x_0-k)=p^{\frac{N-1}{2}}\sum_{1\leq k\leq p-1}\tilde{\varphi}^1_{N,k,0}(\xi), \qquad \xi\in S_N.$$

Let $|\sigma|_p = p^{1-N}$. Then for $\xi \in S_N$ we have

$$\chi_p(\xi\sigma) = \chi_p\left(\frac{\xi_0\sigma_0}{p}\right) = \sum_{1\le k\le p-1} \delta(\xi_0 - k)\chi_p\left(\frac{\xi_0\sigma_0}{p}\right)$$

$$= \sum_{1\le k\le p-1} \chi_p\left(\frac{k\sigma_0}{p}\right)\delta(\xi_0 - k)$$

$$= p^{\frac{N-1}{2}} \sum_{1\le k\le p-1} \chi_p\left(\frac{k}{p}\sigma_0\right)\bar{\varphi}^1_{N,k,0}(\xi).$$

Let $|\sigma|_p = p^{n-N}$, $n = 2, 3, \ldots$. Then for $\xi \in S_N$ we have

$$C^l_{N,k,\epsilon_l} = \left(\chi_p(\xi\sigma), \bar{\varphi}^l_{N,k,\epsilon_l}\right) = \int_{\mathbb{Q}_p} \chi_p(\xi\sigma)\bar{\bar{\varphi}}^l_{N,k,\epsilon_l}(\xi)d\xi$$

$$= \int_{\mathbb{Q}_p} \overline{\bar{\varphi}^1_{N,k,\epsilon_l}(\xi)\chi_p(-\xi\sigma)}d\xi = \bar{\varphi}^l_{N,k,\epsilon_l}(\sigma).$$

Hence owing to (5.16) $C^l_{N,k,\epsilon_l} = 0$ if $l \neq n$ and for $l = n$

$$|C^n_{N,k,\epsilon_n}|^2 = |\bar{\varphi}^n_{N,k,\epsilon_n}(\sigma)|^2 = p^{N+1-n}\delta(\sigma_0 - k).$$

For fixed $l = n \geq 2$ and $k = 1, 2, \ldots, p-1$ a number of eigen-functions (5.16′) is equal to $(p-1)p^{n-2}$. Therefore

$$\|\chi_p(\xi\sigma)\|^2_{L^2(S_N)} = p^N\left(1 - \frac{1}{p}\right) = (p-1)p^{n-2}p^{N+1-n} \sum_{1\le k\le p-1} \delta(\sigma_0 - k)$$

$$= \sum_{1\le l<\infty} \sum_{1\le k\le p-1} \sum_{\epsilon_l} |C^l_{N,k,\epsilon_l}|^2,$$

so that the Parseval-Steklov equality is realized.

Let $p = 2$. For $|\sigma|_2 = 2^{-N}$ or $|\sigma|_2 = 2^{1-N}$ we have $\chi_2(\xi\sigma) = 1$ on S_N and hence owing to (5.4″) $\chi_2(\xi\sigma) \sim \bar{\psi}^1_{N,0,0}(\xi)$ on S_N. For $|\sigma|_2 = 2^{n-N}$, $n = 3, 4, \ldots$ as in the case $p \neq 2$ we have

$$C^l_{N,k,\epsilon_l} = \left(\chi_2(\xi\sigma), \bar{\psi}^l_{N,k,\epsilon_l}\right) = \bar{\psi}^l_{N,k,\epsilon_l}(\sigma).$$

Therefore owing to (5.6) $C^l_{N,k,\varepsilon_l} = 0$ if $l+1 \neq n$ and for $l+1 = n$

$$|C^{n-1}_{N,k,\varepsilon_{n-1}}|^2 = |\bar{\psi}^l_{N,k,\varepsilon_l}(\sigma)|^2 = 2^{N-n+1}.$$

For fixed $l = n - 1 \geq 2$ a number of eigen-functions (5.6′) is equal to $2 \cdot 2^{n-2} = 2^{n-1}$. Therefore

$$\|\chi_2(\xi\sigma)\|^2_{L^2(S_N)} = 2^N\left(1 - \frac{1}{2}\right) = 2^{N-l}2^{l-1}$$

$$= \sum_{1\leq l<\infty} \sum_{k=0,1} \sum_{\varepsilon_l} |C^l_{N,k,\varepsilon_l}|^2 . \qquad \blacksquare$$

Finally we note that functions of the form

$$\{\chi_p(x\sigma),\ |\sigma|_p = p^N\} \tag{5.18}$$

satisfy the equation (4.3) for $\lambda = \lambda_N = p^{\alpha N}$ owing to (1.7).

Thus we have proved the following

Theorem. *The spectrum of the operator D^α is essential* and it consists of a countable number of eigen-values $\lambda_N = p^{\alpha N}$, $N \in \mathbb{Z}$ each of which is infinite multiplicity, and the point $\lambda = 0$ (the limit point of the eigen-values). The functions (5.16) for $p \neq 2$ and (5.6) for $p = 2$ form an orthonormal basis of eigen-functions of the operator $\mathcal{D}^\alpha$. These eigen-functions belong to $\mathcal{D}(\mathbb{Q}_p)$ and satisfy the conditions (5.7) and (5.8). The generalized eigen-function $\psi(x) = 1$ is the only one corresponding to the point $\lambda = 0$, and the eigenvalues λ_N correspond to generalized eigenfunctions of the form (5.18).*

6. *Expansions on Eigen-Functions*

For uniformity we shall redenote the eigen-functions (5.6) $\psi^l_{N,k,\varepsilon_l}$ $(p = 2)$ by $\varphi^l_{N,k,\varepsilon_l}$. From the results of Sec. 9.4–9.5 it follows the that

Theorem on expansion in eigen-functions $\varphi^l_{N,k,\varepsilon_l}$ of the operator D^α,

$$D^\alpha \varphi^l_{N,k,\varepsilon_l}(x) = p^{\alpha N}\varphi^l_{N,k,\varepsilon_l}(x),$$
$$N \in \mathbb{Z}, \quad l = 1, 2, \ldots, \quad k = 1, 2, \ldots, p-1, \quad \varepsilon_l. \tag{6.1}$$

* The definition of essential spectrum is found in Sec. 10.1.

D^α,

$$D^\alpha \varphi^l_{N,k,\varepsilon_l}(x) = p^{\alpha N} \varphi^l_{N,k,\varepsilon_l}(x),$$
$$N \in \mathbb{Z}, \quad l = 1, 2, \ldots, \quad k = 1, 2, \ldots, p-1, \quad \varepsilon_l. \tag{6.1}$$

Theorem. *Every function $f \in L^2(\mathbb{Q}_p)$ is expanded in the Fourier-series in eigen-functions $\{\varphi^l_{N,k,\varepsilon_l}\}$ of the operator D^α:*

$$f(x) = \sum_{N\in\mathbb{Z}} \sum_{1\le l<\infty} \sum_{1\le k\le p-1} \sum_{\varepsilon_l} f^l_{N,k,\varepsilon_l} \varphi^l_{N,k,\varepsilon_l}(x) \tag{6.2}$$

where

$$f^l_{N,k,\varepsilon_l} = \int_{\mathbb{Q}_p} f(x) \bar{\varphi}^l_{N,k,\varepsilon_l}(x) dx \, . \tag{6.3}$$

The series (6.2) converges in $L^2(\mathbb{Q}_p)$, and the Parseval-Steklov equality is valid

$$\|f\|^2 = \sum_{N\in\mathbb{Z}} \sum_{1\le l<\infty} \sum_{1\le k\le p-1} \sum_{\varepsilon} |f^l_{N,k,\varepsilon_l}|^2. \tag{6.4}$$

In the other words the space $L^2(\mathbb{Q}_p)$ is expanded in a direct sum of finite-demensional eigen-subspaces (series) $\mathcal{H}^l_N$, $N \in \mathbb{Z}$, $l = 1, 2, \ldots$,

$$L^2(\mathbb{Q}_p) = \sum_{N\in\mathbb{Z}} \sum_{1\le l<\infty} \oplus \mathcal{H}^l_N. \tag{6.5}$$

For $p \neq 2$ the subspace $\mathcal{H}^l_N$, $l = 2, 3, \ldots$ are spanned on the eigen-functions of I kind (5.16′) so that $\dim \mathcal{H}^l_N = (p-1)^2 p^{l-2}$; $\mathcal{H}^1_N$ are spanned on the eigen-functions of II kind (5.16″) so that $\dim \mathcal{H}^1_N = p - 1$. For $p = 2$, $\mathcal{H}^l_N$, $l = 2, 3, \ldots$ are spanned on the eigen-functions of I kind (5.6′), $\dim \mathcal{H}^l_N = 2^{l-1}$; $\mathcal{H}^1_N$ are spanned on (5.6″), $\dim \mathcal{H}^1_N = 2$.

Remark. It is easily seen that just constructed eigen-functions of the operator D^α, $\alpha > 0$ (see for example the formulas (5.5) and (5.6)) by complex conjugation convert either in themselves (real) or in other eigen-functions of the some series (complex). Therefore a complex eigen-function (5.5) or (5.6) gives contribution 1 in $\dim \mathcal{H}^l_N$ rather than 2 as it is the case in the classical mathematical physics.

X. p-Adic Schrodinger Operators

A pseudo-differential operator A of the form

$$A\psi = a*\psi + V \cdot \psi$$

we call *the p-adic Schrodinger operator.* Its symbol is the function

$$\tilde{a}(\xi) + V(x)$$

where $\tilde{a}(\xi)$ is the Fourier-transform of a (generalized) function $a(x)$; the function $V(x)$ is called a *potential.*

The simplest example of such operator is the operator $D^\alpha = f_{-\alpha}*$, $\alpha > 0$. Its symbol is $|\xi|_p^\alpha$ (see Sec. 9.).

In presentation of materials in these section we shall follow mainly the work [208].

1. *Bounded from Below Selfadjoint Operators*

Let A be a selfadjoint operator in the Hilbert space H with (dense) domain of definition $\mathcal{D}(A)$. We denote by $\rho(A)$ the resolvent set, by $\sigma(A)$ the spectrum and by $\mathcal{P}(\lambda)$ the projector-valued measure of the operator A,

$$A = \int_{\sigma(A)} \lambda d\mathcal{P}(\lambda), \quad (A\varphi, \psi) = \int_{\sigma(A)} \lambda d(\mathcal{P}(\lambda)\varphi, \psi), \quad \varphi, \psi \in \mathcal{D}(A).$$

Let $\lambda \in \sigma(A)$. In accordance with definitions (see [175], v.1) we say that $\lambda \in \sigma_{\text{ess}}(A)$($\sigma_{\text{ess}}(A)$ is the *essential spectrum of the operator* A) *if the projector*

$$\mathcal{P}(\lambda + \varepsilon) - \mathcal{P}(\lambda - \varepsilon) \tag{1.1}$$

is infinite dimensional for all $\varepsilon > 0$; if the projector (1.1) is finite dimensional for all sufficiently small $\varepsilon > 0$ then $\lambda \in \sigma_{\text{disc}}(A)$($\sigma_{\text{disc}}(A)$ is the *discrete spectrum* of the operator A).

Thus

$$\sigma(A) = \sigma_{\text{ess}}(A) \cup \sigma_{\text{disc}}(A), \quad \sigma_{\text{ess}}(A) \cap \sigma_{\text{disc}}(A) = \phi;$$

$\lambda \in \sigma_{\text{disc}}(A)$ if and only if λ is an isolated point of the spectrum $\sigma(A)$ and λ is the eigen-value of the operator A of finite multiplicity. An operator for which $\sigma(A) = \sigma_{\text{disc}}(A)$ is called operator with *discrete spectrum.*

Another classification of points of the spectrum $\sigma(A)$ is based on expansion of the spectral measure $d\mathcal{P}(\lambda)$ (precisely measures $d(\mathcal{P}(\lambda)\psi,\psi)$, $\psi \in H$) on singular, absolutely continuous, and continuously singular parts. It gives respectively $\sigma_{pp}(A)$-pure pointwise, $\sigma_{\text{ac}}(A)$-absolutely continuous, and $\sigma_{\text{sing}}(A)$-continuously singular parts of the spectrum $\sigma(A)$,

$$\sigma(A) = \sigma_{\text{pp}}(A) \cup \sigma_{\text{ac}}(A) \cup \sigma_{\text{sing}}(A).$$

However these sets may intersect each other (for details see [175], v.1, ch.VII.2).

Now we suppose that a selfadjoint operator A is bounded from below, i.e. there exists a constant C such that

$$(A\psi,\psi) \geq C\|\psi\|^2, \qquad \psi \in \mathcal{D}(A). \tag{1.2}$$

The closed bilinear form generated by the operator A we denote by $A(\varphi,\psi)$; its domain of definition we denote by $\mathcal{Q}(A) \supset \mathcal{D}(A)$ so that

$$A(\varphi,\psi) = (A\varphi,\psi), \quad \varphi,\psi \in \mathcal{D}(A). \tag{1.3}$$

We denote by $\lambda_k(A)$, $k = 1,2,\ldots$ the mini-max numbers which are defined by the mini-max principle

$$\lambda_k(A) = \sup_{\varphi_1,\ldots,\varphi_{k-1}\in H} \inf_{(\psi,\varphi_i)=0, i=1,2,\ldots,k-1} \frac{A(\psi,\psi)}{\|\psi\|^2} \qquad \psi \in \mathcal{Q}(A). \tag{1.4}$$

The following Theorem takes place (see for example [175, v.IV]).

Theorem. *Let A be a bounded from below selfadjoint operator in the Hilbert space H with domain of definition $\mathcal{D}(A)$; let $A(\varphi,\psi)$ be a corresponding bilinear form with domain of definition $\mathcal{Q}(A)$. The following conditions are equivalent.*

(I) *The operator $(A-\lambda)^{-1}$ is a compact for some $\lambda \in \rho(A)$;*
(II) *The operator $(A-\lambda)^{-1}$ is a compact for all $\lambda \in \rho(A)$;*
(III) *The set of functions*

$$[\psi \in \mathcal{D}(A): \ \|\psi\| \leq a, \ \|A\psi\| \leq b]$$

is compact for all $a > 0$ and $b > 0$;

(IV) *The set of functions*

$$[\psi \in \mathcal{Q}(A): \ \|\psi\| \le a, \ A(\psi,\psi) \le b]$$

is compact for all $a > 0$ *and* $b > 0$;

(V) *There exists an orthonormal basis* $\{\varphi_k \in \mathcal{D}(A), \ k = 1,2,\dots\}$ *of eigen-functions* φ_k *of the operator* A *which correspond to eigenvalues* λ_k,

$$A\varphi_k = \lambda_k \varphi_k, \qquad k = 1,2,\dots, \tag{1.5}$$

in addition λ_k *is of finite multiplicity, and*

$$\lambda_1 \le \lambda_2 \le \dots, \qquad \lambda_k \to \infty, \quad k \to \infty; \tag{1.6}$$

(VI)
$$\lambda_k(A) \to +\infty, \qquad k \to \infty, \tag{1.7}$$

where numbers $\lambda_k(A)$ *are defined by the mini-max principle* (1.4).

In other words by realization one of the conditions (I)–(VI) the others conditions will be fulfilled, and the spectrum $\sigma(A) = [\lambda = \lambda_k, \ k = 1,2,\dots]$ is discrete, in addition $\lambda_k = \lambda_k(A) \to \infty, \ k \to \infty$.

Example 1. Let a potential $V(|x|_p)$ be bounded from below and locally finite in $\mathbb{Q}_p : -C \le V(|x|_p) \in L^{\infty}_{\text{loc}}(\mathbb{Q}_p)$.The p-adic Schrodinger operator

$$A\psi = D^{\alpha}\psi + V(|x|_p)\psi, \quad \alpha > -1 \tag{1.8}$$

has symbol

$$|\xi|_p^{\alpha} + V(|x|_p)$$

and the domain of definition

$$\mathcal{D}(A) = [\varphi \in L^2 : \ |\xi|_p^{\alpha}\tilde{\varphi} \in L^2(\mathbb{Q}_p), \ V(|x|_p)\varphi \in L^2] \,.$$

It is bounded from below and symmetric in the Hilbert space $L^2(\mathbb{Q}_p) = L^2$. It admits the Friedrichs selfadjoint extension (see for example [175], v.2) with the help of the closed bounded from below bilinear form

$$\begin{aligned} A(\varphi,\psi) &= \int_{\mathbb{Q}_p} |\xi|_p^{\alpha} \tilde{\varphi}(\xi)\bar{\tilde{\psi}}(\xi)d\xi + \int_{\mathbb{Q}_p} V(|x|_p)\varphi(x)\bar{\psi}(x)dx \\ &= (D^{\alpha/2}\varphi, D^{\alpha/2}\psi) + (V\varphi,\psi) \end{aligned} \tag{1.9}$$

with the domain of definition

$$Q(A) = [\varphi \in L^2(\mathbb{Q}_p) : |\xi|_p^{\alpha/2}\tilde{\varphi} \in L^2(\mathbb{Q}_p), \sqrt{V(|x|_p) + |C| + 1}\varphi \in L^2]. \tag{1.10}$$

It is clear that

$$\mathcal{D}(A) \subset \mathcal{D}(\tilde{A}) \subset Q(A).$$

Example 2. Multidimensional p-adic Schrodinger operators with symbols

$$|\xi_1|_p^2 + \dots |\xi_n|_p^2 + V(x), \qquad |(\xi, \xi)|_p + V(x),$$

where a potential $V \in L^\infty_{\text{loc}}(\mathbb{Q}_p^n)$ and $V(x) \geq -C$, $x \in \mathbb{Q}_p^n$, give examples of bounded from below symmetric operators in $L^2(\mathbb{Q}_p^n)$.

2. *Compactness in $L_p^2(\mathbb{Q}^n)$*

As the space $\mathbb{Q}_p^n$ is locally-compact then some criterions of compactness of functions of real arguments are directly carried over to complex-valued functions of p-adic arguments, for example the Ascoli-Arzela Lemma (see [175], v.1, [237]).

If M is an infinite set of continuous functions on a compact $K \subset \mathbb{Q}_p^n$ which is bounded in $C(K)$ and it consists of equicontinuous functions on K, then a sequence may be chosen from M which converges in $C(K)$.

Let G be an clopen set in $\mathbb{Q}_p^n$ (for instance, a ball B_r, a sphere S_r, the exterior of a ball B_r : $\mathbb{Q}_p^n \backslash B_r$, the exterior of a sphere S_r : $\mathbb{Q}_p^n \backslash S_r$, whole space $\mathbb{Q}_p^n$ and so on). We shall consider the space $L^2(G)$ (see Sec. 4.1) as a set of those functions from $L^2(\mathbb{Q}_p^n)$ whose support is contained in G, i.e. every function f from $L^2(G)$ is defined on the whole space $\mathbb{Q}_p^n$ and vanishes almost everywhere outside of G.

Definitions. 1. We shall say that a set of functions $M \subset L^2(G)$ consists of *equicontinuous on the whole in* $L^2(G)$ functions if for any $\varepsilon > 0$ there exists $n = n_\varepsilon \in \mathbb{Z}$ such that for any function f from M the inequality is valid

$$\int_G |f(x+h) - f(x)|^2 dx < \varepsilon, \quad \forall\, h \in B_n.$$

2. We shall say that a set $M \subset L^2(G)$ consists of functions with *equicontinuous $L^2(G)$-integrable at infinity* if for any $\varepsilon > 0$ there exists $N = N_\varepsilon \in \mathbb{Z}$ such that for any $f \in M$ the inequality is valid

$$\int_{G \setminus B_N} |f(x)|^2 dx < \varepsilon.$$

3. We shall say that a measurable real function $F(x)$ defined on a set $G \subset \mathbb{Q}_p^n$ tends to $+\infty$ at infinity (i.e. $F(x) \to +\infty$, $|x|_p \to \infty$, $x \in G$) if G is an unbounded set and for any $M > 0$ there exists $N = N_\varepsilon \in \mathbb{Z}$ such that $F(x) > M$ for all $|x|_p \geq p^N$, $x \in G$.

Analogy of the Riesz-Kolmogorov criterion on compactness: In order that a set $M \subset L^2(G)$ is compact in $L^2(G)$, it is necessary and sufficient that it satisfies the following conditions:

(I) *is bounded in $L^2(G)$;*
(II) *consists of equicontinuous on the whole in $L^2(G)$ functions;*
(III) *consists of functions with equicontinuous $L^2(G)$-integrals at infinity.*

In terms of the Fourier-transform from the Riesz-Kolmogorov criterion we obtain the following criterion of compactness: *In order that a set $M \subset L^2(G)$ be compact in $L^2(G)$, it is necessary and sufficient that it satisfies the conditions* (I), (III) *and the condition*

(II′) *the set $[g = \tilde{f}, f \in M]$ consists of functions with equicontinuous $L^2(\mathbb{Q}_p^n)$-integrals at infinity.*

From the last criterion it follows directly **the analogy of Rellich's criterion:** *If a set of functions $f \in L^2(G)$ is bounded in $L^2(G)$ and satisfies the conditions*

$$\int_{\mathbb{Q}_p^n} \sigma(\xi)|\tilde{f}(\xi)|^2 d\xi \leq C, \quad f \in M, \tag{2.1}$$

$$\int_G \rho(x)|f(x)|^2 dx \leq C, \quad f \in M, \tag{2.2}$$

where $\sigma(\xi)$ and $\rho(x)$ are positive functions which tend to $+\infty$ at infinity and a number C does not depend on f then M is compact in $L^2(G)$.

Remark. If G is bounded then the conditions (III) and (2.2) naturally are absent.

3. *The Operator $a*+V$*

The operator $a*+V\cdot = A$ is pseudo-differential

$$(A\psi)(x) = \int_{\mathbb{Q}_p^n} \tilde{a}(\xi)\tilde{\psi}(\xi)\chi_p(-(x,\xi))d\xi + V(x)\psi(x). \tag{3.1}$$

We shall consider the operator A in the Hilbert space $L^2(G)$ where G is a clopen set in $\mathbb{Q}_p^n$. Therefore in the equation (3.1) x belongs to G. The symbol of the operator A is the (generalized) function

$$\tilde{a}(\xi) + V(x), \quad \xi \in \mathbb{Q}_p^n, \quad x \in G.$$

We suppose: $\tilde{a}(\xi)$ and $V(x)$ are functions bounded from below, locally bounded and tend to $+\infty$ to at infinity in $\mathbb{Q}_p^n$ and G respectively. Without loss of generality we may assume that functions $\tilde{a}(\xi)$ and $V(x)$ are positive.

As a domain of definition of the operator A we take the set

$$\mathcal{D}(A) = [\psi \in L^2(G) : \ \tilde{a}\tilde{\psi} \in L^2(\mathbb{Q}_p^n), \ V\psi \in L^2(G)].$$

Under these assumptions the operator A is symmetric and positive. The corresponding bilinear form is:

$$A(\psi,\varphi) = \int_{\mathbb{Q}_p^n} \tilde{a}(\xi)\tilde{\psi}(\xi)\bar{\tilde{\varphi}}(\xi)d\xi + \int_G V(x)\psi(x)\bar{\varphi}(x)dx, \tag{3.2}$$

$$A(\psi,\psi) = \int_{\mathbb{Q}_p^n} \tilde{a}(\xi)|\tilde{\psi}(\xi)|^2 d\xi + \int_G V(x)|\psi(x)|^2 dx\ , \tag{3.3}$$

with the domain of definition

$$\mathcal{Q}(A) = [\psi \in L^2(G) : \ \sqrt{\tilde{a}}\tilde{\psi} \in L^2(\mathbb{Q}_p^n), \ \sqrt{V}\psi \in L^2(G)]$$

which is closed and positive. Therefore the operator A admits the (unique) Friedrich selfadjoint extension $\tilde{A}$ with a domain of definition $\mathcal{D}(\tilde{A}) = \mathcal{D}(A^*)$ $\cap \mathcal{Q}(A) \subset \mathcal{Q}(A)$, and

$$A(\psi, \varphi) = (\tilde{A}\psi, \varphi), \quad \varphi, \psi \in \mathcal{D}(\tilde{A}).$$

The "boundary value" problem for the operator A on a set G is posed in the following way: Let $f \in L^2(G)$, find a function φ from $\mathcal{D}(\tilde{A})$ which satisfies the equation

$$(\tilde{A}\varphi)(x) = \lambda\varphi(x) + f(x), \qquad x \in G. \tag{3.4}$$

For $f \equiv 0$ we have the spectrum problem for the operator A.

Remark 1. "Boundary" conditions in the "boundary value" problem (3.4) are hidden in the definition of the quadratic form (3.3) in which it is supposed that functions ψ vanish outside G.

By virtue of the analogy of Rellich's criterion (see Sec. 10.2) any bounded set M in $L^2(G)$ for which for some $b > 0$

$$A(\psi, \psi) \leq b, \quad \psi \in M$$

is compact in $L^2(G)$. From here by using the Theorem of Sec. 10.1 we conclude that the Hilbert-Schmidt theory is valid for the operator A, namely one has the following

Main Theorem *Let locally bounded functions $\tilde{a}(\xi)$ and $V(x)$ be bounded from below and tend to $+\infty$ at infinity in $\mathbb{Q}_p^n$ and G respectively. Then the spectrum of the operator $A = a^* + V.$[1*]) is discrete and it consists of real eigen-value $\lambda_1 \leq \lambda_2 \leq \ldots$, $\lambda_k \to +\infty$, $k \to \infty$; every eigen-value is finite multiplicity. Corresponding eigen-functions $\varphi_k \in \mathcal{D}(\tilde{A})$ form an orthonormal basis in $L^2(G)$,*

$$\tilde{A}\varphi_k = \lambda_k\varphi_k, \quad (\tilde{A}\varphi_k, \varphi_j) = \delta_{kj}, \quad k, j = 1, 2, \ldots. \tag{3.5}$$

The following variational principle is valid:

$$\lambda_k = \min_{(\psi,\varphi_j)=0, j=1,2,\ldots,k-1} \frac{A(\psi, \psi)}{\|\psi\|^2} \qquad \psi_j \in \mathcal{Q}(A) \tag{3.6}$$

[1*] More precisely, the spectrum of its selfadjoint extension $\tilde{A}$.

where the min *is realized on an eigen-function* $\psi = \varphi_k$ *corresponding to the eigen-value* λ_k.

For $\lambda \neq \lambda_k$, $k = 1, 2, \ldots$ *the operator* $(\tilde{A} - \lambda)^{-1}$ *is compact, so the equation* (3.4) *is uniquely solvable in* $\mathcal{D}(\tilde{A})$ *for any* $f \in L^2(G)$, *and its solution is expressed by the formula*

$$\varphi = \sum_{1 \le k < \infty} \frac{(f, \varphi_k)}{\lambda - \lambda_k} \varphi_k . \tag{3.7}$$

For $\lambda = \lambda_k$ *a solution of the equation* (3.4) *exists if, and only if,* f *is orthogonal to all eigen-functions corresponding to the eigen-value* λ_k,

$$(f, \varphi_{k+j}) = 0, \qquad j = 0, 1, \ldots, n_k - 1, \tag{3.8}$$

where n_k *is the multiplicity of* λ_k; *this solution is expressed by the formula*

$$\varphi = \sum_{1 \le j < \infty, \lambda_j \neq \lambda_k} \frac{(f, \varphi_j)}{\lambda_k - \lambda_j} + \sum_{0 \le j \le n_k - 1} c_j \varphi_{k+j} \tag{3.9}$$

where c_j *are arbitrary constants.*

Remark 2. The Main Theorem is valid also for those bounded from below pseudo-differential operators, whose symbol $a(\xi, x)$ is locally integrable in $\mathbb{Q}_p^n \times G$ and every set of functions

$$\left[\psi \in L^2(G) : \ \|\psi\| \le 1, \quad \int_{\mathbb{Q}_p^n \times G} a(\xi, x) \chi_p(-(\xi, x)) \tilde{\psi}(\xi) \bar{\psi}(x) d\xi dx \le 1 \right]$$

is compact in $L^2(G)$.

Conversion of the Main Theorem. *If the Schrodinger-type operator* $A = a * + V.$ *in* $\mathbb{Q}_p^n$ *with a symbol of the form*

$$\mathcal{P}(|\xi_1|_p, \ldots, |\xi_n|_p) + V(|x_1|_p, \ldots, |x_n|_p)$$

where $\mathcal{P}$ *and* V *are locally-bounded and bounded from below functions in* $\mathbb{Q}_p^n$ *such that the mini-max numbers* $\lambda_k(A) \to +\infty$, $k \to \infty$ *then* $\mathcal{P} \to +\infty$, $|\xi|_p \to \infty$ *and* $V \to +\infty$, $|x|_p \to \infty$.

■ At first we consider the case $n = 1$. Let $V(|x|_p)$ not tending to $+\infty$ if $|x|_p \to \infty$. Then there exist $K > 0$ and a sequence $\rho_k \to +\infty$, $k \to \infty$, $\rho_k \in \mathbb{Z}$ such that

$$V(p^{\rho_k}) \leq K, \quad k = 1, 2, \ldots. \tag{3.10}$$

On the eigen-functions of the I kind (see (5.16′) of Sec. 9.5 for $p \neq 2$ and (5.6′) Sec. 9.5 for $p = 2$)

$$\{\varphi^2_{2-\rho_k,1,1}(x), \quad k = 1, 2, \ldots\} \tag{3.11}$$

$$|\varphi^2_{2-\rho_k,1,1}(x)|^2 = \frac{p}{p-1} p^{-\rho_k} \delta(|x|_p - p^{\rho_k}),$$
$$|\tilde{\varphi}^2_{2-\rho_k,1,1}(\xi)|^2 = \frac{p}{p-1} p^{\rho_k - 2} \delta(|\xi|_p - p^{2-\rho_k})$$

the quadratic form

$$A(\varphi, \varphi) = \int_{\mathbb{Q}_p^n} \mathcal{P}(|\xi|_p) |\varphi(\xi)|^2 d\xi + \int_{\mathbb{Q}_p} V(|x|_p) |\varphi(x)|^2 dx$$

takes the values

$$\begin{aligned}
&A(\varphi^2_{2-\rho_k,1,1}, \varphi^2_{2-\rho_k,1,1}) \\
&= \frac{P}{p-1} p^{\rho_k - 2} \int_{\mathbb{Q}_p} \mathcal{P}(|\xi|_p) \delta(|\xi|_p - p^{2-\rho_k}) d\xi \\
&\quad + \frac{p}{p-1} p^{-\rho_k} \int_{\mathbb{Q}_p} V(|x|_p) \delta(|x|_p - p^{\rho_k}) \\
&= \mathcal{P}(p^{2-\rho_k}) + V(p^{\rho_k}),
\end{aligned}$$

and hence owing to (3.10) is bounded on k by the number

$$\sup_{k \geq 1} \mathcal{P}(p^{2-\rho_k}) + K.$$

By the Theorem in Sec. 10.1 the set of functions (3.11) is compact in $L^2(\mathbb{Q}_p)$ which is not possible owing to their orthogonality. The contradiction just obtained proves that $V(|x|_p) \to +\infty$ when $|x|_p \to \infty$.

Similarly it is proved that $\mathcal{P}(|\xi|_p) \to +\infty$, $|\xi|_p \to \infty$. By supposing that it is not the case for a corresponding sequence of integers $\rho_k \to +\infty$, $k \to \infty$ we choose eigen-functions

$$\{\varphi^2_{\rho_k,1,1}(x), \quad k = 1, 2, \dots\} \tag{3.12}$$

similar to (3.11).

For $n > 1$ the proof is similar. As an orthonormal system of functions one must take products of functions (3.11) and (3.12) with distinct arguments. ■

4. *Operator D^α, $\alpha > 0$ in B_r*

Let us calculate in explicit form all eigen-values and eigen-functions of the operator D^α in the disc $B_r \subset \mathbb{Q}_p$, $r \in \mathbb{Z}$.

By the Main Theorem of Sec. 10.3 the spectrum of the operator D^α in the disc B_r is discrete and the eigen-functions form an orthonormal basis in $L^2(B_r)$. According to the definition (see Sec. 10.3) those eigen-functions of the operator D^α in $\mathbb{Q}_p$ whose supports are contained in B_r are eigen-functions of the operator D^α in B_r. Eigen-functions of the operator D^α in $\mathbb{Q}_p$ have been calculated in Sec. 9.5.

Beforehand we prove the following

Lemma. *Let a function $f(|x|_p)$, $x \in \mathbb{Q}_p$ be such that*

$$\sum_{0 \le \gamma < \infty} p^{-\gamma\alpha}|f(p^\gamma)| < \infty, \qquad \sum_{0 \le \gamma < \infty} p^{-\gamma}|f(p^{-\gamma})| < \infty$$

for some $\alpha > 0$. Then

$$\begin{aligned}(D^\alpha f)(x) &= \frac{p^\alpha(p^\alpha + p - 2)}{p^{\alpha+1} - 1}|x|_p^{-\alpha} f(|x|_p) \\ &+ \frac{1}{\Gamma_p(-\alpha)}\left[|x|_p^{-\alpha-1}\int_{|y|_p < |x|_p} f(|y|_p)dy + \int_{|y|_p > |x|_p} |y|_p^{-\alpha-1} f(|y|_p)dy\right], \\ &x \in \mathbb{Q}_p\end{aligned} \tag{4.1}$$

In particular, for $f(|x|_p) = \Omega(p^{-r}|x|_p)$, $r \in \mathbb{Z}$

$$(D^\alpha f)(x) = \frac{p-1}{p^{\alpha+1} - 1}p^{\alpha(1-r)}, \qquad x \in B_r; \tag{4.2}$$

for $f(|x|_p) = \delta(|x|_p - p^r)$, $r \in \mathbb{Z}$

$$(D^\alpha f)(x) = \frac{p^\alpha + p - 2}{p^{\alpha+1} - 1} p^{\alpha(1-r)}, \qquad x \in S_r. \tag{4.3}$$

■ By the definition of the operator D^α (see Sec. 9.1) we have

$$(D^\alpha f)(x) = f_{-\alpha} {}^* f(|x|_p) = \frac{1}{\Gamma_p(-\alpha)} \int_{\mathbb{Q}_p} \frac{f(|y|_p) - f(|x|_p)}{|x-y|^{\alpha+1}} dy$$

$$= \frac{1}{\Gamma_p(-\alpha)} \left[|x|_p^{-\alpha-1} \int_{|y|_p<|x|_p} f(|y|_p) dy + \int_{|y|_p>|x|_p} |y|_p^{-\alpha-1} f(|y|_p) dy \right]$$

$$- \frac{f(|x|_p)}{\Gamma_p(-\alpha)} \left[|x|_p^{-\alpha-1} \int_{|y|_p<|x|_p} dy + \int_{|y|_p>|x|_p} |y|_p^{-\alpha-1} dy \right]. \tag{4.4}$$

But denoting $|x|_p$ by p^N we have

$$- \frac{1}{\Gamma_p(-\alpha)} \left[|x|_p^{-\alpha-1} \int_{|y|_p<|x|_p} dy + \int_{|y|_p>|x|_p} |y|_p^{-\alpha-1} dy \right]$$

$$= - \frac{1}{\Gamma_p(-\alpha)} \left[|x|_p^{-\alpha-1} p^{N-1} + \sum_{N+1 \le \gamma < \infty} p^{-(\alpha+1)\gamma} p^\gamma \left(1 - \frac{1}{p}\right) \right]$$

$$= - \frac{1}{\Gamma_p(-\alpha)} \left[\frac{|x|_p^{-\alpha}}{p} + p^{-\alpha(N+1)} \left(1 - \frac{1}{p}\right) \frac{1}{1 - p^{-\alpha}} \right]$$

$$= |x|_p^{-\alpha} \frac{p^\alpha - 1}{1 - p^{-\alpha-1}} \left[\frac{1}{p} + \left(1 - \frac{1}{p}\right) \frac{p^{-\alpha}}{1 - p^{-\alpha}} \right] = |x|_p^{-\alpha} p^\alpha \frac{p^\alpha + p - 2}{p^{\alpha+1} - 1}$$

From here and (4.4) the equality (4.1) follows.

The equality (4.2) follows simply from (4.4): if $|x|_p \le p^r$ then

$$(D^\alpha f)(x) = - \frac{1}{\Gamma_p(-\alpha)} \int_{|y|_p<|x|_p} |y|_p^{-\alpha-1} dy$$

$$= - \frac{1}{\Gamma_p(-\alpha)} \sum_{r+1 \le \gamma < \infty} p^{-(\alpha+1)\gamma} p^\gamma \left(1 - \frac{1}{p}\right)$$

$$= \frac{p^\alpha - 1}{1 - p^{-\alpha-1}} \left(1 - \frac{1}{p}\right) p^{-\alpha(r+1)} \frac{1}{1 - p^{-\alpha}} = \frac{p-1}{p^{\alpha+1} - 1} p^{\alpha(1-r)}.$$

The equality (4.3) follows from (4.1) if $|x|_p = p^r$. ■

Let $p \neq 2$. The I kind eigen-functions $\varphi^l_{N,j,\epsilon}$ (see (5.16′) of Sec. 9.5) with the indices $l - N \leq r$ have their supports in B_r and with the indices $l - N > r$ vanish in B_r. The II kind eigen-functions $\varphi^1_{N,j,0}$ (see (5.16″) Sec. 9.5) with the indices $1 - N \leq r$ have their supports in B_r and with the indices $1 - N > r$ are equal to 1 in B_r. But by the Lemma (see (4.2)) $\psi_0(x) \equiv 1$, $x \in B_r$ is the eigen-function of the operator D^α in B_r corresponding to the eigen-value

$$\lambda_0 = \frac{p-1}{p^{\alpha+1}-1} p^{\alpha(1-r)}. \tag{4.5}$$

(The equality (4.2) for $\psi_0(x) \equiv 1$ and $x \in B_r$ takes the form $(D^\alpha \psi_0)(x) = \lambda_0 \psi_0(x)$.)

As the eigen-functions $\varphi^l_{N,j,\epsilon}$ of the operator D^α in $\mathbb{Q}_p$ form the orthonormal basis in $L^2(\mathbb{Q}_p)$ then the eigen-functions of the operator D^α in B_r just enumerated form an orthonormal basis in $L^2(B_r)$. In particular, the multiplicity of the eigen-value λ_0 is equal to 1.

Hence we have just obtained the following eigen-values and orthonormal basis of eigen-functions of the operator D^α, $\alpha > 0$ in B_r $(p \neq 2)$:

$$\lambda_0 = \frac{p-1}{p^{\alpha+1}-1} p^{\alpha(1-r)}, \qquad \varphi_0(x) = p^{-r/2}, \quad \text{multiplicity } 1;$$

$$\lambda_1 = p^{\alpha(1-r)}, \qquad \text{II kind } \varphi^1_{1-r,j,0}(x), \quad j = 1, 2, \ldots, p-1,$$

multiplicity $p - 1$;

$$\lambda_k = p^{\alpha(k-r)}, \text{ I kind } \varphi^l_{k-r,j,\epsilon_l}(x), \quad j = 1, 2, \ldots, p-1,$$

$$l = 2, 3, \ldots, k, \epsilon_l,$$

$$\text{II kind } \varphi^1_{k-r,j,0}(x), \quad j = 1, 2, \ldots, p-1,$$

$$\epsilon = 0 \quad \text{multiplicity}$$

$$(p-1)^2 + (p-1)^2 p + \ldots + (p-1)^2 p^{k-2} + p - 1 = (p-1)p^{k-1}, \quad k = 2, 3, \ldots .$$

Let $p = 2$. Supports of the I kind eigen-functions $\varphi^l_{N,j,\epsilon_l}$, $l + 1 - N \leq r$ (see (5.6′) of Sec. 9.5) are contained in B; for indices $l+1-N > r$ they vanish in B_r. Supports of the II kind eigen-functions $\varphi^1_{N,j,0}$, $1 - N \leq r$ for $j = 0$ or

$2-N \leq r$ for $j=1$ (see (5.6″) of Sec. 9.5) are also contained in B_r. (Note that the support of the function $\varphi^1_{N,1,0}$ is contained in S_{2-N}.) Besides, $\varphi^1_{N,0,0}(x) \equiv 1$ in B_r if $1-N > r$, and $\varphi^1_{N,1,0}(x) \equiv 0$ in B_r if $2-N > r$. By the Lemma $\psi_0 \equiv 1$ is an eigen-function of the operator D^α in B_r corresponding to the eigen-value λ_0 (see (4.3)); λ_0 is a simple eigen-value. Hence we have just obtained the following eigen-values and orthonormal basis of eigen-functions of the operator D^α, $\alpha > 0$ in $B_r (p=2)$:

$$\begin{aligned}
\lambda_0 &= \frac{2^{\alpha(1-r)}}{2^{\alpha+1}-1}, && \varphi_0(x) = 2^{-r/2}, \quad \text{multiplicity } 1;\\
\lambda_1 &= 2^{\alpha(1-r)}, && \text{II kind } \varphi^1_{1-r,0,0}(x), \quad \text{multiplicity } 1;\\
\lambda_2 &= 2^{\alpha(2-r)}, && \text{II kind } \varphi^1_{2-r,j,0}(x),\ j=0,1, \quad \text{multiplicity } 2;\\
\lambda_k &= 2^{\alpha(k-r)}, && \text{I kind } \varphi^l_{k-r,j,\varepsilon_l}(x), \quad l = 2,3,\dots,k-1, \quad j = 0,1,\varepsilon_l,\\
& && \text{II kind } \varphi^1_{k-r,j,0}(x), \quad j=0,1 \text{ multiplicity}\\
& && 2(1+2+\dots+2^{k-3})+2 = 2^{k-1}, \quad k = 2,3,\dots.
\end{aligned}$$

5. *Operator* D^α, $\alpha > 0$ *in* S_r

Like in the case of the disc B_r we calculate in explicit form all eigen-values and eigen-functions of the operator D^α on the circle $S_r \subset \mathbb{Q}_p$, $r \in \mathbb{Z}$.

Let $p \neq 2$. The I kind eigen-functions $\varphi^l_{N,j,\varepsilon}$ (see (5.16′) of Sec. 9.5) with the indices $l - N = r$ have their supports in S_r; the others vanish outside S_r. As for the II kind eigen-functions (see (5.16″) of Sec. 9.5) their non-zero traces on S_r are the functions

$$1, v_j(x) = \chi_p(jp^{r-1}x), \quad j = 1,2,\dots,p-1, \quad x \in S_r. \tag{5.1}$$

The functions (5.1) are linearly dependent on S_r as

$$\begin{aligned}
\sum_{1\leq j\leq p-1} v_j(x) &= \sum_{1\leq j\leq p-1} \chi_p(jp^{r-1}x)\\
&= \sum_{1\leq j\leq p-1} \exp\left(2\pi i \frac{j}{p} x\right) = -1, \quad x \in S_r.
\end{aligned}$$

The remaining functions $\{v_j,\ j=1,2,\dots,p-1\}$ are linearly independent on S_r as (see (3.2) of Sec. 4.3)

$$(v_j, v_k) = \int_{S_r} \chi_p((j-k)p^{r-1}x)dx = \begin{cases} p^r\left(1-\frac{1}{p}\right), & j=k, \\ -p^{r-1}, & j \neq k, \end{cases} \tag{5.2}$$

and thus

$$\det \|(v_j, v_k)\| = p^{(r-1)(p-1)} \begin{vmatrix} p-1 & -1 & \dots & -1 \\ -1 & p-1 & \dots & -1 \\ \dots & \dots & \dots & \dots \\ -1 & -1 & & p-1 \end{vmatrix} = p^{rp-r-1} \neq 0.$$

As a basis of the functions (5.1) we choose the functions

$$\psi_0 = 1, \quad \psi_j = \chi_p(jp^{r-1}x) - \chi_p((j+1)p^{r-1}x), \quad j = 1, 2, \dots, p-2. \tag{5.3}$$

By the Lemma of Sec. 10.4 the function $\psi_0(x) \equiv 1$ is an eigen-function of the operator D^α corresponding to the eigen-value (see (4.3))

$$\lambda_0 = \frac{p^\alpha + p - 2}{p^{\alpha+1} - 1} p^{\alpha(1-r)}. \tag{5.4}$$

The functions ψ_j, $j = 1, 2, \dots, p-2$ are proportional on S_r respectively to the functions

$$\varphi_{1-r,j}(x) = \frac{1}{\sqrt{2}}[\varphi^1_{1-r,j,0}(x) - \varphi^1_{1-r,j+1,0}(x)], \quad j = 1, 2, \dots, p-2 \tag{5.5}$$

(see (5.16″) of Sec. 9.5) and therefore they are the eigen-functions of the operator D^α corresponding to the eigen-value $\lambda_1 = p^{\alpha(1-r)}$.

Now we sum up the results. For $p \neq 2$ the operator D^α in S_r, $r \in \mathbb{Z}$ has the following eigen-values and normed basis of eigen-functions:

$$\lambda_0 = \frac{p^\alpha + p - 2}{p^{\alpha+1} - 1} p^{\alpha(1-r)}, \qquad \varphi_0(x) = p^{-r/2}\sqrt{\frac{p}{p-1}}, \quad \text{multiplicity } 1;$$

$$\lambda_1 = p^{\alpha(1-r)}, \qquad \varphi_{1-r,j}(x) = \frac{p^{-r/2}}{\sqrt{2}} \chi_p\left(j\frac{x_0}{p}\right) \cdot \left[1 - \chi_p\left(\frac{x_0}{p}\right)\right], \qquad j = 1, 2, \dots, p-2, \ \text{multiplicity } p-2;$$

$$\lambda_k = p^{\alpha(k-r)}, \quad \varphi^k_{k-r,j,\epsilon_k}(x), \qquad j = 1, 2, \dots, p-1, \epsilon_k,$$

$$\text{multiplicity } (p-1)^2 p^{k-2}, \qquad k = 2, 3, \ldots .$$

The listed eigen-functions are mutually orthogonal except the functions $\{\varphi_{1-r,j},\ j = 1, 2, \ldots, p-2\}$ for which owing to (5.3)

$$(\varphi_{1-r,j}, \varphi_{1-r,k}) = \begin{cases} -1/2 & \text{if } j = k+1, \text{ or } k = j+1, \\ 0 & \text{if } j \neq k,\ j \neq k+1,\ k \neq j+1. \end{cases} \tag{5.6}$$

For $p = 2$ the eigen-values and the orthonormal basis of the eigen-functions of the operator D^α in S_r are:

$$\lambda_0 = \frac{2^{\alpha(2-r)}}{2^{\alpha+1}-1}, \qquad \varphi_0(x) = 2^{\frac{1-r}{2}}, \quad \text{multiplicity } 1;$$

$$\lambda_1 = 2^{\alpha(2-r)}, \qquad \varphi^1_{2-r,1,0}(x), \quad \text{multiplicity } 1;$$

$$\lambda_k = 2^{\alpha(k+1-r)}, \qquad \varphi^k_{k-1-r,j,\varepsilon_k}(x), \quad j = 1, 2, \varepsilon_k,$$

$$\text{multiplicity } 2^{k-1}, \quad k = 2, 3, \ldots .$$

6. *Operator* $D^\alpha + V(|x|_p)$, $\alpha > 0$ *in* $\mathbb{Q}_p$, $p \neq 2$

We consider the Schrodinger-type operator $A = D^\alpha + V(|x|_p)$ (see example 1 Sec. 10.1) by assumptions that the potential $V(|x|_p)$ is a finite bounded from below function and $V(|x|_p) \to +\infty$, $|x|_p \to \infty$. The operator A is bounded from below and selfadjoint if its domain of definition $\mathcal{D}(A)$ consists of those functions ψ from $L^2(\mathbb{Q}_p)$ for which $D^\alpha\psi + V(|x|_p)\psi \in L^2(\mathbb{Q}_p)$. The operator A satisfies the conditions of the Main Theorem of Sec. 10.3. In particular, its spectrum is discrete. Let $\lambda_0, \lambda_1, \lambda_2, \ldots$ be its eigen-values, $\lambda_0 \leq \lambda_1 \leq \lambda_2 \leq \ldots,$ $\varphi_0, \varphi_1, \varphi_2 \ldots$ be its corresponding eigen-functions, so

$$A\varphi_k = \lambda_k \varphi_k, \qquad \varphi_k \in \mathcal{D}(A), \quad k = 0, 1, 2, \ldots .$$

All I kind eigen-functions $\varphi^l_{N,j,\varepsilon}$ (see (5.16′) of Sec. 9.5) are the eigen-functions of the operator A corresponding to the eigen-values

$$\lambda^l_N = p^{\alpha N} + V(p^{l-N}), \qquad l = 2, 3, \ldots, \quad N \in \mathbb{Z}. \tag{6.1}$$

The multiplicity of λ^l_N is no less than

$$(p-1)^2 \sum_y p^{y-2}$$

where (x, y) are (different) solutions of the Diophantine equation

$$p^{\alpha x} + V(p^{y-x}) = p^{\alpha N} + V(p^{l-N}), \qquad y = 2, 3, \ldots, \quad x \in \mathbb{Z}. \tag{6.2}$$

This number is finite by the Main Theorem of Sec. 10.3.

Further the II kind eigen-functions (see (5.16″) of Sec. 9.5) and (5.5))

$$\begin{aligned} \varphi_{N,j}(x) &= \frac{1}{\sqrt{2}}[\varphi^1_{N,j,0}(x) - \varphi^1_{N,j+1,0}(x)] \\ &= \frac{p^{\frac{N-1}{2}}}{\sqrt{2}}\delta(|x|_p - p^{1-N})\chi_p\left(j\frac{x_0}{p}\right)\left[1 - \chi_p\left(\frac{x_0}{p}\right)\right], \\ j &= 1, 2, \ldots p-2 \end{aligned} \tag{6.3}$$

are the eigen-functions of the operator A corresponding to the eigen-values

$$\lambda^1_N = p^{\alpha N} + V(p^{1-N}), \quad N \in \mathbb{Z}. \tag{6.4}$$

The problem is how to find deficient eigen-values and eigen-functions of the operator A?

As it follows from the results of Sec. 10.5 the subspace spanned on those eigen-functions

$$\begin{aligned} \{\varphi^l_{N,j,\varepsilon_l}, \quad & l = 2, 3, \ldots, \quad j = 1, 2, \ldots, p-1, \varepsilon_l; \\ \varphi_{N,j}, \quad & j = 1, 2, \ldots, p-2; \quad N \in \mathbb{Z}\} \end{aligned}$$

which supports are contained in the circle S_r, $r \in \mathbb{Z}$:

$$\begin{aligned} \{\varphi^l_{l-r,j,\varepsilon_l}, \quad & l = 2, 3, \ldots, \quad j = 1, 2, \ldots, p-1, \varepsilon_l; \\ \varphi_{1-r,j}, \quad & j = 1, 2, \ldots, p-2\} \end{aligned} \tag{6.5}$$

has codimension 1 and it is orthogonal to 1 in $L^2(S_r)$. In the other words every function ψ from $L^2(\mathbb{Q}_p)$ which is orthogonal to all eigen-functions (6.5) for all $r \in \mathbb{Z}$ is a constant on every circles S_r, i.e. it is expanded in the orthogonal canonical basis $\{\psi(|x|_p - p^\gamma), \ \gamma \in \mathbb{Z}\}$ in $L^2_0(\mathbb{Q}_p)$

$$\psi(|x|_p) = \sum_{-\infty<\gamma<\infty} \psi_\gamma \delta(|x|_p - p^\gamma), \quad \psi_\gamma = \psi(p^\gamma). \tag{6.6}$$

The set of functions ψ from $L^2(\mathbb{Q}_p)$ of the form (6.6) forms the Hilbert space of $L_0^2(\mathbb{Q}_p)$ which is a (closed) subspace of $L^2(\mathbb{Q}_p)$. It is isomorphic to the Hilbert space l_0^2 of sequences $\Psi = \{\psi_\gamma, \ \gamma \in \mathbb{Z}\}$ with the norm

$$\|\Psi\|_0^2 = \sum_{-\infty<\gamma<\infty} p^\gamma |\psi_\gamma|^2, \tag{6.7}$$

besides

$$\|\psi\|^2 = \left(1 - \frac{1}{p}\right) \|\Psi\|_0^2.$$

7. *Operator* $D^\alpha + V(|x|_p)$, $\alpha > 0$ *in* $L_0^2(\mathbb{Q}_p)$ $(p \neq 2)$

At first we prove the following statement:

The space $L_0^2(\mathbb{Q}_p)$ *is invariant with respect to the Fourier-transform operation.*

■ It follows from the formula (3.3) Sec. 4.3: The Fourier-transform of functions from $\mathcal{D}(\mathbb{Q}_p)$ depending only on $|x|_p$ (dense set in $L_0^2(\mathbb{Q}_p)$) is a function of the same class. As the subspace $L_0^2(\mathbb{Q}_p)$is closed in $L^2(\mathbb{Q}_p)$ and the Fourier-transform operation is continuous in $L^2(\mathbb{Q}_p)$ (see Sec. 7.4) then this statement is valid for all functions from $L_0^2(\mathbb{Q}_p)$. ■

From the proved statement it follows that the operator A maps $\mathcal{D}(A) \cap L_0^2(\mathbb{Q}_p)$ where $\mathcal{D}(A)$ is the domain of definition of the operator A (see Sec. 10.6) into $L_0^2(\mathbb{Q}_p)$ (as its symbol $|\xi|_p^\alpha + V(|x|_p)$ depends only on $|\xi|_p$ and $|x|_p$).

Denote by A_0 the restriction of the operator A on subspace $L_0^2(\mathbb{Q}_p)$. Its domain of definition is $\mathcal{D}(A_0) = \mathcal{D}(A) \cap L_0^2(\mathbb{Q}_p)$. By the Lemma in Sec. 10.4 the operator A_0 on functions ψ from $\mathcal{D}(A_0)$ has the form

$$(A_0\psi)(|x|_p) = (D^\alpha \psi)(|x|_p) + V(|x|_p)\psi(|x|_p) \tag{7.1}$$

where D^α is the positive integral operator

$$(D^\alpha \psi)(|x|_p) = \int_{\mathbb{Q}_p} \mathcal{K}(|x|_p, |y|_p)\psi(|y|_p)dy \tag{7.2}$$

with the real symmetric kernel

$$\mathcal{K}(t,\tau) = \begin{cases} \Gamma_p^{-1}(-\alpha)t^{-\alpha-1}, & \tau < t, \\ \frac{p^\alpha+p-2}{(p-1)(1-p^{-\alpha-1})}t^{-\alpha-1}, & \tau = t. \end{cases} \tag{7.3}$$

In terms of sequences $\Psi = \{\psi_\gamma,\ \gamma \in \mathbb{Z}\}$ the operator A_0 takes the form

$$(A_0\Psi)_\gamma = \sum_{-\infty<\gamma'<\infty} \mathcal{K}_{\gamma\gamma'}\psi_{\gamma'} + V(p^\gamma)\psi_\gamma,\ \gamma \in \mathbb{Z}, \tag{7.4}$$

$$\mathcal{K}_{\gamma\gamma'} = \begin{cases} \dfrac{p-1}{p\Gamma_p(-\alpha)}p^{-\gamma(\alpha+1)+\gamma'}, & \gamma' \le \gamma - 1, \\ \dfrac{p^\alpha(p^\alpha+p-2)}{p^{\alpha+1}-1}p^{-\gamma\alpha}, & \gamma' = \gamma, \\ \dfrac{p-1}{p\Gamma_p(-\alpha)}p^{-\alpha\gamma'}, & \gamma' \ge \gamma + 1. \end{cases} \tag{7.5}$$

The matrix $\mathcal{K}_{\gamma\gamma'}$ is evidently symmetrizable.

Our goal is to investigate the spectral properties of the operator A_0. The operator A_0 is bounded from below and selfadjoint with the domain of definition $\mathcal{D}(A_0) = \mathcal{D}(A) \cap L_0^2(\mathbb{Q}_p)$. It satisfies the conditions of the Theorem in Sec. 10.1 with $H = L_0^2(\mathbb{Q}_p)$. In particular, its spectrum is discrete, i.e. it consists of eigen-values $\{\mu_k,\ k = 0,1,\dots\}$:

$$\mu_0 \le \mu_1 \le \mu_2 \le \dots, \qquad \mu_k \to \infty, \quad k \to \infty,$$

every μ_k is of finite multiplicity n_k; corresponding eigen-functions $\psi_k(|x|_p)$,

$$A_0\psi_k = \mu_k\psi_k, \quad \psi_k \in \mathcal{D}(A_0),\ k = 0,1,\dots \tag{7.6}$$

form an orthonormal basis in $L_0^2(\mathbb{Q}_p)$.

8. *The Lowest Eigenvalue* λ_0

Here we shall prove that *the lowest eigenvalue* λ_0 *of the p-adic Schrödinger operator*

$$A = D^\alpha + V(|x|_p), \quad \alpha > 0$$

is simple, $\lambda_0 = \mu_0$*, satisfies the estimates*

$$\inf_{\gamma\in\mathbb{Z}} V(p^\gamma) < \lambda_0 < \inf_{\gamma\in\mathbb{Z}} \left(1 - \frac{1}{p}\right)\left[\frac{p^{-\alpha\gamma}}{1-p^{-\alpha-1}} + p^{-\gamma}\sum_{-\infty<\gamma'\le\gamma} V(p^{\gamma'})p^{\gamma'}\right], \tag{8.1}$$

and corresponding (unique) eigenfunction $\psi_0(|x|_p)$ is positive.

■ For the lowest eigenvalue λ_0 of the operator A the following variational principles valid (see Sec. 10.1 and 10.3):

$$\lambda_0 = \inf_{\psi \in Q(A)} \frac{(D^\alpha \psi, \psi) + (V\psi, \psi)}{\|\psi\|^2} = (A\varphi_0, \varphi_0)$$
$$= (D^\alpha \varphi_0, \varphi_0) + (V\varphi_0, \varphi_0) \tag{8.2}$$

where $\varphi_0(x)$ is any real eigenfunction of the operator A belonging to λ_0; the kernel $\mathcal{K}(|x|_p, |y|_p)$ of the integral operator D^α posses the property: $\mathcal{K}(t, \tau) < 0$ for $t \neq \tau$ and $\mathcal{K}(t, t) > 0$ (see (7.5)).

At first we prove that $\varphi_0(x) \geq 0$, $x \in \mathbb{Q}_p$.

■ Let, conversely, there exist bounded vicinities $U(x')$ and $V(x'')$ of points x' and x'' such that $\varphi_0(x) > 0$, $x \in U(x')$ and $\varphi_0(x) < 0$, $x \in V(x'')$ so that $|\varphi_0(x)| = \varphi_0(x)$, $x \in U(x')$ and $|\varphi_0(x)| > \varphi_0(x)$, $x \in V(x'')$.

Let us introduce the sequence of functions

$$\{\psi_N(x) = \varphi_0(x)[\Delta_N(x) - \Delta_{-N}(x)], \quad N \to +\infty\}$$

from $\mathcal{D}(\mathbb{Q}_p) \subset \mathcal{D}(A)^*$ with supports in $B_N \backslash B_{-N}$, so that $\psi_N \to \varphi_0$, $N \to \infty$ in L^2. From here and from the closure of the quadratic form $A(\psi, \psi)$ (See Sec. 10.1) and from (8.2) it follows

$$\inf_N \frac{A(\psi_N, \psi_N)}{\|\psi_N\|^2} = \inf_N \frac{(A\psi_N, \psi_N)}{\|\psi_N\|^2} = (A\varphi_0, \varphi_0) = \lambda_0. \tag{8.3}$$

Further for sufficiently big N, $V(x') \cup U(x'') \subset B_N \backslash B_{-N}$ and hence

$$\int_{\substack{\mathbb{Q}_p^2,\\ |x|_p \neq |y|_p}} \mathcal{K}(|x|_p, |y|_p)[\psi_N(x)\psi_N(y) - |\psi_N(x)|\,|\psi_N(y)|]dy \geq \eta \tag{8.4}$$

where $\eta > 0$ which is equal to

$$\eta = -4 \int_{U(x')} \int_{V(x'')} \mathcal{K}(|x|_p, |y|_p)\varphi_0(|x|_p)|\varphi_0(|y|_p)|dxdy.$$

---

* Every eigen-function of the operator A is a locally constant function in $\mathbb{Q}_p \backslash \{0\}$ as the kernel of the operator D^α depends only on $|x|_p$ and $|y|_p$.

From (8.4) we conclude that

$$(A|\psi_N|, |\psi_N|) = \int_{\substack{\mathbb{Q}_p^2, \\ |x|_p \neq |y|_p}} \mathcal{K}(|x|_p, |y|_p)|\psi_N(x)|\, |\psi_N(y)| dx dy$$

$$+ \int_{\mathbb{Q}_p} \mathcal{K}(|x|_p, |x|_p)|\psi_N(x)|^2 dx + \int_{\mathbb{Q}_p} V(|x|_p)|\psi_N(x)|^2 dx$$

$$\leq \int_{\substack{\mathbb{Q}_p^2, \\ |x|_p \neq |y|_p}} \mathcal{K}(|x|_p, |y|_p)\psi_N(x)\psi_N(y) dx dy + \int_{\mathbb{Q}_p} \mathcal{K}(|x|_p, |x|_p)\psi_N(x)^2 dx$$

$$+ \int_{\mathbb{Q}_p} V(|x|_p)\psi_N(x)^2 dx - \eta = (A\psi_N, \psi_N) - \eta.$$

Hence, taking into account that $|||\psi_N||| = ||\psi_N||$ we derive inequality

$$\frac{(A|\psi_N|, |\psi_N|)}{|||\psi_N|||^2} \leq \frac{(A\psi_N, \psi_N) - \eta}{||\psi_N||^2}, \quad N \to \infty,$$

which owing to (8.3) contradicts to the variational principle (8.2). ■

Similarly one can prove that any eigen-function $\psi_0(|x|_p)$ of the operator A_0 corresponding to the lowest eigen-value μ_0 is non-negative in $\mathbb{Q}_p$.

Now we shall prove that $\psi_0(|x|_p) > 0$, $x \in \mathbb{Q}_p$.

■ Let, conversely, there exists $\gamma \in \mathbb{Z}$ such that $\psi_0(p^\gamma) = 0$. Then by integrating the equation

$$A_0\psi_0 \equiv D^\alpha \psi_0 + V\psi_0 = \mu_0\psi_0$$

on the circumference S_γ and denoting

$$\psi = \delta(|x|_p - p^\gamma) \in \mathcal{D}(\mathbb{Q}_p)$$

we obtain the contradiction

$$(A_0\psi_0, \psi) = (\psi_0, A_0\psi) = (\psi_0, D^\alpha \psi) + (\psi_0, V\psi) = \mu_0(\psi_0, \psi) = 0$$

as

$$(\psi_0, V\psi) = 0, \quad (\psi_0, D^\alpha \psi) = \left(1 - \frac{1}{p}\right) p^\gamma \int\limits_{|x|_p \neq p^\alpha} \mathcal{K}(|x|_p, p^\gamma)\psi_0(|x|_p)dx < 0.$$

■

From here it follows that μ_0 *is a simple eigen-value.*

■ In fact if another linear-independent eigenfunction ψ corresponding to μ_0 would exist then what we have just proved is $\psi(|x|_p) > 0$, $x \in \mathbb{Q}_p$. It is possible to construct a linear combination of the eigenfunctions ψ_0 and ψ which takes both positive and negative values which contradicts to the positiveness of every such eigen-functions. ■

From the variation principle (8.2) it follows evidently that $\lambda_0 \leq \mu_0$. But $\lambda_0 < \mu_0$ is not valid otherwise the eigen-functions $\varphi_0(x) \geq 0$ and $\psi_0(|x|_p) > 0$ would be orthogonal in $L^2(\mathbb{Q}_p)$ which is impossible. Hence $\lambda_0 = \mu_0$ and $\varphi_0(x) = \psi_0(|x|_p)$ depend in fact on $|x|_p$. ■

The estimates (8.1) follow from the variational principle (8.2). The lower bound follows evidently

$$\lambda_0 = (D^\alpha \psi_0, \psi_0) + (V\psi_0, \psi_0) > \int\limits_{\mathbb{Q}_p} V(|x|_p)\psi_0^2(|x|_p)dx \geq \inf_{x \in \mathbb{Q}_p} V(|x|_p).$$

To obtain the upper bound we put in (8.2)

$$\psi = \Delta_\gamma(x) = \Omega(p^{-\gamma}|x|_p) \in \mathcal{D}(\mathbb{Q}_p).$$

As

$$||\Delta_\gamma||^2 = p^\gamma, \quad \tilde{\Delta}_\gamma(\xi) = p^\gamma \Omega(p^\gamma |\xi|_p) \qquad \text{(see Sec. 7.2)}$$

and also $\varphi_0 \neq \Delta_\gamma$ then

$$\lambda_0 < \inf_{\gamma\in\mathbb{Z}} p^{-\gamma}[(D^\alpha\Delta_\gamma,\Delta_\gamma)+(V\Delta_\gamma,\Delta_\gamma)]$$

$$= \inf_{\gamma\in\mathbb{Z}} p^{-\gamma}\left[(|\xi|_p^\alpha\tilde{\Delta}_\gamma,\tilde{\Delta}_\gamma)+\int_{B_\gamma} V(|x|_p)dx\right]$$

$$= \inf_{\gamma\in\mathbb{Z}}\left[p^\gamma\int_{B_{-\gamma}}|\xi|_p^\alpha d\xi+\left(1-\frac{1}{p}\right)p^{-\gamma}\sum_{-\infty<\gamma'\le\gamma}V(p^{\gamma'})p^{\gamma'}\right]$$

$$= \inf_{\gamma\in\mathbb{Z}}\left(1-\frac{1}{p}\right)\left[p^{-\gamma}\sum_{-\infty<\gamma'\le-\gamma}p^{(\alpha+1)\gamma'}+p^{-\gamma}\sum_{-\infty<\gamma'\le\gamma}V(p^{\gamma'})p^{\gamma'}\right].$$

■

9. *Operator* $D^\alpha + V(|x|_p)$, $\alpha > 0$ *in* $\mathbb{Q}_p$, $p \neq 2$ *(continuation)*

We sum up the results of Secs. 10.6–10.8 in the following

Theorem. *The spectrum of the operator* $D^\alpha + V(|x|_p)$ *is discrete and it consists of the following eigen-values and eigenfunctions:*

The lowest eigenvalue λ_0 *of multiplicity 1 satisfies the estimates* (8.1); *corresponding eigenfunction is* $\psi_0(|x|_p) > 0$, $x \in \mathbb{Q}_p$.

Eigen-values μ_k, $k = 1, 2, \ldots$, $\mu_k \to +\infty$, $k \to \infty$ *are of multiplicity*

$$n_k + (p-2)\omega_k + (p-1)^2 \sum_{1\le i\le b_k} p^{l_i^k-2} \tag{9.1}$$

where ω_k *is a number of solutions* $\{N_i^k,\ i = 1, 2, \ldots, \omega_k\}$ *of the equation*

$$\mu_k = p^{\alpha N} + V(p^{1-N}), \qquad N \in \mathbb{Z}; \tag{9.2}$$

$\{(N_i^k, l_i^k),\ i = 1, 2, \ldots, b_k\}$ *are all solutions of the equation*

$$\mu_k = p^{\alpha N} + V(p^{l-N}), \qquad N \in \mathbb{Z}, \quad l = 2, 3, \ldots; \tag{9.3}$$

corresponding eigenfunctions are:

$$\psi_{k+i}(|x|_p), \quad i = 0, 1, \ldots, n_k - 1,$$
$$\varphi_{N,j}(x), N = N_i^k, \quad j = 1, 2, \ldots, p-2, \quad i = 1, 2, \ldots, \omega_k,$$
$$\varphi^l_{N,j,\epsilon_l}(x), N = N_i^k, \quad l = l_i^k, \quad j = 1, 2, \ldots, p-1, \quad i = 1, 2, \ldots, b_k.$$

Eigenvalues

$$\lambda_N^1 = p^{\alpha N} + V(p^{1-N}) \neq \mu_k, \qquad N \in \mathbb{Z}, \quad k \in \mathbb{Z}_+ \tag{9.4}$$

are of multiplicity

$$(p-2)\eta_N + (p-1)^2 \sum_{1\leq i\leq c_N} p^{y_i^N - 2} \tag{9.5}$$

where η_N *is a number of solutions* $\{x_i^N,\ i = 1, 2, \dots, \eta_N\}$ *of the equation*

$$\lambda_N^1 = p^{\alpha x} + V(p^{1-x}), \qquad x \in \mathbb{Z}, \tag{9.6}$$

$\{(x_i^N, y_i^N),\ i = 1, 2, \dots, c_N\}$ *are all solutions of the equation*

$$\lambda_N^1 = p^{\alpha x} + V(p^{l-x}), \qquad x \in \mathbb{Z}, \quad l = 2, 3, \dots; \tag{9.7}$$

corresponding eigenfunctions are:

$$\begin{array}{lllll} \varphi_{N,j}(x), & N = x_i^N, & j = 1, 2, \dots, p-2, & i = 1, \dots, \eta_N, & \\ \varphi_{N,j,\epsilon_l}^l(x), & N = x_i^N, & l = y_i^N, & j = 1, 2, \dots, p-1, & i = 1, \dots, c_N. \end{array}$$

Eigenvalues

$$\lambda_N^l = p^{\alpha N} + V(p^{l-N}) \neq \mu_k, \neq \lambda_{N'}^1, \ N \in \mathbb{Z}, \ l = 2, 3, \dots, \ k \in \mathbb{Z}_+, \ N' \in \mathbb{Z} \tag{9.8}$$

are of multiplicity

$$(p-1)^2 \sum_{1\leq i\leq d_{N,l}} p^{y_i^{N,l} - 2} \tag{9.9}$$

where $\{(x_i^{N,l}, y_i^{N,l}),\ i = 1, 2, \dots, d_{N,l}\}$ *are all solutions of the equation*

$$\lambda_N^l = p^{\alpha x} + V(p^{y-x}), \qquad x \in \mathbb{Z}, \ y = 2, 3, \dots; \tag{9.10}$$

corresponding eigen-functions are:

$$\varphi_{N,j,\epsilon_l}^l(x), \quad N = x_i^{N,l}, \quad l = y_i^{N,l}, \qquad j = 1, 2, \dots, p-1, \quad i = 1, \dots, d_{N,l}.$$

The listed eigenfunctions form a normed bases in $L^2(\mathbb{Q}_p)$, they are mutually orthogonal except the functions $\{\varphi_{N,j}, j = 1, 2, \dots, p-2\}$ for which the relations (5.6) take place:

$$(\varphi_{N,j}, \varphi_{N,k}) = \delta_{jk} - \frac{1}{2}\delta_{j,k+1} - \frac{1}{2}\delta_{k,j+1}.$$

Remark. The eigenfunction $\psi_0(|x|_p) > 0$ which corresponds to the lowest eigenvalue λ_0 defines the *ground state* (*"vacuum"*) of the physical system; the "vacuum" is unique. Eigenfunctions which correspond to eigenvalues of multiplicity > 1 define "degenerate" states.

10. *Example. Potential* $|x|_p^\alpha$, $\alpha > 0$ $(p \neq 2)$

For the potential $V(|x|_p) = |x|_p^\alpha$ the results of Sec. 10.9 admit some refinements. The estimates (8.1) take the form

$$0 < \lambda_0 < \frac{2(p-1)p^\alpha}{p^{\alpha+1}-1} < 2. \tag{10.1}$$

■ $$\inf_{\gamma\in\mathbb{Z}} \left(1-\frac{1}{p}\right)\frac{p^{-\alpha\gamma}}{1-p^{-\alpha-1}} + p^{-\gamma} \sum_{-\infty<\gamma'\le\gamma} p^{\gamma'(\alpha+1)}$$
$$= \inf_{\gamma\in\mathbb{Z}} \left(1-\frac{1}{p}\right)\frac{p^{-\alpha\gamma}+p^{\alpha\gamma}}{1-p^{-\alpha-1}} = \frac{2(1-p^{-1})}{1-p^{-\alpha-1}}$$ ■

Now we shall prove the following

Lemma. *All solutions of the Diophantine equation*

$$p^{\alpha N} + p^{\alpha M} = p^{\alpha x} + p^{\alpha y}, \quad M, N, x, y \in \mathbb{Z}$$

if $\alpha \ge \frac{\ln 2}{\ln p}$ *have the form:*

$$\textit{either } x = N,\ y = M \ \textit{ or } \ y = N,\ x = M.$$

■ Without loss of generality we may suppose that

$$N = \max(N, M, x, y).$$

If we suppose that $N \neq x$, $N \neq y$ then $x + 1 \le N$, $y + 1 \le N$ and

$$2p^{\alpha(N-1)} = p^{\alpha(N-1)} + p^{\alpha(N-1)} \ge p^{\alpha x} + p^{\alpha y} = p^{\alpha N} + p^{\alpha M} > p^{\alpha N}$$

which is impossible for $\alpha \geq \frac{\ln 2}{\ln p}$. Hence, either $N = x$ (and then $M = y$) or $N = y$ (and then $M = x$). ∎

From the above Lemma it follows that the equation

$$\lambda_N^l = p^{\alpha x} + p^{\alpha(y-x)}, \qquad N, x \in \mathbb{Z}, \quad l, y \in \mathbb{Z}_+ \tag{10.2}$$

has only two solutions $x = N$, $y = l$ and $x = l - N$, $y = l$ if $l \neq 2N$ and has only one solution $x = N$, $y = l$ if $l = 2N$. From here it follows that in the Theorem in Sec. 10.9 the following cases are possible. If $\alpha \geq \frac{\ln 2}{\ln p}$ then ω_k may take only values 0 and 2, in the last case for solutions N_1^k and N_2^k of the equation (9.2) the relation $N_2^k = 1 - N_1^k$ is valid; b_k may take only values 0 and 1 if $l = 2N$, and values 0 and 2 if $l \neq 2N$, in the last case solutions of the equation (9.3) have the form (N_1^k, l_1^k), $(l_1^k - N_1^k, l_1^k)$; $\eta_k = 2$ and for solutions x_1^N and x_2^N of the equation (9.6) the relation $x_2^N = 1 - x_1^N$ is valid; $c_N = 0$; $d_{N,l}$ may take only value 1 if $l = 2N$ and value 2 if $l \neq 2N$, in the last case solutions of (9.10) have the form

$$(x_1^{N,l}, y_1^{N,l}), \quad (y_1^{N,l} - x_1^{N,l}, y_1^{N,l}).$$

In conclusion we note: *if φ is an eigenfunction of the operator $D^\alpha + |x|_p^\alpha$ corresponding to eigenvalue λ then its Fourier-transform $\tilde{\varphi}$ is also an eigenfunction which corresponds to the same eigenvalue λ so that*

$$\tilde{\varphi}(x) = \sum_{1 \leq k \leq n} c_k \varphi_k(x), \qquad x \in \mathbb{Q}_p \tag{10.3}$$

where $\{\varphi_k,\ k = 1, 2, \ldots, n\}$ are eigenfunctions corresponding to λ, n is the multiplicity of λ and c_k are some constants.

∎ The equation

$$D^\alpha \varphi + |x|_p^\alpha \varphi = \lambda \varphi$$

by the Fourier transform goes to itself (see Sec. 9.4):

$$|\xi|_p^\alpha \tilde{\varphi} + D^\alpha \tilde{\varphi} = \lambda \tilde{\varphi} .$$

∎

Remark. For the eigen-functions of the I kind $\varphi_{N,j,\varepsilon}^l$ (see (5.16′) of Sec. 9.5) the formula (10.3) is simplified

$$\tilde{\varphi}_{N,j,\varepsilon_l}^l(x) = \varphi_{l-N,j',\varepsilon_l'}^l, \quad \varepsilon' = -\frac{1}{4\varepsilon_l}, \quad j' \equiv -2\varepsilon_0 j \pmod p)$$

(see (5.12) of Sec. 9.5) i.e. eigenfunctions of the series $\mathcal{H}_N^l$ $(l \geq 2)$ by the Fourier transform go to eigenfunctions of the series $\mathcal{H}_{l-N}^l$ and they belong to the same eigenvalue $\lambda_N^l = \lambda_{l-N}^l$ (cf. Sec. 9.6).

11. *Operator* $D^\alpha + V(|x|_p)$, $\alpha > 0$ *Outside of a Disc* $(p \neq 2)$

Let us denote by

$$G_s = \mathbb{Q}_p \backslash B_{s-1} = [x \in \mathbb{Q}_p : |x|_p \geq p^s]$$

the exterior of the disc B_{s-1}, $s \in \mathbb{Z}$. We consider the operator $A = D^\alpha + V(|x|_p)$ in the set G_s under the hypotheses of Sec. 11.6 concerning the potential $V(|x|_p)$.

As in Sec. 11.6 we conclude that the basis of eigen-functions of the operator A in $L^2(G_s)$ consists of those eigen-functions $\varphi_{N,j,\epsilon_l}^l$ $(l \geq 2)$ and $\varphi_{N,j}$, whose supports are contained in G_s, and also of eigen-functions $\psi_k(|x|_p)$, $k = 0, 1, \ldots$ of the operator A_0 in G_s,

$$A_0\psi_k = D^\alpha\psi_k + V(|x|_p)\psi_k = \mu_k\psi_k, \quad \psi_k \in \mathcal{D}(A_0) \subset L_0^2(G_s) \tag{11.1}$$

which correspond to eigen-values μ_k.

Let us try to find in an explicit form $\{\mu_k,\ \psi_k(|x|_p),\ k = 0, 1, \ldots\}$. Beforehand we note the following equality which follows from the Lemma in Sec. 10.4 (see also Sec. 10.7)

$$D^\alpha\delta(|x|_p - p^r) = \sum_{-\infty<\gamma<\infty} \mathcal{K}_{r\gamma}\delta(|x|_p - p^\gamma), \quad x \in \mathbb{Q}_p, r \in \mathbb{Z} (\alpha > 0), \tag{11.2}$$

where

$$\mathcal{K}_{r\gamma} = \begin{cases} -\rho p^{-\alpha r}, & \gamma < r, \\ \sigma p^{-\alpha r}, & \gamma = r, \\ -\rho p^{-(\alpha+1)\gamma + r}, & \gamma > r, \end{cases} \tag{11.3}$$

$$\sigma = \frac{p^\alpha + p - 2}{p^{\alpha+1} - 1} p^\alpha, \quad \rho = -\frac{p-1}{p\Gamma_p(-\alpha)} = \frac{(p-1)(p^\alpha - 1)}{p(1 - p^{-\alpha-1})}, \quad \sigma + \rho = p^\alpha. \tag{11.4}$$

Let $\psi(|x|_p)$ be an eigenfunction of the operator A_0 in G_s which corresponds to the eigenvalue μ provided that $\psi(p^s) \neq 0$. We shall see it in the form of a formal series in the orthogonal canonical bases in $L_0^2(G_s)$ (cf. Sec. 11.6)

$$\psi(|x|_p) = \sum_{s \leq r < \infty} d_r\delta(|x|_p - p^r), \quad d_s \neq 0. \tag{11.5}$$

Then owing to (11.2) and (11.4) for $x \in G_s$ we shall have

$$
\begin{aligned}
D^\alpha \psi(|x|_p) &= \sum_{s\le r<\infty} d_r D^\alpha \delta(|x|_p - p^r) \\
&= \sum_{s\le r<\infty} d_r \sum_{-\infty<\gamma<\infty} \mathcal{K}_{r\gamma}\delta(|x|_p - p^\gamma) = \sum_{s\le r<\infty} \delta(|x|_p - p^\gamma) \sum_{s\le r<\infty} \mathcal{K}_{r\gamma} d_r \\
&= \delta(|x|_p - p^s) \sum_{s\le r<\infty} C_s^r d_r + \sum_{s+1\le\gamma<\infty} \delta(|x|_p - p^\gamma) d_r \sum_{s\le r<\infty} C_\gamma^r d_r \\
&= \delta(|x|_p - p^s)\left(\mathcal{K}_{ss} d_s + \sum_{s+1\le r<\infty} \mathcal{K}_{rs} d_r\right) \\
&\quad + \sum_{s+1\le\gamma<\infty} \delta(|x|_p - p^\gamma)\left(\sum_{s\le r\le\gamma-1} \mathcal{K}_{r\gamma} d_r + \mathcal{K}_{\gamma\gamma} d_\gamma + \sum_{\gamma+1\le r<\infty} \mathcal{K}_{r\gamma} d_r\right) \\
&= \delta(|x|_p - p^s)\left((\sigma+\rho)p^{-\alpha s} d_s - \rho \sum_{s\le r<\infty} p^{-\alpha r} d_r\right) + \sum_{s+1\le\gamma<\infty} \delta(|x|_p - p^\gamma) \\
&\quad \cdot \left(-\rho \sum_{s\le r\le\gamma-1} p^{-(\alpha+1)\gamma+r} d_r + \sigma p^{-\alpha\gamma} d_\gamma - \rho \sum_{\gamma+1\le r<\infty} p^{-\alpha r} d_r\right). \qquad (11.6)
\end{aligned}
$$

Denote

$$\sum_{s\le r<\infty} p^{-\alpha r} d_r = \kappa d_s. \qquad (11.7)$$

Then owing to (11.3) the equality (11.6) takes the form

$$
\begin{aligned}
D^\alpha \psi(|x|_p) &= (p^{(1-s)\alpha} - \rho\kappa) d_s \delta(|x|_p - p^s) \\
&+ \sum_{s+1\le\gamma<\infty} \left(-\rho\kappa d_s + p^{(1-\gamma)\alpha} d_\gamma - \rho \sum_{s\le r\le\gamma-1} p^{-\alpha r} C_{\gamma,r} d_r\right) \delta(|x|_p - p^\gamma) \qquad (11.8)
\end{aligned}
$$

where it is denoted

$$C_{\gamma,r} = 1 - p^{(1+\alpha)(r-\gamma)}. \qquad (11.9)$$

If we substitute the expressions (11.8) and (11.5) in the equation (11.1) for $|x|_p \geq p^s$ we obtain

$$
\begin{aligned}
& d_s(p^{(1-s)\alpha} - \rho\kappa)\delta(|x|_p - p^s) \\
& + \sum_{s+1 \leq r < \infty} \left(-\rho\kappa d_s + p^{(1-\gamma)\alpha} d_r + \sum_{s \leq \gamma \leq r-1} p^{-\alpha\gamma} C_{r,\gamma} d_\gamma \right) \delta(|x|_p - p^r) \\
& + \sum_{s \leq r < \infty} d_r V(p^r)\delta(|x|_p - p^r) = \mu \sum_{s \leq r < \infty} d_r \delta(|x|_p - p^r)
\end{aligned}
$$

i.e.

$$
p^{(1-s)\alpha} - \rho\kappa + V(p^s) = \mu \tag{11.10}
$$

$$
\begin{aligned}
-\rho\kappa d_s + p^{(1-r)\alpha} d_r + \rho \sum_{s \leq \gamma \leq r-1} p^{-\alpha\gamma} C_{r,\gamma} d_\gamma & \\
+ d_r V(p^r) = \mu d_r, \quad r = s+1, s+2, \ldots &
\end{aligned} \tag{11.11}
$$

The formula (11.11) gives the recursion relation for determination of the coefficients

$$
d_r = \frac{-\rho\kappa d_s + \rho \sum\limits_{s \leq \gamma \leq r-1} p^{-\alpha\gamma} C_{r,\gamma} d_\gamma}{\mu - p^{(1-r)\alpha} - V(p^r)}, \quad d_s = 1, \ r = s+1, s+2, \ldots \tag{11.12}
$$

provided (see notation (6.4))

$$
\mu \neq p^{(1-r)\alpha} + V(p^r) = \lambda^1_{1-r}, \quad r = s+1, s+2, \ldots. \tag{11.13}
$$

With the help of equality (11.10) we eliminate from the equalities (11.12) and (11.7) an unknown quantity

$$
\mathcal{H} = \frac{1}{\rho}(\lambda^1_{1-s} - \mu),
$$

and as a result we obtain the recursion relation

$$
\begin{aligned}
d_r - \frac{1}{\mu - \lambda^1_{1-r}} \left(\mu - \lambda^1_{1-r} + \rho \sum_{s \leq \gamma \leq r-1} C_{r,\gamma} p^{-\alpha\gamma} d_\gamma \right), & \\
d_s = 1, \quad r = s+1, \quad s+2, \ldots &
\end{aligned} \tag{11.14}
$$

and the transcendental equation

$$\sum_{s+1\leq r<\infty} p^{-\alpha r} d_r(\mu) = \frac{1}{\rho}(\lambda^1_{1-s} - \mu) - p^{-\alpha s}, \quad d_r(\mu) = d_r \tag{11.15}$$

for the determination of eigen-values $\mu = \mu_k$ (provided that the inequalities (11.13) are fulfilled).

The formula (11.8) takes the form

$$D^\alpha \psi(|x|_p) = \sum_{s\leq r<\infty} d'_r \delta(|x|_p - p^r) \tag{11.16}$$

where

$$\begin{aligned} d' &= [\mu - V(p^s)]d_s, \\ d'_r &= p^{(1-r)\alpha} d_r - \rho \sum_{r\leq\gamma<\infty} p^{-\alpha\gamma} d_\gamma - \rho p^{-(1+\alpha)r} \sum_{s\leq\gamma\leq r-1} p^\gamma d_\gamma. \end{aligned} \tag{11.17}$$

These arguments have a formal character as we have not considered here the convergence of the series (11.5), (11.15) and (11.16). Now we sum up our results in the form of the following

Theorem. *Let an eigenvalue μ of the operator A_0 in G_s satisfies the conditions (11.13) and its corresponding eigen-function $\psi(|x|_p)$ satisfies $\psi(p^s) \neq 0$. Then $\psi(|x|_p)$ is represented by the series (11.5) in G_s, and coefficients $d_s \equiv d_s(\mu)$ satisfy the recursion relation (11.14) and μ satisfies the transcendental equation (11.15). Conversely, every (real) solution μ of the equation (11.15) satisfying the condition (11.13) is an eigen-value of the operator A_0 in G_s, and the function $\psi(|x|_p)$ constructed by the formulas (11.5) and (11.14) is a corresponding eigen-functions which satisfies $\psi(p^s) \neq 0$.*

12. *Justification of the Method of Sec.* 10.11

Now let us investigate the convergence of the series (11.5), (11.15) and (11.16). To this end we estimate coefficients d_r for $r \to +\infty$. From the formulas (11.9) and (11.13) the estimates follow:

$$1 - p^{-\alpha-1} \leq C_{r,\gamma} < 1, \quad s \leq \gamma \leq r-1, \ \lambda^1_{1-r} \sim V(p^r), \ r \to +\infty. \tag{12.1}$$

From here and (11.14) (for fixed μ) the recursion estimate follows

$$|d_r| \le \frac{1}{|\mu - \lambda^1_{1-r}|}\left(|\mu - \lambda^1_{1-s}| + \rho \sum_{s \le \gamma \le r-1} p^{-\alpha\gamma}|d_\gamma|\right), \quad d_s = 1, \; r \ge s+1. \tag{12.2}$$

Choose now an integer $s_0 \ge s$ such that

$$\rho \sum_{s_0 \le \gamma} p^{-\alpha\gamma}|\mu - \lambda^1_{1-\gamma}|^{-1} = q < 1. \tag{12.3}$$

(The series (12.3) converges as $\lambda^1_{1-\gamma} \sim V(p^\gamma) \to +\infty$, $\gamma \to +\infty$.) Rewrite the recursion inequality (12.2) in the form

$$|d_r| \le \frac{1}{|\mu - \lambda^1_{1-r}|}\left(C + \rho \sum_{s_0 \le \gamma \le r-1} p^{-\alpha\gamma}|d_\gamma|\right), \quad r \ge s_0 + 1 \tag{12.4}$$

where

$$C = |\mu - \lambda^1_{1-s}| + \rho \sum_{s \le \gamma \le s_0 - 1} p^{-\alpha\gamma}|d_\gamma|.$$

If we introduce the new sequence

$$T_r = |d_r(\mu - \lambda^1_{1-r})|, \qquad r = s_0, \quad s_0 + 1, \ldots \tag{12.5}$$

the recursion inequality (12.4) takes the form

$$T_r \le C + \rho \sum_{s_0 \le \gamma \le r-1} p^{-\alpha\gamma}|\mu - \lambda^1_{1-\gamma}|^{-1}|T_\gamma|,$$
$$T_{s_0} = |d_{s_0}(\mu - \lambda^1_{1-s_0})|, \qquad r \ge s_0 + 1. \tag{12.6}$$

From here we derive the estimate

$$T_r \le \frac{M_1}{1-q}, \quad r \ge s_0, \quad M_1 = \max(C, T_{s_0}). \tag{12.7}$$

■ For inequality (12.7) we prove by induction method with respect to r. This inequality is true for $r = s_0$. Supposing it is true for $s_0 + 1, \ldots, r$ we shall prove it for $r+1$. Then from (12.6) and taking (12.3) into account we have

$$T_{r+1} \le C + \rho \sum_{s_0 \le \gamma \le r} p^{-\alpha\gamma} |\mu - \lambda^1_{1-\gamma}|^{-1} \frac{M_1}{1-q} \le M_1 \left(1 + \frac{q}{1-q}\right) = \frac{M_1}{1-q}.$$

■

From the estimate (12.7) and owing to (12.5) and (12.1) the estimate follows (for some M)

$$|d_r| \le M V^{-1}(p^r), \qquad r \ge s. \tag{12.8}$$

From this estimate it follows that the series (11.15) converges.

Now we suppose that the potential $V(|x|_p)$ satisfies the condition

$$\int_{\mathbb{Q}_p} V^{-2}(|x|_p)dx = \left(1 - \frac{1}{p}\right) \sum_{s \le r < \infty} p^r V^{-2}(p^r) < \infty. \tag{12.9}$$

Under this condition and owing to (12.8) the series (11.5) converges in $L_0^2(G_s)$ as $\sum_{s \le r < \infty} p^r |d_r|^2 < \infty$ so that $\psi \in L_0^2(G_s)$.

Prove now the estimate

$$|d'_r| \le M_2 p^{-(\alpha+1/2)r}, \quad r \ge s \tag{12.10}$$

where the numbers d'_r are defined in (11.17).

■ By using the estimates (12.8) and (12.9) for $r \ge s$ we have

$$\begin{aligned}
|d'_r| &\le p^{(1-r)\alpha}|d_r| + \rho \sum_{r \le \gamma < \infty} p^{-\alpha\gamma}|d_\gamma| + \rho p^{-(\alpha+1)r} \sum_{s \le \gamma \le r-1} p^\gamma |d_\gamma| \\
&\le p^{(1-r)\alpha} M |V^{-1}(p^r)| + \rho M \sum_{r \le \gamma < \infty} p^{-\alpha\gamma} |V^{-1}(p^\gamma)| \\
&\quad + \rho p^{-(\alpha+1)r} M \sum_{s \le \gamma \le r-1} p^\gamma |V^{-1}(p^\gamma)| \\
&\le M_1 \left[p^{-\alpha r}|V^{-1}(p_r)| + \sqrt{\sum_{r \le \gamma < \infty} p^{-(2\alpha+1)\gamma}} \sqrt{\sum_{r \le \gamma < \infty} p^\gamma V^{-2}(p^\gamma)} \right. \\
&\quad \left. + p^{-(\alpha+1)r} \sqrt{\sum_{s \le \gamma \le r-1} p^\gamma} \sqrt{\sum_{s \le \gamma \le r-1} p^\gamma V^{-2}(p^\gamma)} \right] \le M_2 p^{-(\alpha+1/2)r}
\end{aligned}$$

for some M_1 and M_2. Here we have used the inequality

$$|V^{-1}(p^r)| \leq M_3 p^{-r/2}, \qquad r \geq s$$

which follows from (12.9). ■

From the estimate (12.10) it follows that the series $\sum_{s<r<\infty} p^r |d'_r|^2$ converges and hence the series (11.16) converges in $L_0^2(G_s)$ i.e. $D^\alpha \psi \in L_0^2(G_s)$. From here owing to the equation (11.1) it follows that $V\psi \in L_0^2(G_s)$ so that

$$A_0 \psi = D^\alpha \psi + V\psi \in L_0^2(G_s).$$

Thus the function $\psi(|x|_p)$ belongs to the domain of definition $\mathcal{D}(A_0)$ (see Secs. 10.6 and 10.7) and hence it is an eigen-function of the operator A_0. We note also that from the condition $V\psi \in L_0^2(G_s)$ it follows the convergence of the series

$$\sum_{s \leq r < \infty} p^r V^2(p^r)|d_r|^2 < \infty. \tag{12.11}$$

We sum up our results in the form of the following

Theorem. *Let the potential $V(|x|_p)$ satisfy the condition* (12.9), *the eigenvalue μ of the operator A_0 in G_s satisfy the conditions* (11.13) *and its corresponding eigenfunction $\psi(|x|_p)$ be such that $\psi(p^s) \neq 0$. Then the coefficients d_r, calculated by the recursion relation* (11.14), *define this eigenfunction by the formula*

$$\psi(|x|_p) = \sum_{s \leq r < \infty} d_r \delta(|x|_p - p^r),$$

and the eigenvalue μ satisfies the transcendental equation (11.15); *in addition the series* (12.11) *converges.*

Remark. The open question is whether it is possible to obtain all eigen-functions of the operator A_0 in G_s by the stated method.

13. *Further Results on the Spectrum of the Operator* $D^\alpha + V(|x|_p)$

$\alpha > 0$ in $\mathbb{Q}_p (p \neq 2)$ has been obtained by A. N. Kochubey [123] under the hypothesis that the potential $V(|x|_p)$ is real and locally bounded. He used results of Sec. 9.5 by demonstrations. Let us list his results without proofs.

Let us denote by $\mathcal{H}_1$ and $\mathcal{H}_2$ the Hilbert spaces which span the eigen-functions of the I and II kind respectively (see (5.16) of Sec. 9.5) so that $L^2(\mathbb{Q}_p) = \mathcal{H}_1 \oplus \mathcal{H}_2$, and by A_1 and A_2 the restrictions of the operator A on $\mathcal{H}_1$ and $\mathcal{H}_2$ respectively. (It is proved that subspaces $\mathcal{H}_1$ and $\mathcal{H}_2$ reduce the operator A.)

1. *If the sequence $\{V(p^\gamma),\ \gamma = 1,2,\dots\}$ has no finite limit points then the spectrum of the operator A_2 is pure discrete and the operator A has a complete system of the eigen-functions.*

We remind reader that the spectrum of the operator A_1 is known, it is pure discrete and consists of eigen-values (see Sec. 10.6)

$$\lambda_N^l = p^{\alpha N} + V(p^{l-N}), \qquad N \in \mathbb{Z}, \quad l = 2,3,\dots$$

Let $N(\lambda)$ be the distribution function of eigen-values of the operator A_2 i.e. a number of eigen-values (with regard to their multiplicity) smaller than λ.

2. *Let $V(|x|_p) \sim C|x|_p^\beta$, $|x|_p \to \infty$ where $C > 0$ and $\beta > 0$, and there exists N such that $V(p^l) \neq V(p^m)$ if $l \neq m$, $l > N$, $m > N$. Then*

$$N(\lambda) = (p-1)\left(\frac{1}{\alpha} + \frac{1}{\beta}\right) \ln_p \lambda + \mathcal{O}(1), \qquad \lambda \to +\infty.$$

3. *If $V(|x|_p) \to 0$, $|x|_p \to \infty$ then*

$$\sigma_{\text{ess}}(A_2) = \{0\}.$$

4. *If $V(|x|_p) \to 0$, $|x|_p \to 0$, then $\sigma_{\text{ess}}(A_2)$ coincides with the set of finite limit points of the sequence $\{V(p^l),\ l \to +\infty\}$.*

5. *If*

$$\sum_{-\infty<\gamma\leq 0} |V(p^\gamma)| < \infty \tag{13.1}$$

then the spectrum of the operator A_2 (and thus the spectrum of the operator A) is pure singular.

Example. For the potential $V(|x|_p) = \sin(a|x|_p)$, $a \in \mathbb{R}$ the condition (13.1) is fulfilled, and $\sigma_{\text{ess}}(A_2) = [-1,1]$ for almost all a.

14. *Non-Stationary p-Adic Schrodinger Equation*

Non-stationary p-adic Schrodinger equation with the potential $V(|x|_p)$ with respect to a wave function $\psi(t,x)$ has the form

$$D_t\psi = \frac{1}{|4|_p} D_x^2\psi + V(x)\psi. \tag{14.1}$$

For $V(x) = 0$ we get the equation for a free particle

$$\left(D_t - \frac{1}{|4|_p} D_x^2\right)\psi = 0. \tag{14.2}$$

A general solution ψ of the equation (14.2) in the class of the generalized functions $\mathcal{D}'(\mathbb{Q}_p^2)$ is given by the formula

$$\psi(t,x) = \tilde{\Phi}(t,x), \tag{14.3}$$

where $\Phi(k_1,k_2)$ is an arbitrary generalized function from $\mathcal{D}'(\mathbb{Q}_p^2)$ with a support in the manifold $|k_1|_p = |k_2/2|_p^2$.

■ Passing on to the Fourier-transform in the equation (14.2) we get the equation

$$\left(|k_1|_p - \left|\frac{k_2}{2}\right|_p^2\right)\tilde{\psi}(k_1,k_2) = 0, \quad \tilde{\psi} \in \mathcal{D}'(\mathbb{Q}_p^2)$$

from where it follows that

$$\operatorname{supp}\tilde{\psi} \subset \left[(k_1,k_2) \in \mathbb{Q}_p^2 : |k_1|_p - \left|\frac{k_2}{2}\right|_p^2 = 0\right]. \qquad ■$$

In the case when

$$\Phi(k_1,k_2) = \rho(k_2)\delta\left(k_1 - \frac{k_2^2}{2}\right)$$

the formula (14.3) takes the form

$$\psi(t,x) = \int_{\mathbb{Q}_p} \rho(k)\chi_p\left(\frac{k^2}{4}t - kx\right) dk\,. \tag{14.4}$$

The formula (14.4) can be interpreted as the expansion of the solution $\psi(t,x)$ in plane waves

$$\chi_p\left(\frac{k^2}{4}t - kx\right). \tag{14.5}$$

(In order to be convinced that the plane waves (14.5) satisfy the equation (14.2) it is sufficient to use the formula (1.7) of Sec. 9.1 for differentiation of $\chi_p(ax)$.)

For plane waves the adelic formula (see (1.11) of Sec. 3.1)

$$\prod_{2\le p\le\infty} \chi_p\left(\frac{k^2}{4}t - kx\right) = 1, \quad k,t,x \in \mathbb{Q} \tag{14.6}$$

is valid. It connects the plane waves (14.5) with classical ones

$$\chi_\infty\left(\frac{k^2}{4}t - kx\right) = \exp\left[-2\pi i\left(\frac{k^2}{4}t - kx\right)\right]. \tag{14.7}$$

For $\rho \equiv 1$ the formula (14.4) gives the function propagator

$$\mathcal{K}_t^{(p)}(x) = F\left[\chi_p\left(\frac{k^2}{4}t\right)\right] = \lambda_p(t)\left|\frac{2}{t}\right|_p^{1/2}\chi_p\left(-\frac{x^2}{t}\right) \tag{14.8}$$

which is the kernel of the evolution operator for a free particle (see Sec. 11 below).

Owing to the formula (3.8) of Sec. 7.3 the function $\mathcal{K}_t^{(p)}(x)$ satisfies the boundary condition (see (3.9) of Sec. 7.3)

$$\mathcal{K}_t^{(p)}(x) \longrightarrow \delta(x), \quad t \to 0 \text{ in } \mathcal{D}'(\mathbb{Q}_p). \tag{14.9}$$

We consider now the classical Schrodinger equation for a free particle

$$\frac{\partial\psi}{\partial t} - \frac{i}{8\pi}\frac{\partial^2\psi}{\partial x^2} = 0. \tag{14.10}$$

The corresponding propagator is (see (3.2) of Sec. 5.3)

$$\mathcal{K}_t^{(\infty)}(x) = \int_{-\infty}^{\infty} \exp\left[-2\pi i\left(\frac{k^2}{4}t - kx\right)\right] dk$$

$$= \int_{\mathbb{Q}_\infty} \chi_\infty\left(\frac{k^2}{4}t - kx\right) dk = \lambda_\infty(t)\left|\frac{2}{t}\right|^{1/2}\chi_\infty\left(-\frac{x^2}{t}\right) \tag{14.11}$$

$$\mathcal{K}_t^{(\infty)}(x) \longrightarrow \delta(x), \quad t \to 0 \text{ in } \mathcal{D}'(\mathbb{R}). \tag{14.12}$$

Here it is denoted (see (3.3) of Sec. 5.3)

$$\lambda_\infty(t) = \exp\left(-i\frac{\pi}{4}\text{sgn}t\right)^{*}.$$

For the propagators the adelic formula

$$\prod_{p=2}^{p=\infty} \mathcal{K}_t^{(p)}(x) = 1, \quad t, x \in \mathbb{Q}, \ t \neq 0 \tag{14.13}$$

is valid. It follows from the adelic formulas (1.4) of Sec. 1.1, (1.11) of Sec. 3.1 and (4.2) of Sec. 5.4.

Some analogy between "phases" $\lambda_\infty(t)$ and $\lambda_p(t)$ takes place. According to the sgn of t the real time R can be represented as the union of three disjoint sets: two sectors $\mathbb{R}^+$ (the future sector) where $\lambda_\infty(t) = e^{-i\pi/4}$ i.e. $t > 0$,

$$\mathbb{R}^- \text{ (the past sector) where } \lambda_\infty(t) = e^{i\pi/4} \text{ i.e. } t < 0,$$

and the point $t = 0$ (present). Analogously, the p-adic time $\mathbb{Q}_p$ can be represented in the following way (the definition of $\lambda_p(t)$ see in Sec. 5).

For $p \equiv 1 \pmod 4$ it is the union (disjoint) of two sectors: $\mathbb{Q}_p^+$ where $\lambda_p(t) = 1$, $\mathbb{Q}_p^-$ where $\lambda_p(t) = -1$, and the point $t = 0$.

For $p \equiv 3 \pmod 4$ it is the union (disjoint) of three sectors: $\mathbb{Q}_p^+$ where $\lambda_p(t) = 1$, $\mathbb{Q}_p^{\pm i}$ where $\lambda_p(t) = \pm i$, and the point $t = 0$.

For $p \equiv 2$ it is the union (disjoint) of four sectors: $\mathbb{Q}_2^\pm$ where $\lambda_p(t) = e^{\pm i\pi/4}$, $\mathbb{Q}_p^{\pm i}$ where $\lambda_p(t) = ie^{\pm i\pi/4}$, and the point $t = 0$.

The Cauchy problem for the equation (14.10) in the domain $\mathbb{R}^+ \times \mathbb{R}$ with an initial (generalized) function $\psi_0(x)$ from $\mathcal{D}'(\mathbb{R})$ is posed in the following way: Find a solution $\psi(t, x)$ of the equation (14.10) in the domain $(t, x) \in \mathbb{R}^+ \times \mathbb{R}^1$ which satisfies the initial condition

$$\psi(t, x) \longrightarrow \psi_0(x), \qquad t \to +0 \text{ in } \mathcal{D}'(\mathbb{R}).$$

Similarly, the Cauchy problem for the equation (14.2) in a sector, say, $\mathbb{Q}_p^+ \times \mathbb{Q}_p$ with an initial (generalized) function $\psi_0(x)$ from $\mathcal{D}'(\mathbb{Q}_p)$ is posed in the following way: Find a solution $\psi(t, x)$ of the equation (14.10) in the sector $(t, x) \subset \mathbb{Q}_p^+ \times \mathbb{Q}_p$ which satisfies the initial condition

$$\psi(t, x) \longrightarrow \psi_0(x), \qquad t \to 0, \quad t \in \mathbb{Q}_p^+ \text{ in } \mathcal{D}'(\mathbb{Q}_p).$$

---

* sgn$t$ is the Maslov index for the Hamiltonian $p^2/2$ (see[149]).

The solution of the Cauchy problem for the equation (14.10) is given by the formula

$$\psi(t,x) = \mathcal{E}(t,\cdot)*\psi_0(\cdot) = \int_{-\infty}^{\infty} \mathcal{E}(t,x-x')\psi_0(x')dx',$$

where $\mathcal{E}(t,x)$ is the fundamental solution of the equation (14.10)

$$\mathcal{E}(t,x) = \theta(t)\mathcal{K}_t^{(\infty)}(x) = \theta(t)\sqrt{\frac{2}{t}}\exp\left(2\pi i\frac{x^2}{t} - \frac{\pi i}{4}\right)$$

(see [205]). Here $\theta(t)$ is the Heaviside function, $\theta(t) = 1$, $t \geq 0$, $\theta(t) = 0$, $t < 0$.

It is interesting to note that *there does no exist any fundamental solution of the equation* (14.2) *in the space* $\mathcal{D}'(\mathbb{Q}_p^2)$.

■ Let $p \neq 2$ for definiteness, and there exist a solution $\mathcal{E} \in \mathcal{D}'(\mathbb{Q}_p^2)$ of the equation

$$D_t\mathcal{E} - D_x^2\mathcal{E} = \delta(t,x) \text{ in } \mathbb{Q}_p^2. \tag{14.14}$$

Passing on to the Fourier-transform in the equation (14.14) and using the formula (1.7) of Sec. 9.1 for the generalized function $\tilde{\mathcal{E}}(\xi_1,\xi_2)$ we obtain the inconsistent equation

$$(|\xi_1|_p - |\xi_2|_p^2)\tilde{\mathcal{E}}(\xi_1,\xi_2) = 1, \quad (\xi_1,\xi_2) \in \mathbb{Q}_p^2$$

as the left-hand side of this equation vanishes in the open set

$$[(\xi_1,\xi_2) \in \mathbb{Q}_p^2 : \ |\xi_1|_p - |\xi_2|_p^2 = 0, \ (\xi_1,\xi_2) \neq 0] . \qquad ■$$

Chapter 3

p-ADIC QUANTUM THEORIES

XI. p-Adic Quantum Mechanics

This section discusses the quantum mechanics over p-adic number field. The simplest but most important models — free particle and harmonic oscillator are investigated in detail. It turns out that these simplest models have a remarkably rich structure.

Investigation of p-adic quantum mechanics is of great interest from the mathematical point of view as well as from the physical one. As possible physical applications we note a consideration of models with nonarchimedean geometry of space-time at very small distances, and also in a spectral theory of processes in complicated media. Furthermore it seems to us that an extension of the formalism of quantum theory to the field of p-adic numbers is of great interest even independent of possible new physical applications because it can lead to better understanding of the formalism of usual quantum theory. We hope also that the investigation of p-adic quantum mechanics and field theory will be useful in pure mathematical researches in number theory, representation theory and p-adic analysis. Let us recall here that the quantum mechanical Weyl representation (see below) has wide applications in number theory and representation theory. However from the point of view of field theory it corresponds only to the simplest model of the free noninteracting system. No doubt investigation of p-adic nonlinear interacting systems will provide new deep pure mathematical results.

In the above we have considered functions of p-adic argument with values in the p-adic number field $\mathbb{Q}_p$ and also in complex number field $\mathbb{C}$. Correspondingly, different versions of p-adic classical and quantum mechanics are possible. Below we propose various most natural formulations of p-adic classical and quantum mechanics. We begin from the investigation of the formulation, which is defined by a triple $(L_2(\mathbb{Q}_p), W(z), U(t))$ where $W(z)$ is the Weyl representation of commutation relations and $U(t)$ is a unitary representation of an additive subgroup of p-adic number field which defines dynamics. Quantum mechanics will be obtained by means of quantization of classical p-adic mechanics.

1. *Classical Mechanics over* $\mathbb{Q}_p$

Let us start with the consideration of the classical p-adic Hamiltonian equations

$$\dot{p} = -\frac{\partial H}{\partial q}, \quad \dot{q} = \frac{\partial H}{\partial p} \tag{1.1}$$

where all variables: coordinates $q = q(t)$, momentum $p = p(t)$, the Hamiltonian $H = H(p, q)$ and time t take values in $\mathbb{Q}_p$. We shall consider only analytic functions $q(t)$, $p(t)$ and $H(p, q)$. We understand the notion of derivative in the sense of Sec. 2.2.

We consider first the simplest case of a *free particle* with the Hamiltonian

$$H = \frac{1}{2m}p^2, \tag{1.2}$$

here $m \in \mathbb{Q}_p$, $m \neq 0$.

Hamiltonian's equations

$$\dot{p} = 0, \quad \dot{q} = \frac{1}{m}p; \qquad p(0) = p, \quad q(0) = q$$

have a unique analytical solution for $t \in \mathbb{Q}_p$

$$p(t) = p, \qquad q(t) = q + \frac{p}{m} \cdot t. \tag{1.3}$$

Let us also present a solution for the *harmonic oscillator* with the Hamiltonian

$$H = \frac{p^2}{2m} + \frac{m\omega^2}{2}q^2, \tag{1.4}$$

---

* We use the same symbol p for the notation of a prime number and for a momentum. We hope that it does not lead to misunderstanding.

where $m, \omega \in \mathbb{Q}_p$, $m \neq 0$.

The equations of motion

$$\dot{p} = -m\omega^2 q, \quad \dot{q} = \frac{p}{m}; \qquad p(0) = p, \quad q(0) = q$$

have an analytical solution which is analogous to the solution over the field of real numbers

$$\begin{pmatrix} q(t) \\ p(t) \end{pmatrix} = \begin{pmatrix} q\cos\omega t + \frac{1}{m\omega} p \sin\omega t \\ p\cos\omega t - qm\omega \sin\omega t \end{pmatrix} = T_t \begin{pmatrix} q \\ p \end{pmatrix}. \tag{1.5}$$

Here

$$T_t = \begin{pmatrix} \cos\omega t & \frac{1}{m\omega}\sin\omega t \\ -m\omega\sin\omega t & \cos\omega t \end{pmatrix}. \tag{1.6}$$

Properties of the functions $\sin\omega t$ and $\cos\omega t$ were considered in Sec. 2.4, these functions are defined by the series (4.3) and (4.4) of Sec. 2, which converge in the region G_p defined by the inequalities

$$|\omega t|_p \leq \frac{1}{p} \text{ for } p \neq 2 \text{ and } |\omega t|_2 \leq \frac{1}{4} \text{ for } p = 2.$$

Region G_p is an additive group: if $t, t' \in G_p$, then $t + t' \in G_p$. For such t and t' the matrices T_t satisfy the group relation

$$T_t T_{t'} = T_{t+t'} \tag{1.7}$$

On the phase space $V = \mathbb{Q}_p \times \mathbb{Q}_p$ we define a skew-symmetric form

$$B(z, z') = p'q - pq', \tag{1.8}$$

where $z = (q, p) \in V$, $z' = (q', p') \in V$. The pair (V, B) defines a symplectic space.

We then have

$$B(T_t z, T_t z') = B(z, z'), \qquad t \in G_p, \tag{1.9}$$

i.e. the dynamics of the oscillator defines a one-parametric group of symplectic automorphisms of the space (V, B). It is also true for the dynamics of a free particle.

2. *The Weyl Representation*

Here we construct the p-adic quantum mechanics in which states are described by complex-valued wave functions of p-adic arguments.

The standard quantum mechanics starts with a representation of the well-known Heisenberg commutation relation

$$[\hat{q},\hat{p}] = i$$

in the space $L_2(\mathbb{R})$. In the Schrodinger representation the operators $\hat{q}$ and $\hat{p}$ are realized by multiplication and differentiation respectively. However in the p-adic quantum mechanics we have $x \in \mathbb{Q}_p$ and $\psi(x) \in \mathbb{C}$ and therefore the operator $\psi(x) \to x\psi(x)$ of multiplication by x has no meaning. Fortunately in this situation there is a possibility to use the Weyl representation. Recall that in the Weyl representation in the space $L_2(\mathbb{R})$ a pair of unitary operators is considered

$$e^{i\hat{p}q} : \psi(x) \to \psi(x+q); \quad e^{i\hat{q}p} : \psi(x) \to e^{ixp}\psi(x).$$

In this form it is possible to construct the following generalization to the p-adic case. We consider in the space $L_2(\mathbb{Q}_p)$ the unitary operators

$$U_q : \psi(x) \to \psi(x+q), \quad V_p : \psi(x) \to \chi(2px)\psi(x),$$

where $q, p, x \in \mathbb{Q}_p$ and χ is the additive character on $\mathbb{Q}_p$ (see Sec. 3).

A family of unitary operators

$$W(z) = \chi(-qp)U_qV_p, \quad z = (q,p) \in \mathbb{Q}_p^2 \tag{2.1}$$

satisfies the Weyl relation

$$W(z)W(z') = \chi(B(z,z'))W(z+z')\,. \tag{2.2}$$

The operator $W(z)$ acts in the following way

$$W(z)\psi(x) = \chi(2px+pq)\psi(x+q)\,. \tag{2.3}$$

The expression (2.3) is conveniently written in the form

$$W(z)\psi(x) = \int W(z;x,y)\psi(y)dy. \tag{2.4}$$

The family of operators $W(z)$ defines a representation of the Heisenberg-Weyl group consisting of elements (z,α); $z \in V$, $\alpha \in \mathbb{Q}_p$ with the composition law

$$(z,\alpha)\cdot(z',\alpha') = (z+z',\alpha+\alpha'+B(z,z')). \qquad (2.5)$$

A representation of the Heisenberg-Weyl group is defined by the formula

$$(z,\alpha) \longrightarrow \chi(\alpha)W(z)\,. \qquad (2.6)$$

Note that a pair $(L_2(\mathbb{Q}_p), W(z))$, where operators $W(z)$ are defined by the formula (2.4) is a special case of the Weyl system (see Sec. 12.7).

In the standard quantum mechanics the utilization of the Weyl representation is technically convenient. As we saw from the above discussion in p-adic quantum mechanics the use of the Weyl relation is the most appropriate way for constructing canonical commutation relations.

We consider now a question on the description of dynamics in the p-adic quantum mechanics. In the standard quantum mechanics one starts with the quantum Hamiltonian and then one constructs an operator of evolution $U(t)$. From our discussion it is clear that in the p-adic quantum mechanics one needs to construct directly a unitary group $U(t)$. It is understood we shall use a classical p-adic Hamiltonian for heuristic arguments. As it is known the usual quantization procedure is the following. For each function $f(q,p)$ from some class, defined on the phase space, one associates a corresponding operator $\hat{f}$ on $L_2(\mathbb{R})$. This quantization map $f \to \hat{f}$ has to satisfy some natural conditions. In general, the quantization procedure is ambiguous and different quantizations exist.

If the function $f(p,q)$ is the Fourier transform of a function $\varphi(\alpha,\beta)$

$$f(p,q) = \int\limits_{\mathbb{R}^2} e^{i(\alpha p+\beta q)}\varphi(\alpha,\beta)d\alpha d\beta = \tilde{\varphi}(p,q), \qquad (2.7)$$

then the Weyl quantization is the construction of the operator

$$\hat{f} = \int\limits_{\mathbb{R}^2} e^{i(\alpha\hat{p}+\beta\hat{q})}\varphi(\alpha,\beta)d\alpha d\beta,$$

where $\hat{p}$ and $\hat{q}$ are the momentum and position operators. Such quantization theory is closely connected to the theory of pseudo-differential operators. This quantization procedure can be generalized to the p-adic case. Let

$f(q,p)$ be a complex-valued function on the p-adic phase space $V = \mathbb{Q}_p^2$ and from $\mathcal{D}(\mathbb{Q}_p^2)$. It can be represented as the Fourier transform

$$f(p,q) = \int_{\mathbb{Q}_p^2} \chi(\alpha p + \beta q)\varphi(\alpha,\beta)d\alpha d\beta = \tilde{\varphi}(p,q), \quad \varphi \in \mathcal{D}(\mathbb{Q}_p^2).$$

In analogy with (2.4) to any such function one corresponds an operator in $L_2(\mathbb{Q}_p)$

$$\hat{f} = \int_{\mathbb{Q}_p^2} W(\alpha,\beta)\varphi(\alpha,\beta)d\alpha d\beta,$$

where $W(\alpha,\beta) = W(z)$ is the unitary operator (2.1). This function $f(p,q)$ is called the symbol of the operator $\hat{f}$. Note an essential difference of such quantization of the p-adic theory from the standard real theory. In the p-adic theory we cannot quantize polynomial functions $f(p,q)$ since such functions take values in $\mathbb{Q}_p$ but not in $\mathbb{C}$.

In standard quantum mechanics usually one starts with the construction of the Hamiltonian operator and then one proves its selfadjointness. Then one constructs the operator of evolution. In the p-adic quantum mechanics we can proceed in the following way.

As it is known in standard quantum mechanics the symbol $U(t)$ can be given in terms of the Feynman functional integral. It is natural to suspect that in the p-adic quantum mechanics the corresponding kernel will be expressed as the functional integral

$$K_t(x,y) = \int \chi\left(\frac{1}{h}\int_0^t L(q,\dot{q})dt\right)\prod_t dq(t), \tag{2.8}$$

where integration is performed over classical p-adic trajectories with the boundary conditions $q(0) = y$, $q(t) = x$. Here $L(q,\dot{q})$ is a classical p-adic Lagrangian, $L(q,\dot{q}) \in \mathbb{Q}_p$ and $h \in \mathbb{Q}_p$. The integral $\int_0^t Ldt = S(t)$ in the formula (2.8) is understood as a function which is inverse to the operation of differentiation, i.e. $\frac{d}{dt}S(t) = L,\ S(0) = 0,\ S(t) \in \mathbb{Q}_p$.

Here we consider the simplest case of the free particle and harmonic oscillator. In these cases it will be shown that as in the standard quantum mechanics the kernel $K_t(x,y) \sim \chi(S_{\text{cl}}(t))$, where $S_{\text{cl}}(t)$ is the action calculated on the classical p-adic trajectory.

3. *Free Particle*

We construct the dynamics of the free particle which corresponds to the classical Hamiltonian (1.2) by means of the Fourier transformation. Let ψ be from $L_2(\mathbb{Q}_p)$ and $\tilde{\psi}(k)$ is its Fourier transformation.

As it is known (see Sec. 7.4) the Fourier transformation $F : \psi \to \tilde{\psi}$ is an unitary operator in $L_2(\mathbb{Q}_p)$. The evolution operator in momentum representation $\tilde{U}(t)$ is given by the formula

$$\tilde{U}(t)\tilde{\psi}(k) = \chi\left(\frac{k^2}{4m}t\right)\tilde{\psi}(k), \qquad t \in \mathbb{Q}_p, \tag{3.1}$$

and in x-representation by the formula

$$U(t) = F^{-1}\tilde{U}(t)F. \tag{3.2}$$

We get a family of unitary operators $\tilde{U}(t)$, $U(t)$, and the relation is fulfilled

$$\tilde{U}(t)\tilde{U}(t') = \tilde{U}(t+t'), \quad t, t' \in \mathbb{Q}_p, \tag{3.2}$$

$$U(t)U(t') = U(t+t'), \quad t, t' \in \mathbb{Q}_p. \tag{3.3}$$

Let us calculate the kernel K_t of the evolution operator in x-representation. Let us consider the family of regularized operators

$$\tilde{U}_N(t)\tilde{\psi}(k) = \chi_N\left(\frac{k^2}{4m}t\right)\tilde{\psi}(k), \tag{3.4}$$

where

$$\chi_N\left(\frac{k^2}{4m}t\right) = \begin{cases} \chi\left(\frac{k^2}{4m}t\right), & |k|_p \le p^N \\ 0, & |k|_p > p^N. \end{cases}$$

A sequence of operators $\tilde{U}_N(t)$ strongly converges to $\tilde{U}(t)$ when $N \to \infty$,

$$\lim_{N\to\infty} \|\tilde{U}_N(t)\tilde{\psi} - U(t)\tilde{\psi}\| = \lim_{N\to\infty} \int\limits_{|k|_p > p^N} |\tilde{\psi}(k)|^2 dk = 0,$$

$$\tilde{\psi} \in L_2(\mathbb{Q}_p), \qquad t \in \mathbb{Q}_p. \tag{3.5}$$

Correspondingly, a sequence of operators $U_N(t)$ strongly converges to $U(t)$.

Using the theorem on the Fourier transformation of the convolution (3.5) of Sec. 7 let us perform the inverse Fourier transformation in the relation (3.4):

$$U_N(t)\psi(x) = F^{-1}[\tilde{U}_N(t)F\psi(x)] = F^{-1}\left[\chi\left(\frac{k^2}{4m}t\right)F\psi\right]$$
$$= F\left[\chi_N\left(\frac{k^2}{4m}t\right)\right]*\psi = \int\limits_{\mathbb{Q}_p} K_t^{(N)}(x-y)\psi(y)dy, \tag{3.6}$$

where

$$K_t^{(N)} = \dot{F}\left[\chi_N\left(\frac{k^2}{4m}t\right)\right](\xi) = \int\limits_{|k|_p\leq p} \chi_N\left(\frac{k^2}{4m}t + k\xi\right)dk. \tag{3.7}$$

The integral (3.7) have been calculated in Sec. 7 formula (2.3). Note, that in particular the function $K_t^{(N)}(\xi)$ has a compact support. Going to the limit $N \to \infty$ in the formula (3.6) in the space $L_2(\mathbb{Q}_p)$ we have for $t \neq 0$

$$U(t)\psi(x) = \int\limits_{\mathbb{Q}_p} K_t(x-y)\psi(y)dy, \tag{3.8}$$

where integral in (3.3) is singular one converging in $L_2(\mathbb{Q}_p)$.

The kernel $K_t(x-y)$ has the form, see (3.1) of Sec. 7,

$$K_t(x-y) = \lambda_p\left(\frac{t}{m}\right)\left|\frac{m}{t}\right|_p^{1/2}\chi\left(-\frac{m}{t}(x-y)^2\right), \qquad t \neq 0 \tag{3.9}$$

For $t = 0$ we have

$$K_0(x-y) = \delta(x-y). \tag{3.10}$$

It is interesting to note, that the relation (4.1) of Sec. 5 for the function $\lambda_p(a)$ follows from the relations (3.3) and (3.9) without using explicit form of the function $\lambda_p(a)$ from the Legendre symbol. The relation for free evolution of the operator $W(z)$ follows from the formulas:

$$U(t)W(z)U(t)^{-1} = W(z_t), \qquad t \in \mathbb{Q}_p, \tag{3.11}$$

where

$$z_t = (q(t), p(t)) = \left(q + \frac{p}{m}t, p\right) \tag{3.12}$$

is the classical evolution of free particle.

As a summary, quantum mechanics of free particle over the field $\mathbb{Q}_p$ is defined by a triple $(L_2(\mathbb{Q}_p), W(z), U(t))$, where $W(z)$ is a unitary representation of the Heisenberg-Weyl group (2.3), $U(t)$ is the unitary operator of evolution (3.8), and the relations(2.2) and (3.11) are fulfilled.

4. *Harmonic Oscillator*

Quantum mechanics of the harmonic oscillator over the field $\mathbb{Q}_p$ is defined by a triple $(L_2(\mathbb{Q}_p), W(z), U(t))$, where $W(z)$ is a unitary representation of the Heisenberg-Weyl group (2.5), and $U(t)$ is such operator, that the relations

$$U(t+t') = U(t)U(t'), \quad t, t' \in G_p, \tag{4.1}$$

$$U(t)W(z)U(t)^{-1} = W(T_t z), \quad t \in G_p, z \in V, \tag{4.2}$$

are fulfilled, where the classical evolution $T_t z$ is defined by the formulas (1.5) and (1.6). Below we put $m = \omega = 1$.

Let us define the operator $U(t)$ on test functions $\psi \in \mathcal{D}(\mathbb{Q}_p)$ by the formula:

$$U(t)\psi(x) = \int_{\mathbb{Q}_p} K_t(x,y)\psi(y)dy, \quad t \in G_p, \tag{4.3}$$

where the kernel $K_t(x,y)$ has the form

$$K_t(x,y) = \lambda_p(t)\frac{1}{|t|_p^{1/2}}\chi\left(-\frac{x^2+y^2}{\tan t} + \frac{2xy}{\sin t}\right), \quad t \neq 0 \tag{4.4}$$

$$K_0(x,y) = \delta(x-y).$$

The expression (4.4) has no sense when $\sin t = 0$. Note here that $\sin t = 0$ vanishes only for $t = 0$. It follows from the equality (4.19) of Sec. 2, $|\sin t|_p = |t|_p$, $t \in G_p$.

Theorem. *The formula* (4.3) *defines a unitary continuous representation $U(t)$ of the group G_p in the Hilbert space $L_2(\mathbb{Q}_p)$. The operator $U(t)$ satisfies the relations* (4.1) *and* (4.2) *and maps $\mathcal{D}(\mathbb{Q}_p)$ into itself.*

■ It follows from the explicit form of the kernel (4.4), from which we have for $t \neq 0$

$$U(t)\psi(x) = \chi\left(-\frac{x^2}{\tan t}\right)\frac{\lambda_p(t)}{|t|_p^{1/2}}F\left[\psi(y)\chi\left(-\frac{y^2}{\tan t}\right)\right]\left(\frac{2x}{\sin t}\right).$$

It is easy to see from the last formula that the operator $U(t)$ for $t \neq 0$ is a composition of four unitary operators and maps $\mathcal{D}(\mathbb{Q}_p)$ into itself. For $t = 0$ the operator $U(0) = I$.

Let us prove the group property (4.1) on the functions from $\mathcal{D}(\mathbb{Q}_p)$ for $p \geq 3$. Let $\psi(z) = 0$ for $|z| \geq p^N$ and $U(t')\psi(y) = 0$ for $|y|_p \geq p^M (t' \neq 0)$. Then

$$\begin{aligned}
&U(t)U(t')\psi(x) \\
&= \int_{|y|_p \leq p^M} K_t(x,y) \int_{|z|_p \leq p^N} K_{t'}(y,z)\psi(z)dydz \\
&= \int_{|z|_p \leq p^N} \psi(z) \int_{|y|_p \leq p^M} K_t(x,y)K_{t'}(y,z)dydz \\
&= \frac{\lambda_p(t)\lambda_p(t')}{|tt'|_p^{1/2}}\chi\left(-\frac{x^2}{\tan t}\right) \int_{|z|_p \leq p^N} \psi(z) \\
&\quad \cdot \int_{|y|_p \leq p^M} \chi\left(-y^2\left(\frac{1}{\tan t} + \frac{1}{\tan t'}\right) + 2y\left(\frac{x}{\sin t} + \frac{z}{\sin t'}\right)\right) dydz.
\end{aligned} \tag{4.5}$$

In order to calculate the internal integral in (4.5) we use the formula (2.3) of Sec. 5. Let us introduce the notations

$$a = -\frac{1}{\tan t} - \frac{1}{\tan t'}, \qquad b = \frac{2x}{\sin t} + \frac{2z}{\sin t'}. \tag{4.6}$$

Then, assuming that $t + t' \neq 0$,

$$|a|_p = \left|\frac{\tan(t+t')}{(1 - \tan t \cdot \tan t')\tan t \cdot \tan t'}\right|_p = \left|\frac{t+t'}{tt'}\right|_p. \tag{4.7}$$

As the left hand part in (4.5) does not depend on M for sufficiantly large M (and fixed x, t, t') the number M can be chosen as large as is wished and hence we use the upper line in the formula (2.3) of Sec. 5. We have also

$$\left|\frac{b}{2a}p^M\right|_p = p^{-M}\left|\frac{x\sin t' + y\sin t}{t+t'}\right|_p.$$

Therefore

$$\Omega\left(\left|\frac{b}{2a}p^M\right|_p\right) = 1.$$

Thus we have

$$\int_{|y|_p\le p} dy\chi\left(-y^2\left(\frac{1}{\tan t}+\frac{1}{\tan t'}\right)+2y\left(\frac{x}{\sin t}+\frac{z}{\sin t'}\right)\right)$$

$$= \lambda_p\left(-\frac{1}{\tan t}-\frac{1}{\tan t'}\right)\left|\frac{tt'}{t+t'}\right|_p$$

$$\cdot\chi\left(\left(\frac{x}{\sin t}+\frac{z}{\sin t'}\right)^2\left(\frac{1}{\tan t}+\frac{1}{\tan t'}\right)^{-1}\right). \tag{4.8}$$

Taking into account the relations $\lambda_p(ac^2)=\lambda_p(a)$ and (4.1) of Sec. 5 we transform the expression for λ_p:

$$\lambda_p\left(-\frac{1}{\tan t}-\frac{1}{\tan t'}\right)=\lambda_p\left(-\frac{\tan(t+t')}{(1-\tan t\cdot\tan t')\tan t\tan t'}\right)$$

$$=\lambda_p\left(-\frac{t+t'}{tt'}\right)=\lambda_p(t)^{-1}\lambda_p(t')^{-1}\lambda_p(t+t'). \tag{4.9}$$

Let us take into account the relation

$$-\frac{x^2}{\tan t}-\frac{z^2}{\tan t'}+\left(\frac{x}{\sin t}+\frac{z}{\sin t'}\right)^2\left(\frac{1}{\tan t}+\frac{1}{\tan t'}\right)^{-1}$$

$$=-\frac{x^2+z^2}{\tan(t+t')}+\frac{2xz}{\sin(t+t')}. \tag{4.10}$$

Assembling together the relations (4.5), (4.8), (4.9) and (4.10) we get

$$U(t)U(t')\psi(x)$$

$$=\frac{\lambda_p(t+t')}{|t+t'|_p^{1/2}}\int_{\mathbb{Q}_p}\chi\left(-\frac{x^2+z^2}{\tan(t+t')}+\frac{2xz}{\sin(t+t')}\right)\psi(z)dz=U(t+t')\psi(x).$$

Thus the property (4.1) is proved for t, t', $t+t'\neq 0$. The case when one of these parameters vanishes can be proved analogously and simpler because of the relation $K_0(x,y)=\delta(x-y)$. For the case $p=2$ one considers analogously.

As it can be seen from the above discussion in p-adic quantum mechanics, in contrary to that of the usual one, a wave function can remain finite under evolution, for example a wave function from $\mathcal{D}(\mathbb{Q}_p)$. One can estimate a diffusion region. In order to check relation (4.2) one calculates the integral

$$\int K_t(x,y)W(z;u,v)K_{-t}(v,y)dvdt = W(T_tz;x,y), \tag{4.11}$$

where $W(z;u,v)$ is the kernel (2.4).

5. *Lagrangian Formalism*

In Sec. 11.1 the Hamiltonian formalism of p-adic classical mechanics has been discussed. In this section we consider the Lagrangian formalism. We begin our consideration from the definition of the integral of analytic function and from the discussion of its properties.

Let function $f : \mathbb{Q}_p \to \mathbb{Q}_p$ be analytic in the disk B (see Sec. 2) and points a and b belong to the disk B', which is strictly contained in B. The integral of f from a to b is, by the definition, the following p-adic number:

$$\int_a^b f(x)dx = f^{(-1)}(b) - f^{(-1)}(a), \tag{5.1}$$

where the antiderivative $f^{(-1)}$ of the function f was defined in Sec. 2.2. Note, that $f^{(-1)}$ is defined in a and b by virtue of the condition $a, b \in B' \subset B$, see Sec. 2.2. The formula (5.1) defines the p-adic valued functional on the set of analytic functions in the disk B. The properties of this functional are given by the following Lemma.

Lemma. *The integral* (5.1) *has the properties:*

1. $\displaystyle\int_a^b (\lambda f + \mu g)dx = \lambda \int_a^b f dx + \mu \int_a^b g dx, \quad \lambda, \mu \in \mathbb{Q}_p.$

2. $\displaystyle\int_a^c f dx + \int_c^b f dx = \int_a^b f dx.$

3. $\displaystyle\int_a^b f' g dx = fg\Big|_a^b - \int_a^b f g' dx.$

4. *If for any analytic function h in the disk B, which satisfies the conditions*

$$h(a) = h(b) = 0$$

the following condition is valid:

$$\int_a^b fh dx = 0,$$

then $f \equiv 0$.

■ Properties 1 and 2 are direct consequence of the definition (5.1). Property 3 follows directly from the definition (5.1) and from the Leibnitz formula for the derivative of product.

Let us prove property 4. Representing f and h as power series we have:

$$h = \sum_{0 \le n < \infty} h_n (x-a)^n,$$

$$\int_a^b fh dx = \sum_{0 \le n < \infty} (fh)_n \frac{(b-a)^{n+1}}{n+1}, \tag{5.2}$$

where

$$h_0 = h(a) = 0,$$

$$h(b) = \sum_{1 \le n < \infty} h_n (b-a)^n = 0,$$

$$(fh)_n = \sum_{0 \le k \le n} f_k h_{n-k}.$$

Changing summation order in (5.2) and denote $d = b - a \neq 0$, we have

$$\int_a^b fh dx = \sum_{0 \le k < \infty} f_k d^k \sum_{n=1}^{\infty} \frac{h_n d^{n+1}}{n+k+1}. \tag{5.3}$$

We shall define h_n, $n \geq 2$ arbitrary (taking into account the condition $|h_n d^n|_p \to 0$, $n \to \infty$ only), and the coefficient h_1 by the formula:

$$h_1 = - \sum_{2 \le n < \infty} h_n d^{n-1}.$$

Substituting the last formula into (5.3) we get:

$$F = \int_a^b fhdx = \sum_{0 \le k < \infty} \frac{f_k d^k}{k+2} \sum_{2 \le n < \infty} \frac{1-n}{n+k+1} h_n d^{n+1}. \tag{5.4}$$

Taking into account the last formula, we shall reformulate the statement 4 of the lemma in the following way. If for any p-adic number h_n, $n \geq 2$, which satisfy the condition $|h_n d^n|_p \to 0$, $n \to \infty$ the relation $F = 0$ is valid, then $f_k = 0$ for all $k = 0, 1, \ldots$. Let us suppose that lemma is not true. Then there exists such m, that $f_m = 0$. Let us define h_n, $n \geq 0$ by the formula:

$$h_n = \frac{1}{(1-n)d^{n+1}} = \begin{cases} 1, & n = p^M - m - 1, \\ 0, & n \neq p^M - m - 1, \end{cases}$$

where $M \in \mathbb{N}$, $p^M \geq m + 3$. Substituting the last expression in (5.4) and taking into account the notation

$$c_k = \frac{f_k d^k}{k+2}, \qquad k = 0, 1, \ldots,$$

we have

$$F = F(M) = \sum_{0 \le k < \infty} \frac{c_k}{k - m + p^M} = 0. \tag{5.5}$$

It is easy to see by analogy with the discussion of the radius of convergence of antiderivative (see Sec. 2.2), that

$$r(f) = r\left(\sum_{\substack{0 \le k < \infty \\ k \neq m}} \frac{f_k}{(k+2)(k-m)} x^k \right),$$

therefore, the series

$$\sum_{\substack{0 \le k < \infty \\ k \neq m}} \frac{c_k}{k-m} \tag{5.6}$$

converges. As $\lim\limits_{M \to \infty} p^M = 0$, then we have

$$\lim_{M \to \infty} \left| \sum_{\substack{0 \le k \le N \\ k \neq m}} \frac{c_k}{k - m + p^M} \right|_p = \left| \sum_{\substack{0 \le k \le N \\ k \neq m}} \frac{c_k}{k-m} \right|_p$$

for any $N \in \mathbb{N}$, $N > m$. By virtue of convergence of the series (5.6) there exists a constant C, which does not depend on M and N, such that

$$\lim_{M\to\infty}\left|\sum_{\substack{0\le k\le N\\ k\ne m}}\frac{c_k}{k-m+p^M}\right|_p < C.$$

Consequently, beginning from some M we have the inequality:

$$\left|\sum_{\substack{0\le k\le N\\ k\ne m}}\frac{c_k}{k-m+p^M}\right|_p < C' \tag{5.7}$$

for some $C' > C$. Choosing M such that $p^M > \frac{C'}{|c_m|_p}$ and taking into account non-Achimedeanness of the norm and inequality (5.7), we have

$$\left|\sum_{0\le k\le N}\frac{c_k}{k-m+p^M}\right|_p = \left|\sum_{\substack{0\le k\le N\\ k\ne m}}\frac{c_k}{k-m+p^M}+p^{-M}C_m\right|_p = p^M|c_m|_p.$$

If we go over to the limit $N \to \infty$ and take into account (5.5) we get from the last formula:

$$|F(M)|_p = p^M|c_m|_p = 0,$$

therefore $f_m = 0$. The contradiction obtained proves the lemma. ■

Lagrangian formalism for $\mathbb{Q}_p$ can be constructed by analogy to that of for $\mathbb{R}$. As in Sec. 11.1 q and t – coordinate and time variables – take values in $\mathbb{Q}_p$ and $q(t)$ is an analytic function in some disk B, then (see Sec. 2.2) its derivative $\dot{q}(t)$ is analytic in B too. We shall consider only the case, when Lagrangian $L(q,\dot{q})$ is an analytic function (p-adic-valued) on $\mathbb{Q}_p \times \mathbb{Q}_p$. In this case the value of $L(q,\dot{q})$ on the trajectory $q(t)$ is an analytic function $L(t) = L(q(t),\ \dot{q}(t))$ in the disk B.

We shall define an action S as a p-adic-valued functional on the set of trajectories by the formula:

$$S[q] = \int_a^b L(q(t),\dot{q}(t))dt, \tag{5.8}$$

where the integral is understood in the sense of (5.1).

Let us define the variational derivative of the action (5.8) by analogy to that of the real number case:

$$\frac{\delta S[q]}{\delta q} = \left.\frac{dS[q+\varepsilon h]}{d\varepsilon}\right|_{\varepsilon=0},$$

where $\varepsilon \in \mathbb{Q}_p$, $h(t)$ is an arbitrary analytic function in B, that satisfies the conditions $h(a) = h(b) = 0$. The action (5.8) is stationary at the trajectory $q(t)$ if the following condition is valid:

$$\left.\frac{\delta S[\gamma]}{\delta \gamma}\right|_{\gamma=q} = 0.$$

Theorem. *If the action* (5.8) *is stationary at the trajectory* $q(t)$, *then the Euler-Lagrange equation is satisfied on this trajectory:*

$$\frac{d}{dt}\left(\frac{\partial L}{\partial \dot{q}}\right) - \frac{\partial L}{\partial q} = 0. \tag{5.9}$$

■ By the condition of the Theorem, we have:

$$\frac{d}{d\varepsilon} S[q+\varepsilon h]\Big|_{\varepsilon=0} = \int_a^b \left\{\frac{\partial L}{\partial q} h + \frac{\partial L}{\partial \dot{q}} \dot{h}\right\} dt = 0. \tag{5.10}$$

Taking into account statement 3 of the lemma, we get:

$$\int_a^b \frac{\partial L}{\partial \dot{q}} \dot{h} dt = \frac{\partial L}{\partial \dot{q}} h\Big|_a^b - \int_a^b \frac{d}{dt}\left(\frac{\partial L}{\partial \dot{q}}\right) h dt = -\int_a^b \frac{d}{dt}\left(\frac{\partial L}{\partial \dot{q}}\right) h dt.$$

Substituting the last equation in (5.10), we have:

$$\int_a^b \left\{\frac{d}{dt}\left(\frac{\partial L}{\partial \dot{q}}\right) - \frac{\partial L}{\partial q}\right\} h dt = 0.$$

By virtue of the statement 4 of the lemma we get the proof of the theorem. ■

We shall call the solution of the equation (5.9) with boundary conditions $q(t_1) = q_1$, $q(t_2) = q_2$ the classical trajectory passing through the points q_1 and q_2 and denote by $q_{\rm cl}(t)$. The action $S_{\rm cl}[t_1, t_2]$ on this trajectory can be calculated by the formula:

$$S_{\rm cl}[t_1, t_2] = \int_{t_1}^{t_2} L(q_{\rm cl}(t), \dot{q}_{\rm cl}(t))dt.$$

6. *Feynman Path Integral*

One of the possible way to construct Feynman path integral has been suggested in Sec. 11 (formula 2.8). In this section a rigorous construction is given by means of finite approximation method and kernel of evolution operator is calculated for the case of p-adic harmonic oscillator.

Let us consider p-adic system with Lagrangian $L(q, \dot{q})$, which is an analytic function on $\mathbb{Q}_p \times \mathbb{Q}_p$ and let the trajectory $q(t)$ be an analytic function in the disk B_N. We shall choose an integer $n < N$ and construct the covering of B_N by disks $B_n(a_j)$, $j = 0, 1, \ldots, p^{N-n} - 1$ of radius p^n without common points (see Sec. 1.3, example 2). In every disk of this covering we shall choose one point t_j, $j = 0, 1, \ldots, p^{N-n} - 1$, and also we suppose that $t_0 = 0$ and $t_{p^{N-n}-1} = t$. For any pair (t_j, t_{j+1}), $j = 0, 1, \ldots, p^{N-n} - 2$ let us construct the classical trajectory $q_{\rm cl}$ (see Sec. 11.5), which satisfies the conditions

$$q_{\rm cl}(t_j) = q_j, \quad q_{\rm cl}(t_{j+1}) = q_{j+1}.$$

The value of the action on every such trajectory is

$$S_{\rm cl}[t_j, t_{j+1}] = \int_{t_j}^{t_{j+1}} L(q_{\rm cl}(t), \dot{q}_{\rm cl}(t))dt, \qquad (6.1)$$

$$j = 0, 1, \ldots, p^{N-n} - 2.$$

Symbol $\int_0^t L(q(\tau), \dot{q}(\tau))d\tau$ in the formula (2.8) we shall interpret as follows:

$$\int_0^t L(\tau)d\tau = \lim_{n \to -\infty} \sum_{0 \le j \le p^{N-n}-2} \int_{t_j}^{t_{j+1}} L(q_{\rm cl}(\tau), \dot{q}_{\rm cl}(\tau))d\tau$$

$$= \lim_{n \to -\infty} \sum_{0 \le j \le p^{N-n}-2} S_{\rm cl}[t_j, t_{j+1}]. \qquad (6.2)$$

The formula (6.2) can be considered as an approximation of an arbitrary trajectory by segments of classical trajectories. Taking into account the formula (6.2) finite approximation $K_t^{(n)}(x,y)$ of the kernel of evolution operator can be written as follows:

$$K_t^{(n)}(x,y) = C_n \int_{\mathbb{Q}_p} \cdots \int_{\mathbb{Q}_p} \chi_p \left(\sum_{0 \le j \le p^{N-n}-2} S_{\text{cl}}[t_j, t_{j+1}] \right) dq_1 \dots dq_{p^{N-n}-2},$$

where $q_0 = x$, $q_{p^{N-n}-1} = y$, n is the order of approximation, C_n is some normalization factor.

If there exists the limit of $K_t^{(n)}(x,y)$ when $n \to -\infty$, then it defines the kernel of evolution operator:

$$K_t(x,y) = \lim_{n \to -\infty} K_t^{(n)}(x,y).$$

In the case of harmonic oscillator Lagrangian has the form:

$$L(q,\dot{q}) = \frac{1}{2}(\dot{q}^2 - q^2),$$

and Euler-Lagrange equation is

$$\ddot{q} + q = 0. \tag{6.3}$$

The solution of the equation (6.3) with boundary conditions

$$q(t_i) = q_i, \quad q(t_{i+1}) = q_{i+1}$$

as in the case of real numbers is given by the following formula:

$$q_{\text{cl}}(t) = \frac{q_i \sin t_{i+1} - q_{i+1} \sin t_i}{\sin(t_{i+1} - t_i)} \cos t - \frac{q_i \cos t_{i+1} - q_{i+1} \cos t_i}{\sin(t_{i+1} - t_i)} \sin t,$$

where $t_i, t_{i+1}, t \in G_p$ (see Sec. 11.1).

By means of simple, but time consuming calculations, which are analogous to those in the real numbers case one can prove the following equality:

$$S_{\text{cl}}[t_i, t_{i+1}] = \frac{q_i^2 + q_{i+1}^2}{2\tan(t_{i+1} - t_i)} - \frac{q_i q_{i+1}}{\sin(t_{i+1} - t_i)}\,. \tag{6.4}$$

With the help of integrals which have been calculated in Sec. 5.3 we shall prove the following Lemma.

Lemma. *The following formula is valid:*

$$\left(\prod_{0\le i\le k-1}\frac{\lambda_p(-2a_i)}{|a_i|_p^{1/2}}\right)\cdot\int_{\mathbb{Q}_p}\cdots\int_{\mathbb{Q}_p}\chi_p\left(\sum_{0\le i\le k-1}\left[\frac{x_i^2+x_{i+1}^2}{2\tan a_i}-\frac{x_ix_{i+1}}{\sin a_i}\right]\right)dx_1\dots dx_{k-1}$$

$$=\frac{\lambda_p\left(-2\sum_{0\le i\le k-1}a_i\right)}{\left|\sum_{0\le i\le k-1}a_i\right|_p^{1/2}}\chi_p\left(\frac{x_0^2+x_k^2}{2\tan\left(\sum_{0\le i\le k-1}a_i\right)}-\frac{x_0x_k}{\sin\left(\sum_{0\le i\le k-1}a_i\right)}\right),\tag{6.5}$$

where $k\in\mathbb{Z}$, $k\ge 2$, $a_i\in\mathbb{Q}_p$, $|a_i|_p\le\frac{1}{p}$, $i=0,1,\dots,k-1$ *and the function* $\lambda_p(a)$ *has been defined in* Sec. 5.

■ The proof will be carried out by means of induction. When $k=2$ the formula (6.5) can be reduced to that of (3.1) of Sec. 5 by means of elementary transformations. Let us suppose that the formula (6.5) is valid for $k=n$ and prove its validity for $k=n+1$. Denoting the expression in the right-hand part of (6.5) by $F(a_0,\dots,a_{k-1})$ and the sum $\sum_{0\le i\le k-1}a_i$ by A_k we get:

$$F(a_0,a_1,\dots,a_k)$$
$$=\frac{\lambda_p(-2a_k)}{|a_k|_p^{1/2}}\int_{\mathbb{Q}_p}F(a_0,\dots,a_{k-1})\chi_p\left(\frac{x_k^2+x_{k+1}^2}{2\tan a_k}-\frac{x_kx_{k+1}}{\sin a_k}\right)dx_k.$$

By the assumption of the induction we have:

$$F(a_0,\dots,a_n)=\frac{\lambda_p(-2a_n)\lambda_p(-2A_{n-1})}{|a_nA_{n-1}|_p^{1/2}}\chi_p\left(\frac{x_0^2}{\sin A_{n-1}}+\frac{x_{n+1}^2}{2\tan a_n}\right)$$
$$\cdot\int_{\mathbb{Q}_p}\chi_p\left(\frac{x_n^2}{2}\left[\frac{1}{\tan A_{n-1}}+\frac{1}{\tan a_n}\right]-x_n\left[\frac{x_0}{\sin A_{n-1}}+\frac{x_{n+1}}{\sin a_n}\right]\right)dx_n.$$

Taking into account the formula (3.1) of Sec. 5 one can rewrite the last formula in the following form:

$$F(a_0,\ldots,a_n) = \frac{\lambda_p(-2a_n)\lambda_p(-2A_{n-1})\lambda_p\left(\frac{1}{2\tan A_{n-1}} + \frac{1}{2\tan a_n}\right)}{\left|a_n A_{n-1}\left(\frac{1}{\tan A_{n-1}} + \frac{1}{\tan a_n}\right)\right|_p^{1/2}}$$

$$\cdot\chi_p\left(\frac{x_0^2}{2\tan A_{n-1}} + \frac{x_{n+1}^2}{2\tan a_n} - \frac{\left(\frac{x_0}{\sin A_{n-1}} + \frac{x_{n+1}}{\sin a_n}\right)^2}{2\left(\frac{1}{\tan A_{n-1}} + \frac{1}{\tan a_n}\right)}\right). \tag{6.6}$$

Using the properties of the function $\lambda_p(a)$ (see Sec. 5.4) we have:

$$\lambda_p\left(\frac{1}{2\tan A_{n-1}} + \frac{1}{2\tan a_n}\right) = \lambda_p\left(\frac{\sin(a_n + A_{n-1})}{2\sin a_n \sin A_{n-1}}\right)$$
$$= \lambda_p\left(\frac{a_n + A_{n-1}}{2a_n A_{n-1}}\right) = \lambda_p(2a_n)\lambda_p(2A_{n-1})\lambda_p(-2A_n).$$

Analogously, taking into account properties of the functions $\sin x$ and $\cos x$ (see Sec. 2.4), the expression under the norm symbol can be reduced to the following:

$$\left|a_n A_{n-1}\left(\frac{1}{\tan A_{n-1}} + \frac{1}{\tan a_n}\right)\right|_p = |A_n|_p.$$

The further proof can be carried out by means of elementary transformations under the character symbol on the right-hand part of the formula (6.6). ■

For the case of harmonic oscillator let us choose the factor C_n in the following form:

$$C_n = \prod_{0\le j\le p^{N-n}-2} \frac{\lambda_p(-2(t_{j+1}-t_j))}{|t_{j+1}-t_j|_p^{1/2}}. \tag{6.7}$$

Then the approximation of order n of the kernel $K_t(x,y)$ has the form:

$$K_t^{(n)}(x,y) = \prod_{0\le i\le p^{N-n}-2} \frac{\lambda_p(-2(t_{i+1}-t_i))}{|t_{i+1}-t_i|_p^{1/2}} \int_{\mathbb{Q}_p}\cdots$$

$$\cdot\int_{\mathbb{Q}_p}\chi_p\left(\sum_{0\le i\le p^{N-n}-2}\left(\frac{q_i^2+q_{i+1}^2}{2\tan(t_{i+1}-t_i)} - \frac{q_i q_{i+1}}{\sin(t_{i+1}-t_i)}\right)\right) dq_1\ldots dq_{p^{N-n}-2}.$$

By virtue of the Lemma we have:

$$K_t^{(n)}(x,y) = \frac{\lambda_p(-2t)}{|t|_p^{1/2}} \chi_p\left(\frac{x^2+y^2}{2\tan t} - \frac{xy}{\sin t}\right) = K_t(x,y).$$

The formula obtained coincides up to nonessential factors with the kernel of evolution operator of harmonic oscillator, which have been constructed in Sec. 11.4 (see (4.4) of Sec. 11.4).

7. *Quantum Mechanics with p-Adic Valued Functions*

In the previous subsections we considered the formalism for p-adic quantum mechanics with complex valued functions. Here we discuss an approach to p-adic quantum mechanics with p-adic valued functions. Recall the second quantization formulation of the usual quantum mechanics in terms of creation and annihilation operators. Let ℓ_2 be the Hilbert space of sequences $f = (f_0, f_1, \ldots)$ of complex numbers with the inner product

$$(f,g) = \sum_{n=0}^{\infty} \frac{1}{n!} f_n \bar{g}_n. \tag{7.1}$$

The creation and annihilation operators a^* and a act by rules

$$a^* f_n = f_{n+1}, \quad a f_n = n f_{n-1}, \quad n = 0, 1, \ldots \tag{7.2}$$

and satisfy the canonical commutation relations

$$[a, a^*] = aa^* - a^*a = 1 \tag{7.3}$$

for some domain in ℓ_2. Hamiltonian of harmonic oscillator has the form

$$H_0 = \omega a^* a \tag{7.4}$$

where ω is a real number. One can consider a more general Hamiltonian

$$H = H_0 + V \tag{7.5}$$

where V is for example a polynomial expression with respect to operators a^* and a. The time evolution is governed by the Schrödinger equation

$$i\frac{\partial \psi}{\partial t} = H\psi \tag{7.6}$$

where $\psi = \psi(t)$ is a vector from ℓ_2. For a self-adjoint operator H one has

$$\psi(t) = e^{-itH}\psi(0) . \tag{7.7}$$

Now note that Eqs. (7.1)–(7.5) can be immediately extended to the case of p-adic quantum mechanics, if we consider a space of sequences $f = (f_0, f_1, \ldots)$ of p-adic numbers with the inner product

$$(f, g) = \sum_{n=0}^{\infty} f_n g_n \tag{7.8}$$

where the series converges in $\mathbb{Q}_p$. A problem here is that the theory of operators in such p-adic Hilbert space is not developed yet to compare it with the theory of operators in complex Hilbert space.

We describe now briefly a p-adic integral calculus and its application in quantum mechanics with p-adic valued functions which has been developed by A. Yu. Khrennikov. Let $\mathbb{Q}_p(\sqrt{\tau})$ be a quadratic extension of the field $\mathbb{Q}_p$. Take $\rho > 0$ and let A_ρ be the space of analytic functions on

$$U_p = \{x \in \mathbb{Q}_p : |x|_p \leq \rho\}$$

taking values in $\mathbb{Q}_p(\sqrt{\tau})$ and endowed with the topology defined by the norm

$$||f||_\rho = \max_n |f_n|\rho^n ,$$

if

$$f(x) = \sum_{n=0}^{\infty} f_n x^n .$$

The projective limit of the spaces A_ρ

$$A = \lim_{\rho \to \infty} \text{ proj } A_\rho$$

is a non-archimedean Frechet space. We denote the dual space A' and will use a notation

$$\int \varphi(x)\mu(dx) = \langle \varphi, \mu \rangle$$

for an action of $\mu \in A'$ on a function $\varphi \in A$. Any distribution $\mu \in A'$ has a Laplace transform $L(\mu)$ which belongs to an inductive limit of functions analytic in a neighborhood of 0. The Gaussian distribution on $\mathbb{Q}_p$ is a distribution $\nu \in A'$ having the Laplace transform

$$L(\nu)(x) = e^{1/4x^2}.$$

Let us define an inner product on the space A,

$$(f,g) = \int f(x)\overline{g(x)}\nu(dx).$$

The completion of A with respect to a corresponding norm is a p-adic Hilbert space $L_2(\mathbb{Q}_p, \nu(dx))$. A theory of pseudodifferential operators in this space has been developed.

XII. Spectral Theory in p-Adic Quantum Mechanics

Let us discuss the spectral problem for a p-adic harmonic oscillator. In standard quantum mechanics over real numbers field one studies spectral properties of the Hamiltonian operator. In p-adic quantum mechanics there is no Hamiltonian operator, therefore spectral properties should be expressed in terms of the group $U(t)$. At first let us consider a harmonic oscillator in standard quantum mechanics and let $U(t)$ be the corresponding operator of evolution, which defines unitary representation of the additive group of real numbers $\mathbb{R}$. The decomposition of the representation $U(t)$ into irreducible representation has the form

$$L_2(\mathbb{R}) = \bigoplus_{n=0}^{\infty} H_n, \tag{0.1}$$

where invariant subspaces are strained on the Hermite polynomials. The corresponding eigen-function equation has the form

$$U(t)\psi = e^{i\omega_n t}\psi, \quad \psi \in H_n. \tag{0.2}$$

Here ω_n are known eigen-values for the harmonic oscillator, which one interprets as the energy levels. Analogously, the study of spectral properties of p-adic harmonic oscillator is connected with the problem of decomposition into irreducible representations a unitary representation of a group G_p.

The solution of this problem is divided into the following steps:

describe the characters of the group G_p (see Sec. 3);
calculate the dimension of the invariant subspaces H_α;
find explicit formulas for the eigen-functions of the evolution operator $U(t)$.

We shall determine a decomposition, which is analogous to (0.1), (0,2),

$$L_2(\mathbb{Q}_p) = \bigoplus_{\alpha \in I_p} H_\alpha, \tag{0.3}$$

$$U(t)\psi = \chi(\alpha t)\psi, \qquad \psi \in H_\alpha. \tag{0.4}$$

It is well known that in standard quantum mechanics for the harmonic oscillator the invariant subspaces H_n are one-dimensional, that is there is no degeneration. However, the spectral properties of the p-adic harmonic oscillator are considerably more complex. In particular, for $p \equiv 1 \pmod 4$ both the invariant vector (vacuum) and the excited states exhibit an infinite degeneration. For $p \equiv 3 \pmod 4$ there is a unique vacuum vector, and the excited states are degenerate with multiplicity p+1. For $p = 2$ there are two vacuum vectors, and the excited states are degenerate with multiplicity 2 or 4.

In this section we investigate the eigen-functions by making a unitary transformation to a new representation. The case $p \equiv 1 \pmod 4$ admits the most complete analysis.

First we list the prerequisites from the harmonic analysis and the theory of operators.

1. *Harmonic Analysis*

Let G be a locally compact commutative group. Every irreducible unitary representation of G is one-dimensional, hence the description of the representations reduces to describing the characters of G. A character of G is a complex-valued continuous function $\chi : G \to \mathbb{C}$ with the properties $\chi(g + g') = \chi(g)\chi(g')$, $|\chi(g)| = 1$, where g and g' are arbitrary elements of G. Equipped with the operation of pointwise multiplication and the topology of uniform convergence on compact subsets, the set of characters becomes a locally compact commutative group, which we will denote by $\hat{G}$. The group $\hat{G}$ is called the *dual* of G, or the *Pontryagin dual*. The group G is compact if and only if $\hat{G}$ is discrete. We have the Pontryagin duality theorem:

$$\hat{\hat{G}} = G.$$

Let $U(g)$ be a continuous unitary representation of G in a Hilbert space H. Then we have the representation

$$U(g) = \int_{\hat{G}} \chi(g) dE(\chi), \tag{1.1}$$

where $dE(\chi)$ is a spectral measure on $\hat{G}$.

Let the group G be compact. On the group G there exists an invariant measure dg (Haar measure). In this case we will normalize the Haar measure by the condition

$$\int_{G} dg = 1. \tag{1.2}$$

The set of characters $\hat{G} = \{\chi_\alpha(g),\ \alpha \in I\}$, where the abstract index α enumerates the characters, forms a complete orthonormal system in $L_2(G)$,

$$\int_{G} \chi_\alpha(g)\overline{\chi_\beta(g)} dg = \delta_{\alpha\beta}, \tag{1.3}$$

where $\delta_{\alpha\beta}$ is the Kronecker symbol and $L_2(G)$ is the Hilbert space of complex-valued functions square-integrable with respect to Haar measure.

In this case, the Hilbert space H can be decomposed in an orthogonal direct sum

$$H = \bigoplus_{\alpha \in I} H_\alpha, \tag{1.4}$$

where H_α is the largest subspace on which the representation acts as a multiple of $\chi_\alpha(g)$. The Hermitian projection operator on H_α is given by

$$P_\alpha = \int_{G} \overline{\chi_\alpha(g)} U(g) dg, \quad \alpha \in I. \tag{1.5}$$

By (1.4) and (1.5), (1.1) takes the form

$$U(g) = \sum_{\alpha \in I} \chi_\alpha(g) P_\alpha. \tag{1.6}$$

We have

$$U(g) P_\alpha = \chi_\alpha(g) P_\alpha. \tag{1.7}$$

2. *Operator Theory*

Let A be a bounded operator on a Hilbert space H, and $\{\psi_n\}_1^\infty$ an orthonormal basis for H. The *trace* of A is defined by

$$\text{Tr } A = \sum_{1\leq n<\infty} (\psi_n, A\psi_n). \tag{2.1}$$

The trace is not defined for all operators, and, moreover, it may depend on the choice of orthonormal basis. We have the following facts [175,60,237]:

1) Let A be a positive bounded operator on H. Then the sum on the right in (2.1) converges (to a finite or infinite limit) and is independent of the choice of basis.
2) Let A be positive and bounded in H, and let T_n be a sequence of positive bounded operators converging to the identity operator in the strong topology. Then
$$\lim_{n\to\infty} \text{Tr } (T_n A T_n) = \text{ Tr } A.$$
3) Let K be compact, and let dx be a positive measure on K. Let A be an integral operator on $L_2(K)$ with kernel $A(x,y)$ continuous on the compact set $K \times K$. Then the trace of A is well-defined (independent of the choice of basis and finite), and
$$\text{Tr } A = \int_K A(x,x)dx.$$

3. *The Theorem about Dimensions of Invariant Subspaces*

Characters of the group B_γ have been studied in Sec. 3. Remember that the group G_p coincides with B_{-1} for $p \geq 3$ and with B_{-2} for $p = 2$. It follows from Sec. 3.1 that the characters of G_p have the form $\chi(\alpha t)$, where $\alpha \in I_p$. The set I_p for $p \geq 3$ consists of the elements α of the form:

$$\alpha = 0 \text{ or } \alpha = p^{-\gamma}(\alpha_0 + \alpha_1 p + \ldots + \alpha_{\gamma-2}p^{\gamma-2}),$$

where $\gamma = 2,3,4,\ldots,\ 0 \leq \alpha_j \leq p-1,\ \alpha_0 \neq 0,\ \ j = 0,1,\ldots,\gamma-2$. For $p = 2$ the set I_2 consists of the elements

$$\alpha = 0 \text{ or } \alpha = 2^{-\gamma}(1 + \alpha_1 2 + \alpha_2 2^2 + \ldots + \alpha_{\gamma-3}2^{\gamma-3}),$$

where $\gamma = 3, 4, \ldots, \quad 0 \le \alpha_j \le 1, \ j = 1, 2, \ldots, \gamma - 3$.

As noted in Sec. 11.1, the operator

$$P_\alpha = |G_p|^{-1} \int_{G_p} \chi(-\alpha t) U(t) dt,$$

where

$$|G_p| = \int_{G_p} dt = \begin{cases} 1/p & \text{for } p \ge 3, \\ 1/4 & \text{for } p = 2, \end{cases} \tag{3.1}$$

is the projection on the subspace $H_\alpha = P_\alpha H$, $H = L_2(\mathbb{Q}_p)$, H_α is *invariant under* $U(t)$, which acts as a multiple of $\chi(\alpha t)$ on it. We have the following theorem concerning the dimension of H_α.

Theorem. *The invariant subspaces have the following dimensions:*

For $p \equiv 1 \pmod 4$, $\dim H_\alpha = \infty$ *for all* $\alpha \in I_p$.

For $p \equiv 3 \pmod 4$, *if* $\alpha = 0$, *then* $\dim H_\alpha = 1$; *if* $|\alpha|_p = p^\gamma$ *with an even* $\gamma \ge 2$, *then* $\dim H_\alpha = p + 1$.

For $p = 2$, *if* $\alpha = 0$ *or* $|\alpha|_2 = 2^3$, *then* $\dim H_\alpha = 2$; *if* $|\alpha|_2 \ge 2^4$ *and* $\alpha_1 = 1$, *then* $\dim H_\alpha = 4$.

For all other $\alpha \in I_p$, $\dim H_\alpha = 0$.

The following result will be used in the proof.

Proposition 1. *The dimension of the subspace* $H_\alpha, \alpha \in I_p$ *is given in terms of the trace of the projection operator* P_α *by*

$$\dim H_\alpha = \operatorname{Tr} P_\alpha. \tag{3.2}$$

It is easy to prove this proposition by choosing a suitable basis in the space H.

In calculating the trace of P_α we will use statements 2) and 3) of Sec. 12.2. Let us define the bounded operator ω_n in $L_2(\mathbb{Q}_p)$ by the formula

$$\omega_n \psi(x) = \Omega(p^{-n}|x|_p)\psi(x), \quad \psi \in L_2(\mathbb{Q}_p), \tag{3.3}$$

where

$$\Omega(a) = \begin{cases} 1, & 0 \le a \le 1, \\ 0, & a > 1. \end{cases} \tag{3.4}$$

Since $\Omega(p^{-n}|\chi|_p) \to 1$ uniformly on every compact set as $n \to \infty$, we have $\omega_n \to E$ in the strong topology (E is the identity operator). By statement 2) of Sec. 12.2 we have

$$\mathrm{Tr}\ P_\alpha = \lim_{n\to\infty} \mathrm{Tr}\ (\omega_n P_\alpha \omega_n). \tag{3.5}$$

By the definition of the operator P_α (3.1) we have

$$\mathrm{Tr}\ (\omega_n P_\alpha \omega_n) = |G_p|^{-1} \int_{|t|\leq 1/p} \chi(-\alpha t)\ \mathrm{Tr}\ (\omega_n U(t) \omega_n) dt. \tag{3.6}$$

We consider the integrand in (3.6). Using the expression (4.4) of Sec. 11 for the kernel $K_t(x,y)$ of the evolution operator $U(t)$ and statement 3) of Sec. 12.2 we obtain for $t \in G_p$, $t \neq 0$,

$$\begin{aligned}\mathrm{Tr}\ (\omega_n U(t)\omega_n) &= \int_{\mathbb{Q}_p} \Omega(p^{-n}|x|_p) K_t(x,x) \Omega(p^{-n}|x|_p) dx \\ &= \int_{|x|_p\leq p^n} K_t(x,x)dx = \int_{|x|_p\leq p^n} \frac{\lambda_p(t)}{|t|_p^{1/2}} \chi(2x^2 \tan \frac{t}{2}) dx.\end{aligned} \tag{3.7}$$

By the formula (2.3) of Sec. 5 we have

$$\int_{|y|_p\leq 1} \chi(ay^2)dy = \begin{cases} \lambda_p(a)|a|_p^{-1/2} & \text{for } |a|_p \geq 1, \\ 1 & \text{for } |a|_p \leq 1. \end{cases} \tag{3.8}$$

We rewrite the last integral in (3.7) in the form (3.8):

$$\begin{aligned}\int_{|x|_p\leq p^n} \chi\left(2x^2 \tan \frac{t}{2}\right) dx &= p^n \int_{|y|_p \leq 1} \chi\left(2p^{-2n} y^2 \tan \frac{t}{2}\right) dy \\ &= \begin{cases} \lambda_p(t)|t|_p^{-1/2} & \text{for } |t|_p \geq p^{-2n}, \\ p_n & \text{for } |t|_p \leq p^{-2n}. \end{cases}\end{aligned} \tag{3.9}$$

Here we have used the relations

$$\lambda_p\left(2p^{-2n}y^2 \tan \frac{t}{2}\right) = \lambda_p\left(2\frac{t}{2}\right) = \lambda_p(t), \quad |\sin t|_p = |t|_p, |\cos t|_p = 1.$$

Therefore, by (3.7) and (3.9) we have

$$\mathrm{Tr}\,(\omega_n U(t)\omega_n) = \frac{\lambda_p(t)}{|t|_p^{1/2}} \begin{cases} \frac{\lambda_p(t)}{|t|_p^{1/2}} & \text{for } |t|_p \geq p^{-2n}, \\ p^n & \text{for } |t|_p \leq p^{-2n}. \end{cases} \tag{3.10}$$

Let us next consider the case $p \geq 3$. Substituting (3.10) into (3.6), we obtain for $n = 1, 2, \ldots$

$$\mathrm{Tr}\,(\omega_n P_\alpha \omega_n) = p^{n+1} \int_{|t|_p \leq p^{-2n}} \frac{\lambda_p(t)}{|t|_p^{1/2}} \chi(-\alpha t)dt + p \int_{p^{-2n+1} \leq |t|_p \leq p^{-1}} \frac{\lambda_p^2(t)}{|t|_p} \chi(-\alpha t)dt. \tag{3.11}$$

We denote the first term in (3.11) by J_1 and rewrite it in the form

$$J_1 = p \int_{|\tau|_p \leq 1} \frac{\lambda_p(\tau)}{|\tau|_p^{1/2}} \chi(-\alpha p^{2n}\tau)d\tau. \tag{3.12}$$

Since we are interested in the limit as $n \to \infty$, for any fixed α we choose n such that $|\alpha p^{2n}|_p = p^{-2n}|\alpha|_p \leq 1$. Then

$$\chi(-\alpha p^{2n}\tau) = 1, \quad J_1 = p \int_{|\tau|_p \leq 1} \frac{\lambda_p(\tau)}{|\tau|_p^{1/2}} d\tau. \tag{3.13}$$

We evaluate J_1 as follows:

$$\begin{aligned} J_1 &= p \int_{|\tau|_p \leq 1} \frac{\lambda_p(\tau)}{|\tau|_p^{1/2}} d\tau = p \sum_{-\infty < \gamma \leq 0} \frac{1}{p^{\gamma/2}} \int_{|\tau|_p = p^\gamma} \lambda_p(\tau)d\tau \\ &= p \sum_{\substack{-\infty<\gamma\leq 0 \\ \gamma \text{ even}}} \int_{|\tau|_p = p^\gamma} d\tau + \sum_{-\infty<\gamma\leq -1} \frac{1}{p^{\gamma/2}} \epsilon_p \sum_{1 \leq k \leq p-1} \left(\frac{k}{p}\right) \int_{\substack{|\tau|_p = p^\gamma \\ \tau_0 = k}} d\tau \\ &= p \sum_{-\infty<\gamma\leq 0} \frac{1}{p^{\gamma/2}} p^\gamma \left(1 - \frac{1}{p}\right), \end{aligned}$$

since

$$\sum_{1 \leq k \leq p-1} \left(\frac{k}{p}\right) = 0, \qquad \int_{\substack{|\tau|_p = p^\gamma \\ \tau_0 = k}} d\tau = p^{\gamma - 1}. \tag{3.14}$$

Here $\varepsilon_p = 1$ for $p \equiv 1 \pmod 4$ and $\varepsilon_p = i$ for $p \equiv 3 \pmod 4$. Hence

$$J_1 = p \sum_{0 \le n < \infty} \frac{1}{p^n}\left(1 - \frac{1}{p}\right) = p. \tag{3.15}$$

We denote the second term in (3.11) by

$$J_2 = p \int_{p^{-2n+1} \le |t|_p \le p^{-1}} \frac{\lambda_p^2(t)}{|t|_p} \chi(-\alpha t) dt. \tag{3.16}$$

Proposition 2. *Let $n \ge 1$. Then*

$$J_2 = \begin{cases} (p-1)(2n-1) & \text{for } \alpha = 0, p \equiv 1 \pmod 4, \\ (p-1)(2n-N) - 1 & \text{for } |\alpha|_p = p^N, 2 \le N \le 2n, p \equiv 1 \pmod 4, \\ 1 - p & \text{for } \alpha = 0, p \equiv 3 \pmod 4, \\ (-1)^N \frac{p+1}{2} - \frac{p-1}{2} & \text{for } |\alpha|_p = p^N, 2 \le N \le 2n, p \equiv 3 \pmod 4. \end{cases}$$

■ We note that the behavior of the function $\lambda_p^2(t)$ depends on p. For $p \equiv 1 \pmod 4$ we have $\lambda_p^2(t) = 1$ for all t for $p \equiv 3 \pmod 4$ we have $\lambda_p^2(t) = (-1)^\gamma$ if $|t|_p = p^\gamma$.

In the case of $\alpha = 0$ and $p \equiv 1 \pmod 4$ we have

$$J_2 = p \int_{p^{-2n+1} \le |t|_p \le p^{-1}} \frac{dt}{|t|_p} = p \sum_{-2n+1 \le \gamma \le -1} p^{-\gamma} \int_{|\tau|_p = p^\gamma} dt$$
$$= p \sum_{-2n+1 \le \gamma \le -1} \left(1 - \frac{1}{p}\right) = (p-1)(2n-1). \tag{3.17}$$

If $|\alpha|_p = p^N$, $2 \le N \le 2n$ and $p \equiv 1 \pmod 4$ we obtain, using (4.2) of Sec. 4,

$$J_2 = p \int_{p^{-2n+1} \le |t|_p \le p^{-1}} \chi(-\alpha t) \frac{dt}{|t|_p}$$
$$= p \sum_{-2n+1 \le \gamma \le -1} \begin{pmatrix} 1 - 1/p, & \gamma \le -N, \\ -1/p, & \gamma = -N+1, \\ 0, & \gamma \ge -N+2, \end{pmatrix}$$
$$= p \sum_{-2n+1 \le \gamma \le -N} \left(1 - \frac{1}{p}\right) - 1 = (p-1)(2n-N) - 1. \tag{3.18}$$

If $\alpha = 0$ and $p \equiv 3 \pmod 4$, we have

$$J_2 = p \sum_{-2n+1 \le \gamma \le -1} (-1)^\gamma \left(1 - \frac{1}{p}\right) = 1 - p. \tag{3.19}$$

Finally, if $|\alpha|_p = p^N$, $2 \le N \le 2n$, $p \equiv 3 \pmod 4$ we have

$$\begin{aligned} J_2 &= p \sum_{-2n+1 \le \gamma \le -1} (-1)^\gamma \begin{pmatrix} 1 - 1/p, \ \gamma \le -N, \\ -1/p, \ \gamma = -N+1, \\ 0, \ \gamma \ge -N+2, \end{pmatrix} \\ &= p \sum_{-2n+1 \le \gamma \le -N} (-1)^\gamma \left(1 - \frac{1}{p}\right) - (-1)^{N+1}. \end{aligned} \tag{3.20}$$

The last expression is equal to 1 for even N and $-p$ for odd N. Equations (3.17)–(3.20) prove Proposition 2. ■

Since

$$\mathrm{Tr}\,(\omega_n P_\alpha \omega_n) = J_1 + J_2,$$

the next result follows from (3.15) and

Proposition 3. *Let a natural number n and a p-adic number α be given such that $p^{2n} \ge |\alpha|_p$. Then*

$$\mathrm{Tr}\,(\omega_n P_\alpha \omega_n) = \begin{cases} 2n(p-1)+1 & \text{for } \alpha = 0, p \equiv 1 \pmod 4, \\ (p-1)(2n-N+1) & \text{for } |\alpha|_p = p^N, 2 \le N \le 2n, p \equiv 1 \pmod 4, \\ 1 & \text{for } \alpha = 0, p \equiv 3 \pmod 4, \\ 0 & \text{for } |\alpha|_p = p^N, N \text{ odd}, \ 2 \le N \le 2n, \\ & \quad p \equiv 3 \pmod 4, \\ p+1 & \text{for } |\alpha|_p = p^N, N \text{ even}, 2 \le N \le 2n, \\ & \quad p \equiv 3 \pmod 4). \end{cases}$$

Taking the limit as $n \to \infty$ in Proposition 3 and recalling (3.2) and (3.5), we obtain the proof of Theorem for $p \ne 2$.

We now consider the case $p = 2$. In place of (4.11) (for $p \geq 3$), we have for $p = 2$

$$\begin{aligned}
&\mathrm{Tr}\,(\omega_n P_\alpha \omega_n) \\
&= 4 \cdot 2^n \int\limits_{|t|_2 \leq 2^{-2n}} \frac{\lambda_2(t)}{|t|_2^{1/2}} \chi(-\alpha t)dt + 4 \int\limits_{2^{-2n+1} \leq |t|_2 \leq 4^{-1}} \frac{\lambda_2^2(t)}{|t|_2} \chi(-\alpha t)dt \\
&= J_1 + J_2.
\end{aligned} \tag{3.21}$$

As above, one proves that $J_1 = 2$. Next, for $ghb\alpha = 0$ we have

$$J_2 = 4 \int\limits_{2^{-2n+1} \leq |t|_2 \leq 2^{-2}} \frac{dt}{|t|_2} \lambda_2^2(t),$$

where, recalling (0.2) of Sec. 5 for $t = 2^\gamma(1 + 2t_1 + \dots)$ we have

$$\lambda_2^2(t) = \frac{1}{2}(1 + (-1)^{t_1} i)^2 = (-1)^{t_1} i.$$

Therefore,

$$\begin{aligned}
J_2 &= 4i \int\limits_{2^{-2n+1} \leq |t|_2 \leq 2^{-2}} \frac{dt}{|t|_2} (-1)^{t_1} \\
&= 4i \sum_{-2n+1 \leq \gamma \leq -2} 2^\gamma \left[\int\limits_{\substack{|t|=2^\gamma \\ t_1=0}} dt - \int\limits_{\substack{|t|_2=2^\gamma \\ t_1=1}} dt \right] = 0.
\end{aligned} \tag{3.22}$$

Assume now that $|\alpha|_2 = 2^N$ and $N < 2n$. Then

$$\begin{aligned}
J_2 &= 4i \sum_{-2n+1 \leq \gamma \leq -2} 2^\gamma \int\limits_{|t|_2=2^\gamma} (-1)^{t_1} \chi(-\alpha t)dt \\
&= 4i \sum_{-2n+1 \leq \gamma \leq -2} \frac{f(\gamma)}{2^\gamma},
\end{aligned} \tag{3.23}$$

where

$$f(\gamma) = \int\limits_{|t|_2=2^\gamma} (-1)^{t_1} \chi(-\alpha t)dt = f_0(\gamma) - f_1(\gamma),$$

$$f_j = \int\limits_{\substack{|t|_2=2^\gamma \\ t_1=j}} (-1)^{t_1} \chi(-\alpha t)dt, \qquad j = 0, 1.$$

For $|t|_2 \le 2^{-N}$ we have $|\alpha t|_2 \le 1$. Then $f_0(\gamma) = f_1(\gamma)$ and $f(\gamma) = 0$. For $|t|_2 = 2^{-N+1}$ we have $\{\alpha t\} = 1/2$. Then again $f_0(\gamma) = f_1(\gamma)$ and $f(\gamma) = 0$. For $|t|_2 = 2^{-N+2}$ we have $\{\alpha t\} = 1/4 + (\alpha_1 + t_1)/2$. Then

$$f_0(\gamma) = \exp\left[2\pi i\left(\frac{1}{4} + \frac{\alpha_1}{2} + \frac{t_1}{2}\right)\right] \int_{\substack{|t|_2 = 2^\gamma \\ t_1 = 0}} dt = i(-1)^{\alpha_1} 2^{\gamma-2}.$$

Similarly, $f_1(\gamma) = -i(-1)^{\alpha_1}2^{\gamma-2}$. Hence for $|t|_2 = 2^{-N+2}$, i.e. for $\gamma = -N+2$ we have $f(-N+2) = 2^{-N+1}i(-1)^{\alpha_1}$. Finally, for $|t|_2 \ge 2^{-N+3}$, we have $f_0(\gamma) = f_1(\gamma)$ and $f(\gamma) = 0$. We have thus proved the following result.

Proposition 4. *For $p = 2$*

$$J_2 = \begin{cases} 0, & \alpha = 0_2 \ or \ |\alpha| = 2^3, \\ -2(-1)^{\alpha_1}, & |\alpha|_2 = 2^N, N \ge 4. \end{cases}$$

This Proposition completes the proof of Theorem 3.

Remark. By Theorem, the $\chi(\alpha t)$ are eigen-values of the evolution operator $U(t)$ if and only if the number α is of the form:

$$\alpha = 0 \text{ or } \alpha = p^{-\gamma}(\alpha_0 + \alpha_1 p + \ldots + \alpha_{\gamma-2}p^{\gamma-2}),$$

$$0 \le \alpha_j \le p-1, \quad \alpha_0 \ne 0; \quad j = 0, 1, \ldots, \gamma-2,$$

where $\gamma = 2, 3, 4, 5, \ldots$ for $p \equiv 1 \pmod 4$, while $\gamma = 2, 4, 6, \ldots$ for $p \equiv 3 \pmod 4$; for $p = 2$ we have

$$\alpha = 0 \text{ or } \alpha = 2^{-\gamma}(1 + 2 + \alpha_2 2^2 + \ldots + \alpha_{\gamma-3}2^{\gamma-2}),$$

$$\gamma = 4, 5, \ldots; \quad \alpha_j = 0, 1; \quad j = 2, 3, \ldots, \gamma-3.$$

We denote this set of indices by J_p; the numbers in J_p are analogous to the "energy levels" in standard quantum mechanics.

4. *Study of the Eigenfunctions*

Up to now we have worked in the space $L_2(\mathbb{Q}_p)$, however, the rather complicated action (4.4) of Sec. 11 of the evolution operator $U(t)$ in this space makes the analysis of the spectrum difficult.

We first consider the case $p \equiv 1 \pmod 4$. The field $\mathbb{Q}_p$ then contains the square root of -1, i.e., there exists an element $\tau \in \mathbb{Q}_p$ such that $\tau^2 = -1$. We will make a unitary transformation to a new representation (called the $\mathfrak{I}$-representation), in which the evolution operator acts in a very simple way. This representation will be used to derive explicit expression for the eigen-functions of the evolution operator. For functions $f \in L_2(\mathbb{Q}_p)$ we introduce the integral operator $\mathfrak{I}$ with Gaussian-type kernel by the formula

$$\mathfrak{I}[f](x) = \int_{\mathbb{Q}_p} \chi\left(\tau x^2 - \frac{\tau}{2} z^2 + 2xz\right) f(z) dz. \tag{4.1}$$

Proposition 1. *The operator $\mathfrak{I}$ given by* (4.1) *is unitary in $L_2(\mathbb{Q}_p)$, takes $\mathcal{D}(\mathbb{Q}_p)$ onto itself, and the inversion formula is valid:*

$$f(z) = \int_{\mathbb{Q}_p} \chi\left(\frac{\tau}{2} z^2 - \tau x - 2xz\right) \mathfrak{I}[f](x) dx. \tag{4.2}$$

■ This result follows from the expression

$$\mathfrak{I}[f](x) = \chi(\tau x^2) F\left[f(z)\chi\left(-\frac{\tau}{2} z^2\right)\right](2x), \tag{4.3}$$

which is a composition of four unitary operators. ■

We say that (4.1) defines the transition to the $\mathfrak{I}$-representation. In the $\mathfrak{I}$-representation, the dynamics is described by the theorem.

Theorem 1. *Let $p \equiv 1 \pmod 4$. Then, for any function in $L_2(\mathbb{Q}_p)$ we have*

$$U(t)\mathfrak{I}[f](x) = \mathfrak{I}[f(e^{-\tau t} z)](x), \qquad |t|_p \leq 1/p. \tag{4.4}$$

■ First assume that $f \in \mathcal{D}(\mathbb{Q}_p)$, so that, by Proposition 1, $\mathfrak{I}[f] \in \mathcal{D}(\mathbb{Q}_p)$ and suppose that $f(z) = 0$ for $|z|_p > p^N$ and $\mathfrak{I}[f](y) = 0$ for $|y|_p \geq p^M$.

By (3.8) and (3.9) of Sec. 11 after interchanging the order of integration we have

$$U(t)\mathfrak{I}[f](x) = \frac{\lambda_p(t)}{|t|_p^{1/2}}\chi\left(\frac{x^2}{\tan t}\right)\int\limits_{|t|_p\le p^N} f(z)\chi\left(-\frac{\tau}{2}z^2\right)$$
$$\cdot\int\limits_{|z|_p\le p^N}\chi\left(\left(\tau-\frac{1}{\tan t}\right)y^2+\left(\frac{2x}{\sin t}+2z\right)y\right)dydz. \tag{4.5}$$

The formula (2.1) of Sec. 5 can be used to evaluate the inner integral in (4.5). We set

$$a=\tau-\frac{1}{\tan t},\quad b=\frac{2x}{\sin t}+2z. \tag{4.6}$$

Then

$$|a|_p=\left|\frac{\tau\sin t-\cos t}{\sin t}\right|=\frac{1}{|t|_p}>1 \tag{4.7}$$

for $|t|_p\le 1/p$. Note that since the left-hand side in (4.5) is independent of M as $M\to\infty$, we can take M arbitrarily large; the top formula in (2.1)of Sec. 5 can then be used to evaluate the integral. We have, furthermore,

$$\left|\frac{b}{2a}p^M\right|_p=\left|\frac{x+z\sin t}{\tau\sin t-\cos t}p^M\right|_p=p^{-M}|x+z\sin t|_p. \tag{4.8}$$

Since the variable z lies in the disk $|z|\le p^N$, we have $p^{-M}|z\sin t|_p\le 1$ for sufficiently large M and $|t|_p\le 1/p$. Also, the variable x lies in a bounded disk, since it appears in the argument (4.8) of the function $\Omega\left(\left|\frac{b}{2a}p^M\right|_p\right)$, which is nonzero only for $p^{-M}|x+z\sin t|_p\le 1$. Hence, for these values of parameters,

$$\Omega\left(\left|\frac{b}{2a}p^M\right|_p\right)=1. \tag{4.9}$$

We now observe that $\cos t-\tau\sin t=c^2$ for some $c\in\mathbb{Q}_p$. Therefore, using (0.3) of Sec. 5 we have

$$\lambda_p(a)=\lambda_p\left(-\frac{\cos t-\tau\sin t}{\sin t}\right)=\lambda_p\left(-\frac{1}{\sin t}\right)$$
$$=\lambda_p\left(-\frac{1}{t}\right)=\lambda_p(-t). \tag{4.10}$$

Using (4.7)-(4.10), for these parameter values we thus obtain

$$\int\limits_{|y|_p \le p^M} dy\chi\left(\left(\tau - \frac{1}{\tan t}\right)y^2 + \left(\frac{2x}{\sin t} + 2z\right)y\right)$$

$$= \frac{\lambda_p(-t)}{|t|_p^{-1/2}}\chi\left(-\frac{\left(z + \frac{x}{\sin t}\right)^2 \sin t}{\tau \sin t - \cos t}\right). \tag{4.11}$$

Consequently, (4.5) takes the form

$$U(t)\mathfrak{I}[f](x) = \chi\left(x^2\left(-\frac{1}{\tan t} + \frac{1}{\sin t(\cos t - \tau \sin t)}\right)\right)$$

$$\cdot \int\limits_{\mathbb{Q}_p} f(z)\chi\left(-\frac{\tau}{2}z^2 + \frac{z^2 \sin t + 2zx}{\cos t - \tau \sin t}\right) dz. \tag{4.12}$$

We now exploit the relation

$$-\frac{1}{\tan t} + \frac{1}{\sin t(\cos t - \tau \sin t)} \cdot \frac{\tau\left(\cos t + \frac{1}{\tau}\sin t\right)}{\cos t - \tau \sin t} = \tau,$$

$$-\frac{\tau}{2} + \frac{\sin t}{\cos t - \tau \sin t} = \frac{-\tau \cos t + \sin t}{2(\cos t - \tau \sin t)} = -\frac{\tau}{2}e^{2\tau t},$$

$$e^{\tau t} = \cos t + \tau \sin t.$$

Then (4.5) takes the form

$$U(t)\mathfrak{I}[f](x) = \chi(\tau x^2)\int\limits_{\mathbb{Q}_p} dz(e^{-\tau t}z)\chi\left(-\frac{\tau}{2}z^2 + 2zx\right), \tag{4.13}$$

which coincides with (4.4) on functions $f \in \mathcal{D}(\mathbb{Q}_p)$. By continuity, (4.4) extends to the entire space $L_2(\mathbb{Q}_p)$. ∎

By (4.4), the dynamics in the $\mathfrak{I}$-representation is thus given by the simple formula

$$f(z) \longrightarrow f(e^{-\tau t}z), \quad f \in L_2(\mathbb{Q}_p). \tag{4.14}$$

The $\mathfrak{I}$-representation may be regarded as a distinctive p-adic analog of the second quantization representation known in standard quantum mechanics.

We construct explicit formulas for all eigen-functions of the evolution operator in $\mathfrak{J}$-representation.

Let us find, for example, an invariant vector ("vacuum"), i.e., the element $\psi \in L_2(\mathbb{Q}_p)$ satisfying

$$U(t)\psi = \psi, \quad |t|_p \leq 1/p. \tag{4.15}$$

To do this we find the invariant vectors in the $\mathfrak{J}$-representation, i.e., the functions $f \in L_2(\mathbb{Q}_p)$ satisfying

$$f(e^{-\tau t}z) = f(z), \quad |t|_p \leq 1/p. \tag{4.16}$$

We note that every nonzero p-adic number $z \in \mathbb{Q}_p^*$ can be uniquely represented in the canonical form $z = p^\gamma \varepsilon^k e^a$, where $|a|_p \leq 1/p$. The dynamics, given by (4.14), reduces to the substitution

$$z = p^\gamma \varepsilon^k e^a \longrightarrow e^{-\tau t}z = p^\gamma \varepsilon^k e^{a-\tau t}, \tag{4.17}$$

i.e., the numbers γ and k in the canonical expression do not change. This means that any function $f(z)$ in $L_2(\mathbb{Q}_p)$ depending only on γ and k is a general solution of (4.16), i.e. an invariant vector. Equivalently, if $z \in \mathbb{Q}_p$ has the canonical representation $z = p^\gamma(z_0 + z_1 p + \ldots)$ then *any function* $f(z) = f(|z|_p, z_0)$ *in* $L_2(\mathbb{Q}_p)$ *is an invariant vector, and, conversely, any such vector is of this form.*

Let us give the explicit form of the invariant vectors in the original representation. By (4.1), we have

$$\begin{aligned}\psi(x) &= \int_{\mathbb{Q}_p} \chi\left(\tau x^2 - \frac{\tau}{2}z^2 + 2xz\right) f(|z|_p, z_0)dz \\ &= \sum_{-\infty<\gamma<\infty} \; \sum_{1\leq k\leq p-1} f(p^\gamma, k)\psi_{\gamma,k}(x),\end{aligned} \tag{4.18}$$

where

$$\psi_{\gamma,k}(x) = \int_{\substack{|z|_p = p^\gamma \\ z = k}} dz \chi\left(\tau x^2 - \frac{\tau}{2}z^2 + 2xz\right).$$

It can be seen by computation that

$$\psi_{\gamma,k}(x) = \chi(\tau x^2 + 2p^{-\gamma}kx)p^{\gamma-1}\Omega(p^{\gamma-1}|x|_p), \quad \gamma \leq 0.$$

In particular, if $f(|z|_p, z_0) = \Omega(|z|_p)$, $\delta(p^\gamma - |z|_p)$, $\gamma = 1, 2, \ldots$, from (4.18) we obtain the vacuum vectors:

$$\psi_0(x) = \Omega(|x|_p), \quad \psi_\gamma(x) = \chi(\tau x^2)\delta(p^\gamma - |x|_p), \quad \gamma = 1, 2, \ldots$$

The dimension of the vacuum subspace is thus seen to be infinite, in agreement with the theorem of Sec. 12.3.

For the excited states, i.e., for vectors ψ_α in H_α the equation

$$U(t)\psi_\alpha = \chi(\alpha t)\psi_\alpha, \qquad \alpha \in J_p, \quad \alpha \neq 0, \quad |t|_p \leq 1/p, \tag{4.19}$$

reduces to

$$f_\alpha(e^{-\tau t}z) = \chi(\alpha t) f_\alpha(z), \quad |t|_p \leq 1/p, \tag{4.20}$$

in the $\mathfrak{J}$-representation $\psi_\alpha = \mathfrak{J}[f_\alpha]$. Taking into account (4.17), we find that the general solution of (4.20) is

$$f_\alpha(z) = \varphi(|z|_p, z_0)\chi(-\alpha\tau a), \tag{4.21}$$

where φ is an arbitrary function in $L_2(\mathbb{Q}_p)$ depending on the parameter α. Substituting (4.21) into (4.1) we obtain an explicit formula for all the eigen-functions of problem (4.19)

$$\psi_\alpha(x) = \int_{\mathbb{Q}_p} \chi\left(\tau x^2 - \frac{\tau}{2}z^2 + 2xz - \alpha\tau a\right)\varphi(|z|_p, z_0)dz,$$

where $z = p^\gamma \varepsilon^k e^a$.

5. *Weyl Systems and Coherent States*

The Weyl representation of commutation relations plays an important role in the formalism of Secs. 11.2–11.4. Here we study the general properties of this representation. In particular, one of the questions is to describe such representations up to unitary equivalence.

Let us start with a discussion of geometry of symplectic space. Let $V = \mathbb{Q}_p^{2n}$, $n \geq 1$ and B is a nondegenerate symplectic form on V, then the pair (V, B) is called the symplectic space over $\mathbb{Q}_p$. The subspace of V is said to be nondegenerate if the restriction of the form B to this subspace is nondegenerate. If (V, B) is some symplectic space, then V can be represented as the direct orthogonal sum of its two-dimensional subspaces h_i, $i = 1, 2, \ldots, n$ (so-called hyperbolic planes):

$$V = \bigoplus_{i=1}^{N} h_i.$$

In every such plane h_i we shall choose a basis (e_i, f_i), $i = 1, 2, \ldots, n$ with the property:

$$B(e_i, f_i) = 1.$$

The basis $\{(e_i, f_i),\ i = 1, 2, \ldots, n\}$ of the space V satisfies the conditions

$$(e_i, e_j) = (f_i, f_j) = 0,$$

$$(e_i, f_j) = -(f_i, e_j) = \delta_{ij}, \qquad i, j = 1, 2, \ldots, n$$

and is called the symplectic basis of V. The matrix of B in this basis has the canonical form:

$$B = \begin{pmatrix} 0 & E \\ -E & 0 \end{pmatrix}.$$

Later on we shall suppose that in V some symplectic basis is chosen and the form B has the canonical form. The following inequality is valid:

$$|B(z, z')|_p \leq \|z\|\,\|z'\|, \tag{5.1}$$

$z, z' \in V$ and the norm $\|\cdot\|$ was defined in Sec. 1.7.

Let now (V, B) be a two-dimensional symplectic space over $\mathbb{Q}$. A Weyl system over (V, B) is a pair (H, W), where H is a Hilbert space and W is a map from V to the set of unitary operators on H which satisfies the relation (so-called Weyl relation):

$$W(x)W(y) = \chi_p(B(x, y))W(x + y), \tag{5.2}$$

where $x, y \in V$ and $\chi_p(\xi)$ is the additive character of $\mathbb{Q}_p$ that satisfies the condition

$$\chi_p(\xi) \equiv 1 \text{ if and only if } |\xi|_p \leq 1 \text{ (see} \approx 3.1)$$

and the map W is continuous in strong topology on the set of unitary operators on H. One can use operations of direct sum and tensor product to construct new Weyl systems from existing ones.

Example 1. Weyl system $(L_2(\mathbb{Q}_p), W)$ over the space $(V = \mathbb{Q}_p \times \mathbb{Q}_p, B)$, where operators $W(z)$, $z \in V$ are defined by the formula (2.3) of Sec. 11.

Example 2. Tensor product $\overset{n}{\otimes} (L_2(\mathbb{Q}_p), W)$ of n Weyl system from Example 1 is the Weyl system $(L_2(\mathbb{Q}_p^n), W^{(n)})$ over the space $(\mathbb{Q}_p^{2n}, B)$, where

$$W^{(n)}(z) = \bigotimes_{i=1}^{n} W(z_i), \quad z = (z_1, \ldots, z_n) \in \mathbb{Q}_p^{2n}.$$

Let us denote by V_0 the following compact subgroup of the additive group of the space V:

$$V_0 = \{x \in V : \ \|x\| \leq 1\}. \tag{5.3}$$

Example 3. Let (V, B) be an arbitrary finite-dimensional symplectic space over $\mathbb{Q}_p$. We shall define the Hilbert space L_2^χ as the following closed subspace of $L_2(V)$

$$L_2^\chi = \{\phi \in L_2(V) : \ \phi(x + x') = \chi_p(B(x, x'))\phi(x), \ x' \in V_0\}, \tag{5.4}$$

and the set of operators $\tilde{W}(z)$, $z \in V$ by the formula:

$$\tilde{W}(z)\phi(x) = \chi_p(B(z, x))\phi(x - z), \quad z \in V, \ \phi \in L_2^\chi. \tag{5.5}$$

then the pair $(L_2^\chi, \tilde{W})$ is Weyl system over (V, B).

■ Unitarity of operators $\tilde{W}(z)$, $z \in V$ is obvious. It is sufficient to check Weyl relation (5.2). One has

$$\begin{aligned}
\tilde{W}(z)\tilde{W}(z')\phi(x) &= \tilde{W}(z)[\chi_p(B(z', x))\phi(x - z')] \\
&= \chi_p(B(z, x))\chi_p(B(z', x - z))\phi(x - z' - z) \\
&= \chi_p(B(z, z'))\chi_p(B(z + z', x))\phi(x - (z + z')) \\
&= \chi_p(B(z, z'))\tilde{W}(z + z')\phi(x) \ .
\end{aligned}$$ ■

The investigation of Weyl systems over p-adic symplectic space is essentially based on the notion of vacuum vector which does not have apparently a natural analog in real number case. Let us prove the following important theorem.

Theorem 1. *For any Weyl system (H, W) over p-adic symplectic space (V, B) there exists a vector $\phi_0 \in H$ such that the following relation*

$$W(x)\phi_0 = \phi_0 \tag{5.6}$$

is valid for all $x \in V_0$. This vector ϕ_0 we shall call the vacuum vector of (H, W).

■ First we note that V_0 (as well as V) is a commutative additive group. Let $x = (x_1, \dots, x_n)$, $y = (y_1, \dots, y_n) \in V$. By using scalar product

$$(x, y) = \sum_{1 \le i \le 2n} x_i y_i$$

and notation

$$\bar{x} = (-x_{n+1}, \dots, -x_{2n}, x_1, \dots, x_n)$$

the value of the form B on the vectors $x, y \in V$ can be represented in the form:

$$B(x, y) = (\bar{x}, y).$$

Let now $x, y \in V_0$, then we have (see (5.1))

$$|B(x, y)|_p \le \|x\| \cdot \|y\| \le 1.$$

Taking into account the property of the character χ_p:

$$\chi_p(\xi) \equiv 1, \qquad |\xi|_p \le 1, \tag{5.7}$$

we get that the restriction (H, W_0) of the Weyl system (H, W) to V_0 ($W_0 = W|_{V_0}$) is a unitary representation of the group V_0 in H. Naturally, Weyl relation (5.2) under the restriction on (H, W_0) has the form

$$W_0(x)W_0(y) = W_0(x + y), \qquad x, y \in V_0.$$

Thus the restriction of Weyl system (H, W_0) on V_0 is a unitary representation of the compact commutative group V.

By the well-known theorem from representation theory any irreducible representation of compact commutative group is one-dimensional (that is be some character of this group). The group of character of V_0 (Pontryagin dual, denoted by $\hat{V}_0$) is isomorphic to quotient group V/V_0, $\hat{V}_0 \simeq V/V_0$ (see 3.1) and any character of V_0 has the form

$$\lambda_{\hat{\alpha}}(x) = \chi_p((\alpha, x)), \quad x \in V_0,$$

where α is an arbitrary element in co-set $\hat{\alpha} \in V/V_0$. By virtue of (5.7) $\lambda_{\hat{\alpha}}$ does not depend on the choice of element α in co-set $\hat{\alpha}$. By Peter-Weyl theorem the space of representation H can be represented as the direct orthogonal sum

$$H = \bigotimes_{\hat{\alpha} \in V/V_0} H_{\hat{\alpha}}, \tag{5.8}$$

where H_α is a maximal subspace in which the representation is divisible to $\lambda_{\hat{\alpha}}(x)$. By virtue of (5.8) we can find $\hat{\alpha} \in V/V_0$ such that H_α is nontrivial. Let us denote by $\|\cdot\|_H$ the norm in H and choose $\psi \in H_{\hat{\alpha}}$ such that $\|\psi\|_H = 1$, then the vector

$$\phi_0 = W\left(\frac{1}{2}\bar{\alpha}\right)\psi, \qquad \alpha \in \hat{\alpha} \in V/V_0. \tag{5.9}$$

can be chosen as required vacuum vector. In fact by means of Weyl relation (5.2) and the condition $\psi \in H_{\hat{\alpha}}$ for $x \in V_0$ we have

$$\begin{aligned} W(x)\phi_0 &= W(x)W\left(\frac{1}{2}\bar{\alpha}\right)\psi = \chi_p((x, \bar{\alpha}))W\left(\frac{1}{2}\bar{\alpha}\right)W(x)\psi \\ &= \chi_p(-(\alpha, x))W\left(\frac{1}{2}\bar{\alpha}\right)\chi_p((\alpha, x)\psi = W\left(\frac{1}{2}\bar{\alpha}\right)\psi = \phi_0. \quad \blacksquare \end{aligned}$$

The next important notion is the notion of the system of coherent states. Let $\phi_0 \in H$ be a vacuum vector of the Weyl system (H, W). Let us choose an element α from any co-set $\hat{\alpha} \in V/V_0$ and denote the family of such elements by J_0. Let us construct the following set of vectors $\Phi \subset H$:

$$\Phi = \{\phi_\alpha = W(\alpha)\phi_0, \quad \alpha \in J_0\}. \tag{5.10}$$

Let α_1 and α_2 belong to the same co-set $\hat{\alpha} \in V/V_0$. Then from (5.2), (5.6) and (5.7) we get

$$\begin{aligned} \phi_{\alpha_1} &= W(\alpha_1)\phi_0 = W(\alpha_2 + (\alpha_1 - \alpha_2))\phi_0 \\ &= \chi_p(-B(\alpha_2, \alpha_1 - \alpha_2))W(\alpha_2)W(\alpha_2 - \alpha_1)\phi_0 \\ &= \chi_p(B(\alpha_1, \alpha_2))W(\alpha_2)\phi_0 = \chi_p(B(\alpha_1, \alpha_2))\phi_{\alpha_2}. \end{aligned}$$

Hence $|\phi_\alpha|$ does not depend on the choice of the element α in co-set $\hat{\alpha} \in V/V_0$. The set of vectors (5.10) we call the system of *coherent states* of Weyl system (H, W). The main property of coherent states is given by the following theorem.

Theorem 2. *If Weyl system (H, W) has unique (up to multiplication to a constant) vacuum vector ϕ_0, then the system of coherent states of (H, W) forms an orthonormal basis in H.*

■ Let us prove that the subspace $H_{2\bar{\alpha}}$ is strained on the vector $\phi_\alpha = W(\alpha)\phi_0$. In fact, if $x \in V_0$ we have

$$W(x)\phi_\alpha = W(x)W(\alpha)\phi_0 = \chi_p(2B(x, \alpha))W(\alpha)\phi_0 = \chi_p(B(-2\bar{\alpha}, x))\phi_\alpha,$$

thus $\phi_\alpha \in H_{2\bar{\alpha}}$. Conversely, let $\phi \in H_{2\bar{\alpha}}$, $\|\phi\|_H = 1$. Then by virtue of the formula (5.9) and uniqueness of the vacuum vector we get:

$$\phi_0 = W(-\alpha)\phi,$$

therefore $\phi = W(\alpha)\phi_0$. Further proof follows immediately from decomposition (5.8).

Remark. We see that the defined system of coherent states forms orthonormal basis in contrast to real numbers case where coherent states form overfilled system.

For further consideration the following definitions are needed.

Weyl system (H, W) is called irreducible if there are no nontrivial subspaces of H, which are invariant under the action of operators $W(x)$, $x \in V$. We shall say that (H, W) can be represented as a direct sum of Weyl systems (H_i, W)

$$(H, W) = \bigoplus_i (H_i, W),$$

if the following condition is valid:

$$H = \bigoplus_i H_i$$

and subspaces H_i are invariant under the action of operators $W(x)$, $x \in V$.

The set of vacuum vectors of the Weyl system (H, W) forms the subspace H_0 which is called the *vacuum subspace* of (H, W). In this subspace H_0 we shall choose some orthonormal basis $\{\phi_0^i,\ i \in I\}$. The following theorem is valid.

Theorem 3. *Weyl system (H, W) is irreducible if and only if the vacuum subspace H_0 of this system is one-dimensional. Otherwise (H, W) can be represented as direct sum of the following type:*

$$(H, W) = \bigoplus_{i \in I} (H_i, W),$$

where subspace H_i is the span of the basis

$$\{\phi_\alpha^i = W(\alpha)\phi_0^i,\ i \in I\}.$$

■ Let H_0 be one-dimensional and $\phi_0 \in H_0$ is vacuum vector. Let us suppose that (H, W) is reducible. Then there exists a nontrivial subspace H' of the space H which is invariant under the action of the operators $W(x)$, $x \in V$. Let us consider the Weyl system (H', W). By virtue of Theorem 1 there exists a vacuum vector ϕ_0' of this Weyl system. By virtue of uniqueness of vacuum vector we can choose ϕ_0' such that

$$\phi_0 = \phi_0'.$$

According to the Theorem 2 the set of vectors

$$\{\phi_\alpha = W(\alpha)\phi_0, \quad \alpha \in J_0\}$$

forms orthonormal basis in H, but $\phi_\alpha \in H'$, hence $H' = H$. The contradiction obtained proves the irreducibility of the Weyl system (H, W).

Let now H_0 is not one-dimensional. We shall prove that subspaces H_i and H_j are orthogonal when $i \neq j$, $i, j \in I$. From definition of these subspaces it follows that it is sufficient to prove the relation:

$$(W(x)\phi_0^i, W(y)\phi_0^j) = 0, \qquad x, y \in V,\ i, j \in I,\ i \neq j. \tag{5.11}$$

Let $z \in V_0$. By virtue of unitarity of $W(z)$ and formulas (5.2) and (5.6) we have:

$$\begin{aligned}(W(x)\phi_0^i, W(y)\phi_0^j) &= (W(z)W(x)\phi_0^i, W(z)W(y)\phi_0^j) \\ &= (\chi_p(2B(z,x))W(x)\phi_0^i, \chi_p(2B(z,y))W(y)\phi_0^j) \\ &= \chi_p(2B(z, x-y))(W(x)\phi_0^i, W(y)\phi_0^j).\end{aligned} \tag{5.12}$$

For any $x, y \in V$, $x - y \notin V_0$ we can always find $z \in V_0$ such that $\chi_p(2B(z, x - y)) \neq 1$ and in this case we get (5.11) from (5.12). In the case of $x - y \in V_0$ and $i \neq j$ we have

$$(W(x)\phi_0^i, W(y)\phi_0^j) = (W(-y)W(x)\phi_0^i, \phi_0^j) = \chi_p(B(-y, x))(\phi_0^i, \phi_0^j) = 0.$$

The formula (5.11) is proved.

Let us consider now the space $\tilde{H} = \bigoplus_{i \in I} H_i$ and prove that $\tilde{H} = H$. Let us suppose that it is not true and consider the orthogonal complement $\tilde{H}^\perp$ of $\tilde{H}$ in H. Because of unitarity of operators $W(x)$, $x \in V$ the space $\tilde{H}^\perp$ is invariant under the action of all these operators. Let us consider the Weyl system $(\tilde{H}^\perp, W)$. According to Theorem 1 there exists a vacuum vector $\tilde{\phi}_0$ of this system which satisfies the relation

$$W(x)\tilde{\phi}_0 = \tilde{\phi}_0, \quad x \in V_0,$$

therefore $\tilde{\phi}_0 \in H_0$ which is impossible by virtue of the condition $H_0 \subset \tilde{H}$. The contradiction obtained proves the relation

$$H = \tilde{H} = \bigoplus_{i \in I} H_i.$$

Invariantness of subspaces H_i, $i \in I$ under the action of operators $W(x)$, $x \in V$ follows directly from the definition of these subspaces. ■

As an application of the Theorem 3 we shall prove the irreducibility of Weyl systems from Examples 1–3. For this, according to this Theorem it is sufficient to prove that vacuum subspaces of these systems are one-dimensional.

■ Let $\phi_0 \in L_2(\mathbb{Q}_p)$ be the vacuum vector of the Weyl system $(L_2(\mathbb{Q}_p), W)$ from Example 1. Then it satisfies the relation (see (2.3) of Sec. 11)

$$\chi_p(2px + pq)\phi_0(x + q) = \phi_0(x), \tag{5.13}$$

where $z = (q, p) \in V_0 = B_0 \times B_0$, $x \in \mathbb{Q}$. If we put $q = 0$ in (5.13) we get the formula

$$\chi_p(2px)\phi_0(x) = \phi_0(x),$$

from which it follows that

$$\operatorname{supp} \phi_0 \subset \{x \in \mathbb{Q}_p : \quad \chi_p(2px) = 1, \ p \in B_0\} = B_0.$$

If we put $p = 0$ in (5.13) we get

$$\phi_0(x) = \phi_0(x+q), \quad x, q \in B_0.$$

Therefore, $\phi_0(x) = C\Omega(|x|_p)$, where $\Omega(|x|_p)$ is the characteristic functions of B_0 (see Sec. 6.2), C is an arbitrary nonzero constant and thus the vacuum subspace H_0 of the Weyl system $(L_2(\mathbb{Q}_p), W)$ is one-dimensional. Irreducibility of the Weyl system $(L_2(\mathbb{Q}_p^n), W^{(n)})$ follows directly from the irreducibility of $(L_2(\mathbb{Q}_p), W)$ and the definition of tensor product of Weyl systems. Vacuum vector of this Weyl system has the form:

$$\phi_0^{(n)}(x) = \Omega(\|x\|) = \prod_{1 \leq i \leq n} \Omega(|x_i|_p, \quad x = (x_1, \ldots, x_n) \in \mathbb{Q}_p^n.$$

Irreducibility of the Weyl system $(L_2^\chi, \tilde{W})$ from Example 3 can be proved by analogy with the previous case. Vacuum vector has the form

$$\phi_0(x) = \Omega(\|x\|). \qquad \blacksquare$$

The following Corollary follows directly from the last Theorem.

Corollary 1. *Any Weyl system can be represented as the direct sum of irreducible Weyl systems.*

Two Weyl systems (H, W) and $(\tilde{H}, \tilde{W})$ over space $(V, B(\cdot,\cdot))$ are *equivalent* by the definition if there exists a unitary operator $U : H \to \tilde{H}$ that satisfies the condition

$$UW(x) = \tilde{W}(x)U, \qquad x \in V. \tag{5.14}$$

Let us prove one more Corollary from Theorem 3.

Corollary 2. *Any two irreducible Weyl systems over the space (V, B) are equivalent.*

■ Let (H, W), $(\tilde{H}, \tilde{W})$ be irreducible Weyl systems over the space (V, B). By virtue of Theorems 1–3 for any of these systems there exist unique vacuum vector $\phi_0 \in H$ and $\tilde{\phi}_0 \in \tilde{H}$ and spaces H and $\tilde{H}$ are strained on the basis of coherent states

$$\begin{aligned} \Phi &= \{\phi_\alpha = W(\alpha)\phi_0, \ \ \alpha \in J_0\}, \\ \tilde{\Phi} &= \{\tilde{\phi}_\alpha = \tilde{W}(\alpha)\tilde{\phi}_0, \ \ \alpha \in J_0\}. \end{aligned} \tag{5.15}$$

Let us construct a unitary operator $U : H \to \tilde{H}$ by the formula:

$$U\phi_\alpha = \tilde{\phi}_\alpha, \qquad \alpha \in J_0. \tag{5.16}$$

It is easy to see that the operator (5.13) satisfies the condition (5.14). In fact, it is sufficient to check (5.14) for basis vectors ϕ_α and $\tilde{\phi}_\alpha$, $\alpha \in J_0$. Using formulas (5.12), (5.15) and (5.16) we get:

$$\begin{aligned} UW(x)\phi_\alpha &= UW(x)W(\alpha)\phi_0 = \chi_p(B(x,\alpha))UW(x+\alpha)\phi_0 \\ &= \chi_p(B(x,\alpha))U\phi_{x+\alpha} = \chi_p(B(x,\alpha))\tilde{\phi}_{x+\alpha} \\ &= \chi_p(B(x,\alpha))\tilde{W}(x+\alpha)\tilde{\phi}_0 = \tilde{W}(x)\tilde{W}(\alpha)\tilde{\phi}_0 = \tilde{W}(x)U\phi_\alpha. \end{aligned}$$

■

6. *Symplectic Group*

The symplectic group $Sp(2n, \mathbb{Q}_p)$ is the group of linear automorphisms of the space $V = \mathbb{Q}_p^{2n}$ which preserve the symplectic form. Any element $g \in Sp(2n, \mathbb{Q}_p)$ can be represented as a matrix (g_{ij}), $1 \le i, j \le 2n$ in some basis. Norm of matrix g means the following

$$||g|| = \max_{1 \le i,j \le 2n} |g_{ij}|_p. \tag{6.1}$$

It is known that $\det g = 1$ for any $g \in Sp(2n, \mathbb{Q}_p)$. We shall be mainly interested in the subgroup of $Sp(2n, \mathbb{Q}_p)$ defined by the following Lemma.

Lemma. *The set of matrices $G = \{g \in Sp(2n, \mathbb{Q}_p) : ||g|| = 1\}$ forms a subgroup of $Sp(2n, \mathbb{Q}_p)$.*

■ Let us prove that for any element $g \in Sp(2n, \mathbb{Q}_p)$ the following inequality is valid

$$||g|| \ge 1. \tag{6.2}$$

In fact, taking into account the definition of det g and properties of the norm (6.1) we get:

$$1 = |\det g|_p \leq \max_{1 \leq i_1, \dots, i_{2n} \leq 2n} |g_{1i_1} \cdots g_{2ni_{2n}}|_p \leq \|g\|^{2n}.$$

Let now $g, h \in G$. Let us prove that $gh \in G$. Taking into account formulas (6.1) and (6.2) we have:

$$1 \leq \|gh\| \leq \max_{1 \leq i,j \leq 2n} \left| \sum_{1 \leq k \leq 2n} g_{ik} h_{kj} \right|_p \leq \|g\| \, \|h\| = 1.$$

By analogy we get that if $g \in G$, then $g^{-1} \in G$:

$$1 \leq \|g^{-1}\| \leq \max_{1 \leq i,j \leq 2n} |(-1)^{i+j} M_{ij}|_p \leq \|g\|^{2n-1} = 1,$$

where M_{ij} is the complement minor of element g_{ij} in g. ∎

Remark 1. The constructed group G is a symplectic group over ring $\mathbb{Z}_p$ of p-adic integers, it is not commutative and is maximal compact subgroup of the group $Sp(2n, \mathbb{Z}_p)$.

Remark 2. In the case of $n = 1$ the group of matrices $\{T_t, \ t \in G_p\}$ constructed in Sec. 11 and which defines the evolution of harmonic oscillator is a subgroup of the group G.

Let $g \in Sp(2n, \mathbb{Q}_p)$, $x \in V$ and gx denotes the following element of the space V:

$$(gx)_i = \sum_{j=1}^{2n} g_{ij} x_j. \tag{6.3}$$

Formula (6.3) defines the action of $Sp(2n, \mathbb{Q}_p)$ on V.

Let us consider an arbitrary irreducible Weyl system (H, W) over the space (V, B). Because $Sp(2n, \mathbb{Q}_p)$ acts on V transitively and preserves the symplectic form we get that (H, W_g), $W_g(x) = W(gx)$ is an irreducible Weyl system over (V, B) too. Therefore, according to Corollary 2 from Theorem 3 of Sec. 12.5 there exists a unitary operator $U(g) : H \to H$ that satisfies the condition

$$U(g)W(x) = W(gx)U(g), \qquad g \in Sp(2n, \mathbb{Q}_p), \quad x \in V. \tag{6.4}$$

It is known (see [225]) that the constructed set of operators $\{U(g),\ g \in Sp(2n, \mathbb{Q}_p)\}$ forms the projective representation of $Sp(2n, \mathbb{Q}_p)$ in H, which is a unitary representation on two-fold covering of $Sp(2n, \mathbb{Q}_p)$ (so-called metaplectic group). We shall be interested in the restriction of this representation to the subgroup G. Remark also that by virtue of irreducibility of (H, W), formula (6.4) defines $U(g)$ uniquely up to a factor. If $n = 1$, then the evolution operator $U(t)$, $t \in G_p$ of quantum p-adic harmonic oscillator constructed in Sec. 11.4 is the representation of subgroup $\{T_t,\ t \in G_p\}$ of the group G defined by the formula (6.4) by means of the Weyl system $(L_2(\mathbb{Q}_p), W)$ from Example 1 of Sec. 12.5.

If $n > 1$ the similar operator can be constructed by means of tensor product of n operators $U(t)$ of one-dimensional oscillator using the Weyl system $(L_2(\mathbb{Q}_p^n), W^{(n)})$ from Example 2 of Sec. 12.5. In both cases representations defined by the formula (6.4) are unitary (not projective). This is true for the whole group G. In fact, the following Theorem is valid.

Theorem 1. *The set of operators $\{U(g), g \in G\}$ which satisfies the relation (6.4) for some irreducible Weyl system (H, W) over the space (V, B) gives a unitary representation of G in H.*

■ By virtue of equivalence of irreducible Weyl systems it is sufficient to prove the Theorem for arbitrary chosen irreducible Weyl system. We choose the Weyl system $(L_2^\chi, \tilde{W})$ from Example 3 of Sec. 12.5 as such Weyl system.

In the space L_2^χ we define the set of operators $\{\tilde{U}(g), g \in G\}$ by the formula:

$$\tilde{U}(g)f(z) = f_g(z) = f(g^{-1}z), \quad f \in L_2^\chi. \tag{6.5}$$

It is easy to see that $\tilde{U}(g)$ is the operator from L_2^χ to L_2^χ. In fact, if $f \in L_2^\chi$, $z' \in V_0$, $g \in G$ the following relation is valid

$$\begin{aligned} f_g(z + z') &= f(g^{-1}z + g^{-1}z') \\ &= \chi_p(B(g^{-1}z, g^{-1}z'))f(g^{-1}z) = \chi_p(B(z, z'))f_g(z). \end{aligned}$$

Taking into account the definitions of $\tilde{W}(z)$ and $\tilde{U}(g)$ it is easy to check the relation (6.4) for Weyl system $(L_2^\chi, \tilde{W})$ and operators $\tilde{U}(g)$, $g \in G$. In fact, if $f \in L_2^\chi$ then we have:

$$\begin{aligned} \tilde{U}(g)\tilde{W}(z)f(x) &= \tilde{U}(g)[\chi_p(B(z, x))f(x - z)] = \chi_p(B(gz, x))f(g^{-1}(x - gz)) \\ &= \tilde{W}(gz)f(g^{-1}x) = \tilde{W}(gz)\tilde{U}(g)f(x). \end{aligned}$$

Obviously, the set of operators $\{\tilde{U}(g), g \in G\}$ forms unitary representation of the group G in L_2^χ. ■

Study of the Weyl system gives us an opportunity to get some information about the properties of the representation $\{U(g), g \in G\}$ defined by (6.4).

We shall say that vector $\phi_0 \in H$ is an eigen-vector of representation $\{U(g), g \in G\}$ if it satisfies the condition:

$$U(g)\phi_0 = \lambda(g)\phi_0,$$

where $\lambda(g)$ is a complex number, $|\lambda(g)| = 1$.

The following theorem is valid.

Theorem 2. *Let (H, W) be an irreducible Weyl system over (V, B) and $\{U(g),\ g \in G\}$ be unitary representation of group G in H defined by* (6.4). *Then vacuum vector ϕ_0 of the Weyl system (H, W) is an eigen-vector of representation $\{U(g),\ g \in G\}$.*

■ By the condition, vector ϕ_0 satisfies the relation:

$$W(z)\phi_0 = \phi_0, \quad z \in V_0.$$

The following equality is valid because of invariantness V_0 under the action of G:

$$W_g(z)\phi_0 = \phi_0$$

and hence ϕ_0 is the vacuum vector of irreducible Weyl system (H, W_g). On the other hand from (6.4) it follows that $U(g)\phi_0$, $g \in G$ is vacuum vector of (H, W_g) too. Taking into account formula (6.6) and Theorem 3 of Sec. 12.5 we get the required statement. ■

7. *Investigation of Eigen-Functions for* $p \equiv 3 \pmod 4$

As it has been noted in Sec. 12, the investigation of spectral properties of harmonic oscillator constructed in Sec. 11.4 is equivalent to that of representation of the group T of matrices of the type

$$T_t = \begin{pmatrix} \cos t & \sin t \\ -\sin t & \cos t \end{pmatrix}, \quad t \in G_p \text{ (see (1.6) of Sec. 11)},$$

which is defined by formulas (4.3)–(4.4) of Sec. 11. This problem has been solved completely for the case of $p \equiv 1 \pmod 4$ (see Sec. 12.4), but for the case of $p \equiv 3 \pmod 4$ only the dimensions of eigen-subspaces have been calculated.

Analysis of the representation of the group T is closely connected with that of larger group S of matrices of the type:

$$\begin{pmatrix} a & b \\ -b & a \end{pmatrix}, \qquad a, b \in \mathbb{Q}_p, \quad a^2 + b^2 = 1.$$

In the case of $p \equiv 3 \pmod 4$ the group S is compact (see Sec. 1.5) and is the subgroup of $Sp(2, \mathbb{Z}_p)$.

In order to study eigen-functions it is convenient to consider the phase space $V = \mathbb{Q}_p \times \mathbb{Q}_p$ of classical system as quadratic extension of $\mathbb{Q}_p$: $V \cong \mathbb{Q}_p(\sqrt{-1})$ (see Sec. 1.4). Using the notation $i = \sqrt{-1}$ any element $z \in \mathbb{Q}_p(\sqrt{-1})$ can be uniquely represented in the form $z = x + iy$, $\bar{z}$ denotes the conjugate element from $\mathbb{Q}_p(\sqrt{-1})$ $\bar{z} = x - iy$, $x, y \in \mathbb{Q}_p$. The group S is isomorphic to the subgroup of the multiplicative group $\mathbb{Q}_p^*(\sqrt{-1})$ of $\mathbb{Q}_p(\sqrt{-1})$ of the following type:

$$S \cong \{z \in \mathbb{Q}_p^*(\sqrt{-1}) : z\bar{z} = 1\}.$$

Let us also define the function $e^{it} : G_p \to \mathbb{Q}_p^*(\sqrt{-1})$ by the formula:

$$e^{it} = \cos t + i \sin t, \quad t \in G_p.$$

This function satisfies the relation:

$$e^{it} e^{it'} = e^{i(t+t')}.$$

The group T is isomorphic to a subgroup of $\mathbb{Q}_p^*(\sqrt{-1})$ of the following type:

$$T \cong \{e^{it}, t \in G_p\}.$$

T is the subgroup of S and the following lemma is valid:

Lemma 1. *Group S is isomorphic (for $p \equiv 3 \pmod 4$ to the direct product of the cyclic group Z_{p+1} of order $p+1$ and the group T:*

$$S \simeq Z_{p+1} \times T, \tag{7.1}$$

(see [82]).

■ Let us give the sketch of the proof. It follows from the equation $a^2 + b^2 = 1$, $a, b \in \mathbb{Q}_p$, that for $p \equiv 3 \pmod 4$ either $|a|_p = 1$, $|b|_p \leq 1$ or $|a|_p \leq 1$ $|b|_p = 1$ (see Sec. 1.4). Therefore, taking into account that $\sin t$ maps G_p to G_p in a one-to-one manner, we get that any $a + ib \in S$ can be represented in the form:

$$a + ib = (a_0 + ib_0) \cdot (\cos t + i \sin t) = (a_0 + ib_0)e^{it}$$

for some $t \in G_p$ and a_0 and b_0 which satisfy the relation:

$$a_0^2 + b_0^2 \equiv 1 \text{ (mod p)}. \tag{7.2}$$

The set of pairs (a_0, b_0), which satisfy (7.2), is isomorphic to some subgroup of multiplicative group $\mathbb{F}_p^*(\sqrt{-1})$ of quadratic extension of the finite field $\mathbb{F}_p$ which is cyclic (as any subgroup of the multiplicative group of the finite field).

The order of this group is equal to the number of solutions of (7.2) which can be calculated by means of Gauss sums and equal $p + 1$. ■

Before the study of eigen-functions let us prove the following lemma, ε denotes some element from $\mathbb{Q}_p(\sqrt{-1})$ with the property $\varepsilon\bar{\varepsilon} = -1$.

Lemma 2. *Any element $z \in \mathbb{Q}_p^*(\sqrt{-1})$ (for $p \equiv 3 \pmod 4$) can be represented in the following form:*

$$z = r\varepsilon^k c^n e^{i\tau}, \tag{7.3}$$

where $r = r(z) \in \mathbb{Q}_p^{*2}$; $k = k(z) = 0, 1$; c is the generator of $Z_{p+1} \cong S/T$; $n = n(z) = 0, 1, \ldots, p$; $\tau = \tau(z) \in G_p$.

■ Let $z \in \mathbb{Q}_p^*(\sqrt{-1})$. Only two cases are possible: either $z\bar{z} = a^2$ or $z\bar{z} = -a^2$, $a \in \mathbb{Q}_p$ (see Sec. 1.4). In the first case we shall choose r as the square root of $z\bar{z}$ which belongs to $\mathbb{Q}_p^{*2}$: $r = \sqrt{z\bar{z}}$. Then $\frac{1}{r}z \in S$, $k = 0$ and (7.3) follows from (7.1). In the second case $r = \sqrt{-z\bar{z}} \in \mathbb{Q}_p^{*2}$ and $\frac{1}{r\varepsilon}z \in S$, $k = 1$. Further proof is evident. ■

The representation (7.3) we shall call the *polar decomposition* of $z \in \mathbb{Q}_p^*(\sqrt{-1})$. As it follows from Lemma 2 this polar decomposition defines on

$\mathbb{Q}_p^*(\sqrt{-1})$ four functions:

$$r(z) : \mathbb{Q}_p^*(\sqrt{-1}) \longrightarrow \mathbb{Q}_p^{*2}; \quad k(z) : \mathbb{Q}_p^*(\sqrt{-1}) \longrightarrow \{0,1\};$$
$$n(z) : \mathbb{Q}_p^*(\sqrt{-1}) \longrightarrow \{0,1,\dots,p\} \text{ and } \tau(z) : \mathbb{Q}_p^*(\sqrt{-1}) \longrightarrow G_p.$$

By the definition these functions have the properties:

$$\begin{gathered} r(e^{it}z) = r(z), \quad k(e^{it}z) = k(z), \\ n(e^{it}z) = n(z), \quad \tau(e^{it}z) = \tau(z) + t, \\ z \in \mathbb{Q}_p^*(\sqrt{-1}), \quad t \in G_p. \end{gathered} \tag{7.4}$$

Less obvious properties of these functions are given by the following lemma.

Lemma 3. *Let $z, z' \in \mathbb{Q}_p^*(\sqrt{-1})$ and $\|z\| \geq p$, $\|z'\| \leq 1$ ($z' \in V_0$). Then the following relations are valid:*

1) $|r(z+z') - r(z)|_p \leq 1$,
2) $k(z+z') = k(z)$,
3) $n(z+z') = n(z)$.

■ Let $z = x + iy$ and $z' = x' + iy'$ satisfy the conditions of the Lemma. 1) By means of elementary calculations we get the formula:

$$r(z+z') - r(z) = \frac{\gamma^2(z') + 2(xx' + yy')}{\gamma(z+z') + \gamma(z)} .$$

Since $r \in \mathbb{Q}_p^{*2}$ then $(r(z+z'))_0 + (r(z))_0 \equiv 0 \pmod{p}$ and hence

$$\begin{aligned} |r(z+z') + r(z)|_p &= \max\{|r(z+z')|_p, |r(z)|_p\} \\ &= \max\{\|z+z'\|, \|z\|\} = |r(z)|_p. \end{aligned} \tag{7.5}$$

From equation $\|z\| = |r(z)|_p = \max\{|x|_p, |y|_p\}$ it follows

$$|x|_p \leq |r(z)|_p, \quad |y|_p \leq |r(z)|_p. \tag{7.6}$$

From (7.5) and (7.6) we have:

$$|r(z+z') - r(z)|_p \leq \frac{1}{|\gamma(z)|_p} \max\{|r(z')|_p^2, |xx'|_p, |yy'|_p\} \leq 1.$$

2) From the definition of the function $k(z)$ it follows that it depends only on the first term in canonical decomposition of p-adic number $z\bar{z}$:

$$k(z) = \tilde{k}((z\bar{z}_0).$$

The relation $((z+z')(\overline{z+z'}))_0 = (z\bar{z})_0$ follows from the equality:

$$|(z+z')(\overline{z+z'}) - z\bar{z}|_p = |z'\bar{z}' + 2(z\bar{z}' + z'\bar{z})|_p < |z\bar{z}|_p.$$

3) By virtue of relations $\|z\| = |r(z)|_p = \max\{|x|_p, |y|_p\}$ we have that either $|x|_p = |r(z)|_p$, $|y|_p \le |r(z)|_p$, or $|y|_p = |r(z)|_p$, $|x|_p \le |r(z)|_p$. Let us consider the first case. By the definition of $n(z)$ and Lemma 1, it follows that if $|y|_p < |r(z)|_p$ then $n(z) = \tilde{n}\left(\left(\frac{x}{\gamma}\right)_0\right)$, if $|y|_p = |r(z)|_p$, then $n(z) = n\left(\left(\frac{x}{\gamma}\right)_0, \left(\frac{y}{\gamma}\right)_0\right)$. Let us consider the second case (the first one can be considered analogously). By virtue of statement 1) of this lemma, relations $|x|_p = |y|_p = |r(z)|_p \ge p$ and inequality $|x'|_p \le |r(z')|_p \le 1$ we have:

$$n(z+z') = \tilde{n}\left(\left(\frac{x+x'}{\gamma(z+z')}\right)_0, \left(\frac{y+y'}{\gamma(z+z')}\right)_0\right) = \tilde{n}\left(\left(\frac{x}{\gamma}\right)_0, \left(\frac{y}{\gamma}\right)_0\right) = n(z)$$

■

Denoting by $\delta_{k,n}$ the Kroneker symbol on $\mathbb{Q}_p^*(\sqrt{-1})$ we define:

$$\delta_m^\varepsilon(z) = \delta_{m,k(z)}, \quad m = 0, 1;$$

$$\delta_n^c(z) = \delta_{n,n(z)}, \quad n = 0, 1, \ldots, p.$$

From the definition of these functions and properties of functions $n(z)$ and $k(z)$ we have:

$$\text{supp}\ \delta_m^{\varepsilon,c}(z) \cap \text{supp}\ \delta_n^{\varepsilon,c} = \emptyset, \qquad \text{if } n \ne m; \tag{7.7}$$

$$\delta_m^{\varepsilon,c}(z+z') = \delta_m^{\varepsilon,c}(z), \qquad \|z\| \ge p, \quad \|z'\| \le 1; \tag{7.8}$$

$$\delta_m^{\varepsilon,c}(e^{it}z) = \delta_m^{\varepsilon,c}(z), \qquad t \in G_p. \tag{7.9}$$

As it has been noted in Sec. 11.7 each of them is connected with some irreducible Weyl system over the space $(V = \mathbb{Q}_p \times \mathbb{Q}_p, B)$. For example in Secs. 11.4 and 12.3 oscillator has been considered in representation which corresponds to the Weyl system from Example 1 of Sec. 12.5. In the case of $p \equiv 3 \pmod 4$ the investigation of eigen-functions is carried out easily in representation which is connected to the Weyl system $(L_2^\chi, \tilde{W})$ from

Example 3 of Sec. 12.5. Remember that in this case the representation space has the form:

$$L_2^\chi = \{f \in L_2(\mathbb{Q}_p \times \mathbb{Q}_p) : f(x + x') = \chi_p(B(x, x'))f(x), \|x'\| \le 1\},$$

representation $U(t)$ of the group T is defined by the formula:

$$U(t)f(x) = f(e^{-it}x), \quad t \in G_p, f \in L_2^\chi,$$

and representation $V(z)$ of S is defined by the formula:

$$V(z)f(x) = f(\bar{z}x), \quad z \in S, f \in L_2^\chi.$$

In this representation the problem of determination of eigen-functions of the oscillator is equivalent to the finding of any $\alpha \in I_p$ (see Sec. 12.3) a complete system of linear independent solutions of the following system of equations in $L_2(V)$:

$$\begin{cases} f(e^{-it}z) & = \chi(\alpha t)f(z), \\ f(z + z') & = \chi(B(z, z'))f(z), \quad \|z'\| \le 1. \end{cases} \tag{7.10}$$

Taking into account the theorem about dimension of invariant subspaces (Sec. 12.3) for the case of $p \equiv 3 \pmod 4$, we have that, if $\alpha \in I_p$, $|\alpha|_p = p^{2k+1}$, $k = 0, 1, \ldots$ the system (7.10) has no solutions; if $\alpha = 0$, then there exists a unique linear independent solution of (7.10) and in the case of $\alpha \in J_p$ there are $p + 1$ linear independent solutions.

The explicit formulas for eigen-functions are given by the following theorem.

Theorem *Let $p \equiv 3 \pmod 4$. If $\alpha = 0$, then any solution of the system (7.10) is proportional to the function*

$$\phi_0(z) = \Omega(\|z\|).$$

If $\alpha \in J_p$, $\alpha \ne 0$, then $p + 1$ functions

$$\phi_\alpha^n(z) = \delta_m^\varepsilon(z)\delta_n^c(z)\Omega(|r(z) - a|_p)\chi_p(-\alpha\tau(z)), \tag{7.11}$$

are linear independent and give us the solutions of (7.10), where

$$m = \frac{1}{2}\left(1 + \left(\frac{\alpha_0}{p}\right)\right), \quad a = \sqrt{(-1)^{m+1}\alpha}, \quad n = 0, 1, \ldots, p.$$

■ As it can be checked by direct substitution, $\phi_0(z)$ is the solution of (7.10). Let now $\alpha \in J_p$, $\alpha \neq 0$. From (7.7) we have that the functions (7.11) are orthogonal for different n and hence are linear independent. Besides that, from (7.9) it follows that the functions (7.11) satisfy the first equation of the system (7.10). For $\alpha \in J_p$, $\alpha \neq 0$ we have $|\alpha|_p \geq p^2$ and therefore $|a|_p \geq p$. Taking into account this inequality, substituting functions (7.11) to the second equation of the system (7.10) and using the property 1) of the function $r(z)$ (see Lemma 3) and formula (7.8) we get:

$$\begin{aligned}&\delta_m^{\varepsilon}(z)\delta_n^{c}(z)\Omega(|r(z)-a|_p)\chi_p(-\alpha\tau(z+z'))\\ &= \chi_p(B(z,z'))\delta_m^{\varepsilon}(z)\delta_n^{c}(z)\Omega(|r(z)-a|_p)\chi_p(-\alpha\tau(z)).\end{aligned}$$

Taking into account (7.7), the last formula is equivalent to the following one:

$$\chi_p(\alpha\Delta\tau + B(z,z')) = 1, \tag{7.12}$$

$$\begin{aligned}\text{where } \Delta\tau = \tau(z+z') - \tau(z), \quad & z = r(z)\varepsilon^m c^n e^{i\tau(z)},\\ z' = r(z')\varepsilon^m c^n e^{i\tau(z')}, \quad & z+z' = r(z+z')\varepsilon^m c^n e^{i\tau(z+z')}.\end{aligned}$$

By virtue of the last equations, the expression for $B(z,z')$ can be transformed to the following form:

$$\begin{aligned}B(z,z') &= B(z,z+z')\\ &= r(z)r(z+z')(\varepsilon\bar{\varepsilon})^m B(e^{i\tau(z)}, e^{i\tau(z+z')})\\ &= r(z)r(z+z')(-1)^m \sin\Delta t.\end{aligned} \tag{7.13}$$

For further proof the following equality is required: (for $p \equiv 3 \pmod 4$)

$$\|e^{ia} - e^{ib}\| = |a-b|_p, \quad a, b \in G_p. \tag{7.14}$$

In fact:

$$\begin{aligned}\|e^{ia} - e^{ib}\| &= \|e^{i(a-b)} - 1\|\\ &= \max\{|\cos(a-b) - 1|_p, |\sin(a-b)|_p\}\\ &= |\sin(a-b)|_p = |a-b|_p.\end{aligned}$$

Taking into account (7.14) we can prove the following inequality:

$$|a\Delta\tau|_p \leq 1. \tag{7.15}$$

In fact:

$$\begin{aligned}|a\Delta\tau|_p &= |a|_p\|e^{i\tau(z+z')} - e^{i\tau(z)}\| \\ &= \|(r(z+z') + (a - r(z+z'))e^{i\tau(z+z')} - (r(z) + (a - r(z)))e^{i\tau(z)}\| \\ &= \|r(z+z')e^{i\tau(z+z')} - r(z)e^{i\tau(z)}\| = \|z'\| \le 1.\end{aligned}$$

Taking into account formulas (7.13) and (7.15) and inequalities $|r(z)-a|_p \le 1$, $|r(z+z')-a|_p \le 1$, let us rewrite the expression (7.12) in equivalent form:

$$\chi_p(\alpha\Delta\tau + (-1)^m a^2 \sin\Delta\tau) = 1. \tag{7.16}$$

From the properties of function $\sin x$, which have been pointed out in Sec. 2.4, the following inequality (for $p \equiv 3 \pmod 4$) follows easily:

$$|\sin\Delta\tau - \Delta\tau|_p \le p|\Delta\tau|_p^3. \tag{7.17}$$

By means of (7.15) and (7.17) the expression (7.17) can be transformed to the form:

$$\chi_p((\alpha + (-1)^m a^2)\Delta\tau) = 1. \tag{7.18}$$

Thus, the functions (7.11) satisfy the second equation of the system (7.10), if a, α and m satisfy the relation (7.18) for any $\Delta\tau \in G_p$. It is easy to see that a and m from the condition of the Theorem satisfy the equality:

$$\alpha + (-1)^m a^2 = 0,$$

from which formula (7.18) follows. ∎

The theorem gives us the opportunity to construct eigen-functions of oscillator for $p \equiv 3 \pmod 4$ in an arbitrary representation. Namely, the following corollary is valid.

Corollary. *Let (H, W) be some irreducible Weyl system over the space $(\mathbb{Q}_p \times \mathbb{Q}_p, B)$ with vacuum vector ψ_0. Then eigen-functions of oscillator in representation (H, W) are given by the formula:*

$$\psi_\alpha^n = \int_{G_p} \chi_p(-\alpha t) W(a\varepsilon^m c^n e^{it})\psi_0 dt, \tag{7.19}$$

where $\alpha \in J_p$, $n = 0, 1, \ldots, p$, a and m are as in the Theorem.

■ Let us consider the Weyl system $(L_2^\chi, \tilde{W})$ with vacuum vector $\phi_0 = \Omega(||z||)$ and corresponding representation for harmonic oscillator.

It is easy to see that the operator $S : L_2^\chi \to H$ defined by the formula

$$S\phi = \int_V (\phi, \tilde{W}(z))_{L_2(V)} W(z)\psi_0 dz, \quad \phi \in L_2^\chi, \tag{7.20}$$

is a unitary operator (see Sec. 11.7). Substituting in (7.20) eigen-functions of oscillator in representation $(L, W)\phi(z)$, we get:

$$\begin{aligned}\psi_\alpha^n &= \int_V dz W(z)\psi_0 \int_V dy \phi_\alpha^n(y)\chi_p(B(-z,y))\Omega(||y-z||) \\ &= \int_V dz W(z)\psi_0 \int_V dy \delta_m^\varepsilon(y)\delta_n^c(y)\Omega(|r(y)-a|_p) \\ &\quad \cdot \chi_p(-\alpha\tau(y))\chi_p(B(-z,y))\Omega(||y-z||).\end{aligned}$$

Since integration in the last formula is actually carried out on bounded domain $||y|| = ||z|| = |a|_p$, then we can change the integration order. Taking into account the formula

$$\int_V dz \chi_p(B(-z,y))\Omega(||y-z||)W(z)\psi_0 = W(y)\psi_0,$$

we get

$$\psi_\alpha^n = \int_V dy \delta_m^\varepsilon(y)\delta_n^c(y)\Omega(|r(y)-a|_p)\chi_p(-\alpha\tau(y))W(y)\psi_0.$$

Using the properties of functions $\delta_m^\varepsilon(y)$, $\delta_n^c(y)$ and the polar decomposition, from the last formula we have:

$$\psi_\alpha^n = \int_{\mathbb{Q}_p^{*2}} |r|_p \Omega(|r-a|_p) dr \int_{|t|_p \le 1/p} \chi_p(-\alpha t) W(a\varepsilon^m c^n e^{it})\psi_0 dt.$$

After rejection of unessential nonzero coefficient we get the required expression for eigen-functions.

Remark 1. By analogy to representation $U(t)$, $t \in G_p$ of the group T we can construct unitary representation $V(g)$, $g \in S$ of the group S. In this case, taking into account the theorem and formula (7.1) it is easy to prove, that eigen-subspaces of $V(g)$ are one-dimensional and eigen-functions coincide with eigen-functions ϕ_α^n of the representation $U(t)$ (but in this case the functions ϕ_α^n for different n correspond to different subspaces).

Remark 2. Investigation of eigen-functions for $p \equiv 3 \pmod 4$ (that is the solution of the system (7.10)) have been carried out without the theorem about dimensions of invariant subspaces by means of direct analysis of the system (7.10).

XIII. Weyl Systems. Infinite Dimensional Case

Let (V, B) be an infinite dimensional symplectic space over $\mathbb{Q}_p$. A subspace $U \subset V$ is nondegenerate if the restriction of B on U is a nondegenerate symplectic form on U. Let F denotes the set of all finite dimensional nondegenerate subspace of V.

A *Weyl system* over (V, B) is a pair (H, W), where H is a complex Hilbert space and W is a map from V to the set of unitary operators on H, which satisfies the property

$$W(x)W(y) = \chi_p(B(x, y))W(x + y)$$

for all $x, y \in V$ and the restriction of W to any $U \in F$ is continuous in strong topology.

Let us be reminded that in the case of dim $V < \infty$ (see Sec. 12.4) all irreducible Weyl systems are unitary equivalent (Stone-von Neumann uniqueness theorem) and any Weyl system can be represented as a direct orthogonal sum of irreducible ones. But it is not the case if dim $V = \infty$.

1. *Weyl Algebras*

Let (H, W) be a Weyl system over (V, B), $U \in F$ and let $\mathfrak{M}_u(H, W)$ denote the W^*-algebra generated by the set of operators $\{W(x), x \in U\}$. As dim $U < \infty$ using the Stone-von Neumann uniqueness theorem (see Sec. 12.4) it is easy to prove the following lemma.

Lemma 1. *For any two Weyl systems (H_1, W_1) and (H_2, W_2) over (V, B) and any $U \in F$ there exists a unique $*$-isomorphism α of the algebras $\mathfrak{M}_u(H_1, W_1)$ and $\mathfrak{M}_u(H_2, W_2)$ which maps $W_1(x)$ into $W_2(x)$ for all $x \in U$:*

$$\alpha(W_1(x)) = W_2(x).$$

The *Weyl algebra* of the Weyl system (H, W) over (V, B) is defined as a C^*-algebra $\mathfrak{U}(H, W)$ which is the uniform closure of the union of algebras $\mathfrak{M}_u(H, W)$, when U runs all subspaces from F:

$$\mathfrak{U}(H, W) = \overline{\bigcup_{U \in F} \mathfrak{M}_u(H, W)}.$$

As it was mentioned, we do not have the Stone-von Neumann uniqueness theorem for the case of dim $V = \infty$. But in this case the so-called C^*-algebraic uniqueness theorem is valid:

Theorem 1. *For any two Weyl systems (H_1, W_1) and (H_2, W_2) over (V, B) there exists a unique $*$-isomorphism α of the algebras $\mathfrak{U}(H_1, W_1)$ and $\mathfrak{U}(H_2, W_2)$ which maps $W_1(x)$ into $W_2(x)$ for all $x \in V$:*

$$\alpha(W_1(x)) = W_2(x).$$

■ Algebras $\mathfrak{M}_u(H, W)$, $U \in F$ form the partially ordered set under imbedding (this ordering is induced by the natural ordering on F) and by virtue of the Lemma 1 and the Zorn Lemma there exists a unique $*$-isomorphism $\tilde{\alpha}$ of the $*$-algebras $\bigcup_{U \in F} \mathfrak{M}_u(H_2, W_2)$. The isomorphism $\tilde{\alpha}$ is continuous in the uniform topology (as a $*$-morphism of $*$-algebras) and thus can be uniquely extended to the needed $*$-isomorphism α ■

2. *Positive Functionals*

A complex valued functional $\mu : V \to \mathbb{C}$ on the symplectic p-adic space (V, B) is *positive*, if

(I) $\mu(0) = 1$,
(II) for any finite sets $\lambda_1, \ldots, \lambda_n \in \mathbb{C}$ and $x_1, \ldots, x_n \in V$ we have

$$\sum_{1 \leq i,j \leq n} \lambda_i \bar{\lambda}_j \mu(x_j - x_i) \chi_p(B(x_i, x_j)) \geq 0,$$

(III) μ is continuous on any $U \in F$.

Positive functionals on (V, B) give us an opportunity to study cyclic Weyl systems. A *cyclic Weyl system* over (V, B) is a triple (H, W, φ), where

(H, W) is a Weyl system over (V, B), $\varphi \in H$, $\|\varphi\|_H = 1$ and the closure of linear span of the set $\{W(x)\varphi, x \in V\}$ in H coincides with H. Two cyclic Weyl systems (H_1, W_1, φ_1) and (H_2, W_2, φ_2) are equivalent if there exists a unitary operator $U : H_1 \to H_2$, which satisfies the conditions: $U\varphi_1 = \varphi_2$ and $UW_1(x)U^{-1} = W_2(x)$ for any $x \in V$. Study of an arbitrary Weyl system over (V, B) reduces to that of cyclic, because any Weyl system can be represented as a direct orthogonal sum of cyclic ones.

On the other hand, positive functionals on (V, B) and cyclic Weyl systems over (V, B) are closely connected. In fact, it is easy to see that if (H, W, φ) is a cyclic Weyl system, then the map μ defined by the formula

$$\mu(x) = (\varphi, W(x)\varphi) \tag{13.1}$$

defines a positive functional on (V, B). The inverse statement is less obvious, but it is true.

Theorem 2. *For any positive functional μ on (V, B) there exists a cyclic Weyl system (H, W, φ) over (V, B), such that the relation (13.1) is valid for all $x \in V$. This Weyl system is unique up to equivalence.*

■ We shall prove the *existence* by means of direct construction.

The space H. Let K denotes the vector space of complex valued functions on V, which are nonzero in not more then finite number of points of V. The formulas

$$\langle f, g\rangle = \sum_{u,v\in V} f(u)\bar{g}(v)\mu(v-u)\chi_p(B(u,v)), \qquad t, g \in K,$$

$$\|f\|^2 = \langle f, f\rangle$$

define nonnegative Hermitian form and seminorm on K respectively. Let N be the closed subspace of K consist of f with zero seminorm. Then on K/N the form $\langle\cdot,\cdot\rangle$ naturally induces the positive Hermitian form and K/N is provided by the prehilbertian structure. The required space H is the closure of K/N with respect to the scalar product mentioned above.

The map W. On the space K we shall define the following set of operators $W(x)$, $x \in V$:

$$\tilde{W}(x)f(u) = \chi_p((Bx, u))f(u - x), \qquad f \in K.$$

These operators satisfy the Weyl relation (this fact has been proved in Sec. 12.4, Example 3). It is easy to see that $\tilde{W}(x)$, $x \in V$ are isometric. Hence, we have correctly defined isometric operators $W(x)$, $x \in V$ on K/N:

$$W(x)[f+N] = \tilde{W}(x)f + N, \qquad x \in V, \quad f \in K,$$

which are uniquely extended to unitary operators on H with needed properties.

The cyclic vector φ can be chosen as follows:

$$\varphi = \tilde{\varphi} + N, \qquad \tilde{\varphi}(u) = \begin{cases} 1, & u = 0 \\ 0, & u \neq 0. \end{cases}$$

It is a simple exercise to verify the relation (13.1) for the constructed Weyl system.

Uniqueness. Let (H_1, W_1, φ_1) and (H_2, W_2, φ_2) be two cyclic Weyl systems corresponding to the same positive functional μ . Let $\tilde{H}_1$ (resp. $\tilde{H}_2$) denote linear span of the set of vectors $\{W_1(x)\varphi_1,\ x \in V\}$ in H_1 (resp. $\{W_2(x)\varphi_2,\ x \in H_2\}$. The formula

$$UW_1(x)\varphi_1 = W_2(x)\varphi_2, \qquad x \in V$$

defines an operator $U : \tilde{H}_1 \to \tilde{H}_2$. It is easy to see that U is an isometric operator:

$$\begin{aligned}
&(UW_1(x)\varphi_1, UW_1(y)\varphi_1) \\
&= (W_2(x)\varphi_2, W_2(y)\varphi_2) = \chi_p(B(x,y))\varphi_2, W_2(y-x)\varphi_2) \\
&= \chi_p(B(x,y))\mu(y-x) = (W_1(x)\varphi_1, W_1(y)\varphi_1).
\end{aligned}$$

As $\tilde{H}_1$ and $\tilde{H}_2$ are dense in H_1 and H_2 respectively, then U is uniquely extended to the unitary operator $H_1 \to H_2$ with needed properties. ■

Remark. The statements of Sec. 13.1 and 13.2 and their proofs coincide with that of for the real number case.

3. *Fock Representation*

Representations of commutation relations (or Weyl systems) are proved to be closely connected with the notion of a *lattice* in p-adic vector space.

Let V be a p-adic vector space (finite- or infinite-dimensional). A lattice L in V is a Z_p-submodule of V which does not contain any nonzero subspace of V and absorbs V (that is for any $x \in V$ there exists $\lambda \in \mathbb{Q}_p\setminus\{0\}$ such that $\lambda x \in L$). In the case of $\dim V < \infty$ this definition coincides with the ordinary one (that is L is a finitely generated Z_p-submodule of V which contains a basis of V). On the other hand, this notion coincides with that of absolutely convex absorbing set (a nonempty subset A of V is absolutely convex it $x, y \in A$, $\lambda, \mu \in Z_p$ implies $\lambda x + \mu y \in A$) without nonzero subspaces.

Let L be a lattice in V. We shall define the following R_+-valued functional ρ_l on V, $x \in V$:

$$\rho_l(x) = \operatorname*{int}_{s\in\mathbb{Q}^*, sx\in L} |s|_p^{-1} .$$

(This is an analog of the Minkowski functional.) It is not hard to prove, that for any lattice L the functional ρ_l is a non-Archimedian norm on V, the topology on V generated by ρ_l we call L-topology.

Let now (V, B) be a p-adic symplectic space and L be a lattice in V. The subset L^* of V defined by the formula

$$L^* = \{x \in V : B(x, y) \in Z_p, \ \forall y \in L\}$$

is a lattice in V and is called a *dual* lattice. If $L = L^*$, then L is *self-dual*. Properties of this duality are rather similar to that of orthogonal complement and therefore are given without proof.

Lemma 2. *Let L, L_1, L_2 be lattices in (V, B). Then we have:*

$$\text{(I) } (L^*)^* = L,$$
$$\text{(II) } (L_1 + L_2)^* = L_1^* \cap L_2^*,$$
$$\text{(III) } (L_1 \cap L_2)^* = L_1^* + L_2^*.$$

A connection of selfdual lattices and Weyl systems can be easily seen from the following lemma.

Lemma 3. *Let L be a selfdual lattice in (V, B). Then the functional $\mu_l : V \to \mathbb{C}$*

$$\mu_l(x) = \begin{cases} 1, & x \in L, \\ 0, & x \notin L \end{cases}$$

is positive.

■ It is sufficient to prove, that for any $\lambda_1, \ldots, \lambda_n \in \mathbb{C}$ and $x_1, \ldots, x_n \in V$ the following inequality is valid:

$$\sum_{1 \le i,j \le n} \lambda_i \bar{\lambda}_j \mu_l(x_j - x_i) \chi_p(B(x_i; x_j)) \ge 0.$$

As $\mu_l(x) = 0$ for all $x \notin L$ the above inequality should be proved only for the case when all x_i, $i = 1, \ldots, n$ have the form

$$x_i = \alpha + u_i, \quad u_i \in L$$

for some $\alpha \in V$. Thus we have:

$$\sum_{1 \le i,j \le n} \lambda_i \bar{\lambda}_j \mu_l(u_j - u_i) \chi_p(B(\alpha + u_i; \alpha + u_j))$$
$$= \sum_{1 \le i,j \le n} \lambda_i \bar{\lambda}_j \chi_p(B(u_i, \alpha)) \bar{\chi}_p(B(u_j, \alpha))$$
$$= \left| \sum_{1 \le i \le n} \bar{\lambda}_i \chi_p(B(u_i, \alpha)) \right|^2 \ge 0 \qquad ■$$

Let L be a selfdual lattice in (V, B). By the definition Fock (or L-Fock) representation of commutation relations is described by the cyclic Weyl system (H, W, φ) which corresponds to the functional μ_l (in the sence of the Theorem 2 of this Section). The vector φ is called vacuum vector. As in the finite-dimensional case we shall define the set of coherent states. From any coset $\alpha \in V/L$ we choose one element and form the set T of all such elements. The family of vectors in H:

$$\phi_\alpha = \{W(\alpha)\varphi, \ \alpha \in T\}$$

is the set of coherent states in H. The Weyl system (H, W, φ) mentioned is called the L-Weyl system for brevity. It satisfies the following properties.

Theorem 3. *Let L be a selfdual lattice in (V, B) and (H, W, φ) be the corresponding L-Weyl system. Then*

(I) $W(x)\varphi = \varphi$ *for all* $x \in L$;

(II) *the set of coherent states forms an orthonormal basis in* H;

(III) *the Weyl system* (H, W, φ) *is irreducible;*

(IV) *the map* W *is continuous in* L*-topology on* V *and strong topology on the set of unitary operators on* H.

■ (I) Because of the relation

$$\mu_l(x) = (\varphi, W(x)\varphi) = \begin{cases} 1, & x \in L, \\ 0, & x \notin L, \end{cases} \tag{13.2}$$

it is sufficient to prove that the closure k of the linear span of the set of vectors $\{W(x)\varphi, x \in L\}$ in H is one-dimensional. Let us suppose that it is not true. Then there exists a nonzero vector ψ from K of the form

$$\psi = \sum_{\alpha \in L} C_\alpha W(\alpha)\phi,$$

which is orthogonal to φ. Thus we have

$$(\varphi, \psi) = \sum_{\alpha \in L} \bar{C}_\alpha = 0,$$

but in this case

$$\|\psi\|^2 = \sum_{\alpha, p \in L} C_\alpha \bar{C}_\beta (W(\alpha)\phi, W(\beta)\phi) = \sum_{\alpha, \beta \in L} C_\alpha \bar{C}_\beta = 0.$$

The contradiction obtained completes the proof of (I).

(II) From (13.2) we can see that the set $\phi_\alpha = \{W(\alpha)\varphi, \ \alpha \in T\}$ of coherent states forms an orthonormal of (H, W, φ), H is the closure of the linear span K of the set $\{W(x)\varphi, x \in V\}$. Let ψ be a vector from K, that is

$$\psi = \sum_\beta C_\beta W(\beta)\varphi.$$

For any $\beta \in V$ there exist $\alpha \in T$ and $u \in L$, such that $\beta = \alpha + u$. Hence, taking into account the property (I) from this theorem and the Weyl relation we have

$$\psi = \sum_{\alpha \in T} \tilde{C}_\alpha W(\alpha)\varphi,$$

where $\tilde{C}_\alpha = \tilde{C}_{\alpha+u}\chi_p(-B(\alpha, u))$ and therefore ϕ_α is a basis in H.

(III) From (II) we easily get that any vector from H is cyclic, then (H, W, φ) is irreducible.

(IV) Let ψ be an arbitrary vector from H. Then

$$\psi = \sum_{\alpha \in T} \tilde{C}_\alpha W(\alpha)\varphi, \qquad \sum_{\alpha \in T} |C_\alpha|^2 < \infty$$

and for $x \in L$ (that is $\rho_l(x) \leq 1$) we have

$$\begin{aligned} \|W(x)\psi - \psi\| &= \|\sum_{\alpha \in T}(1 - \chi_p(2B(\alpha, x))C_\alpha W(\alpha)\varphi\| \\ &= \sum_{\alpha \in T} |C_\alpha|^2 |1 - \chi_p(2B(\alpha, x))|^2. \end{aligned} \tag{13.3}$$

By virtue of convergence of the series $\sum\limits_{\alpha \in T} |C_\alpha|^2$ for any $\varepsilon > 0$ there exists such $\Delta > 0$, such that

$$\sum_{\alpha \in T, \rho_l(\alpha) > \Delta} C_\alpha|^2 < \frac{\varepsilon}{2}. \tag{13.4}$$

For all $\rho_l(x) < \frac{1}{\Delta}$ and $\rho_l(\alpha) \leq \Delta$ we have

$$|B(x, \alpha)|_p \leq \rho_l(x)\rho_l(\alpha) \leq 1$$

and hence $\chi_p(2B(\alpha, x)) = 1$. Therefore for all $\rho_l(x) < \min\{1, \frac{1}{\Delta}\} = \delta$ we have from (13.3) and (13.4)

$$\begin{aligned} \|W(x)\psi - \psi\| &= \left(\sum_{\alpha \in T, \rho_l(\alpha) \leq \Delta} + \sum_{\alpha \in T, \rho_l(\alpha) > \Delta} \right) \cdot |C_\alpha|^2 |1 - \chi_p(2B(\alpha, x))|^2 \\ &\leq 2 \sum_{\alpha \in T, \rho_l(\alpha) > \Delta} |C_\alpha|^2 < \varepsilon. \end{aligned}$$

Thus the map W is continuous in a neighborhood of $x = 0$, the continuity at any other point follows from the continuity at $x = 0$ and the Weyl relations. ■

The theorem proved shows that the properties of the L-Weyl systems are similar to that of Weyl systems over a finite-dimensional symplectic space. This justifies the name L-Fock for the corresponding representation of commutation relations.

4. *Equivalence of L-Fock Representations*

Let us be reminded, that in the case of dim $V < \infty$ all irreducible Weyl systems are unitary equivalent but it is not true if dim $V = \infty$. Therefore a question about an equivalence of two Fock representations for different lattices L is very natural in the last case.

Let L_1 and L_2 be selfdual lattices in (V, B). We shall say that L_1 and L_2 coincide almost everywhere if there exists a nongenerate subspace U of the space V of finite condimension such that $L_1 \cap U = L_2 \cap U$. The following theorem gives us a solution of the problem mentioned above.

Theorem 4. *Let L_1 and L_2 be selfdual lattices in (V, B). Then L_1- and L_2-Fock representation of commutation relations are unitary equivalent if and only if L_1 and L_2 coincide almost everywhere.*

■ *Necessity.* Let (H_1, W_1, φ_1) and (H_2, W_2, φ_2) denote L_1- and L_2-Weyl systems respectively. Since (H_1, W_1) and (H_2, W_2) are unitary equivalent, then there exists a unitary operator $U : H_1 \to H_2$ such that the relation

$$UW_1(x)U^{-1} = W_2(x)$$

is valid for all $x \in V$. Let $\nu : V \to \mathbb{R}$ denote the map defined by the formula

$$\nu(x) = |(U\varphi_1, W_2(x)\varphi_2)|. \tag{13.5}$$

Let us prove the relation

$$\nu(x) = \begin{cases} \nu(0), & x \in L_1 + L_2, \\ 0, & x \notin L_1 + L_2 . \end{cases} \tag{13.6}$$

In fact, by virtue of the Weyl relation and the Theorem 3(i) of Sec. 13.3, for $x_1 \in L_1$ and $x_2 \in L_2$ we have

$$\begin{aligned} \nu(x_1 + x_2) &= |(U\varphi_1, W_2(x_1 + x_2)\varphi_2)| = |W_2(-x_1)U\varphi_1, W_2(x_2)\varphi_2)| \\ &= |(UW_1(-x_1)\varphi_1, W_2(x_2)\varphi_2)| = |(U\varphi_1\varphi_2)| = \nu(0). \end{aligned}$$

For $x \notin L_1 + L_2$ by virtue of the relation $(L_1+L_2)^* = L_1 \cap L_2$ (see Sec. 13.3) there exists $y \in L_1 \cap L_2$ such that $B(x,y) \notin \mathbb{Z}_p$ and $\chi_p(2B(x,y)) \neq 1$. Thus we have

$$(U\varphi_1, W_2(x)\varphi_2) = (W_2(y)U\varphi_1, W_2(y)W_2(x)\varphi_2)$$
$$= \chi_p(2B(x,y))(U\varphi_1, W_2(x)\varphi_2),$$

therefore $\nu(x) = 0$.

Let now $d(L_1, L_2)$ denote the order of the group $(L_1, L_2)/L_2$. Taking into account Theorem 3(ii) of Sec. 13.3, the formula (13.6) and the Parseval-Steklov equality we have

$$1 = \|\varphi_1\|^2_{H_1} = \|U\varphi_1\|^2_{H_2} = \sum_{\alpha \in V/L_2} |(U\varphi_1, W_2(\alpha)\varphi_2)|^2$$
$$= \sum_{\alpha \in (L_1+L_2)/L_2} \nu^2(0) = \nu^2(0)d(L_1, L_2),$$

and from the last formula we get $d(L_1, L_2) < \infty$. It is easy to see that $d(L_1, L_2) < \infty$ if and only if L_1 and L_2 coincide almost everywhere. This finishes the proof of necessity.

Sufficiency. Let L_1 and L_2 coincide almost everywhere and so $d(L_1, L_2) < \infty$. We shall construct the vector $\psi_2 \in H$ by the formula

$$\psi_2 = d^{-1}(L_1, L_2) \sum_{\alpha \in L_1/(L_1 \cap L_2)} W_2(\alpha)\varphi_2 \,. \tag{13.7}$$

Note that the expression under the sum symbol in the formula (13.7) does not depend on the choice of representative in the coset $\alpha \in L_1/(L_1 \cap L_2)$. From the Weyl relations and the last formula we get for all $x \in L_1$:

$$W_2(x_1)\psi_2 = d^{-1}(L_1, L_2) \sum_{\alpha \in L_1/(L_1 \cap L_2)} \chi_p(B(x,\alpha))W_2(x+\alpha)\varphi_2$$
$$= d^{-1}(L_1, L_2) \sum_{\beta \in L_1/(L_1 \cap L_2)} W_2(\beta)\varphi_2 = \psi_2 \,. \tag{13.8}$$

Besides that, by virtue of the Theorem 3(ii) of Sec. 13.3 and the Parseval-Steklov equation we have

$$\|\psi_2\|^2_{H_2} = d^{-1}(L_1, L_2) \sum_{\alpha \in L_1/(L_1 \cap L_2)} \|W_2(\alpha)\varphi_2\|^2_{H_2} = 1, \tag{13.9}$$

because groups $L_1/(L_1 \cap L_2)$ and $(L_1+L_2)/L_2$ are isomorphic and $d(L_1, L_2)$ is equal to the order of the group $L_1/(L_1 \cap L_2)$.

From the formulas (13.8) and (13.9) it follows that (H_2, W_2, ψ_2) is the L_1-Weyl system. Therefore (H_1, W_1, φ_1) and (H_2, W_2, ψ_2) are equivalent by the Theorem 2 of Sec. 13.2 and the Weyl systems (H_1, W_1) and (H_2, W_2) are unitary equivalent. ∎

XIV. p-Adic Strings

In this section elements of p-adic string theory are given. First, expressions for so-called dual amplitudes are introduced. After that p-adic analogues of these expressions are considered and their properties are investigated.

1. *Dual Amplitides*

The origins of modern string theory, as it is known, originate from dual theory of strong interaction of elementary particles. Strong interactions of hadrons must be described by functions (scattering amplitudes), which satisfy some general requirements, such as Lorentz invariance, unitarity, duality and others. It is a rather nontrivial problem to construct functions with such properties. In 1968 G. Veneziano proposed the following function describing the interaction of four particles

$$A(s,t) = \int_0^1 dx x^{-\alpha(s)-1}(1-x)^{-\alpha(t)-1} . \qquad (1.1)$$

Let k_1, k_2, k_3, k_4 be relativistic n-dimensional momenta of scattering particles. Their squares must be negative (so called tachyons)

$$k_i^2 = (k_i^0)^2 - (k_i^1)^2 - \ldots - (k_i^{n-1})^2 = -2 . \qquad (1.2)$$

The variables s and t have the form

$$s = (k_1 + k_2)^2, \qquad t = (k_2 + k_3)^2, \qquad (1.3)$$

one introduces also the variable $u = (k_1 + k_3)^2$, and besides we have $s + t + u = -8$. The energy-momentum conservation law is valid,

$$k_1 + k_2 + k_3 + k_4 = 0. \qquad (1.4)$$

At last, the function $\alpha(s)$ in (1.1) is linear

$$\alpha(s) = 1 + \frac{1}{2}s . \tag{1.5}$$

The beta function (1.1) can be expressed in terms of the Euler gamma function

$$A(s,t) = \frac{\Gamma(-\alpha(s))\Gamma(-\alpha(t))}{\Gamma(-\alpha(s) - \alpha(t))} \tag{1.6}$$

where

$$\Gamma(a) = \int_0^\infty x^{a-1} e^{-x} dx. \tag{1.7}$$

The Veneziano amplitude (1.1) has the symmetry property

$$A(s,t) = A(t,s) \tag{1.8}$$

(so called crossing-symmetry). It can be represented in the form

$$A(s,t) = \sum_{n=0}^{\infty} \frac{(\alpha(t)+1)(\alpha(t)+2)\dots(\alpha(t)+n)}{n!} \cdot \frac{1}{\alpha(s) - n} . \tag{1.9}$$

The equality (1.8) and the representation (1.9) ensure the duality property: one and the same amplitude can be represented either as a sum over poles in s-channel (formula (1.9)), or over poles in t-channel.

The asymptotic of $A(s,t)$ at large s and fixed t has the form

$$A(s,t) \sim s^{\alpha(t)}, \tag{1.10}$$

so called Regge behavior. It was proved that for validity of the unitary condition the dimension of space-time must be $n = 26$.

The generalization of the Veneziano amplitude to the case of scattering of N particles with momenta $(k_1, \dots, k_N)$ has the form

$$A_N = \int_{0<x_2<x_3<\dots<x_{N-2}<1} dx_2 dx_3 \dots dx_{N-2} \tag{1.11}$$

$$\prod_{j=2}^{N-2} |x_j|^{-k_1 \cdot k_j} |1 - x_j|^{-k_2 \cdot k_j} \prod_{2 \le l < m \le N-2} |x_l - x_m|^{-k_l \cdot k_m} .$$

Or in more symmetric form

$$A_N = \int \prod_{i=1}^{N} dx_i (dV_3)^{-1} \prod_{1 \leq l < m \leq N} |x_l - x_m|^{-k_l \cdot k_m}, \qquad (1.12)$$

where

$$dV_3 = dx_a dx_b dx_c (x_a - x_b)^{-1} (x_b - x_c)^{-1} (x_c - x_a)^{-1}$$

and x_a, x_b, x_c are arbitrarily chosen from $x_1, \ldots, x_N$ different variables. Division on dV_3 in (1.12) corresponds to a separation of the volume of the group $SL(2, R)$ from the expression

$$A = \int \prod_{i=1}^{N} dx_i \prod_{1 \leq l < m \leq N} |x_l - x_m|^{-k_l \cdot k_m} \qquad (1.13)$$

which is invariant under $SL(2, R)$ fractional-linear transformation of variables x_i.

There exists another collection of dual amplitudes (Virasoro-Shapiro), which corresponds to integration over the complex plain

$$B_N = \int \prod_{l=4}^{N} d^2 z_l \prod_{4 \leq i,j \leq N} |z_i - z_j|^{1/2 k_i \cdot k_j} \qquad (1.14)$$

The formula (1.14) is obtained from the expression

$$B = \int \prod_{i=1}^{N} d^2 z_i \prod_{1 \leq l < m \leq N} |z_l - z_m|^{1/2 k_l \cdot k_m} \qquad (1.15)$$

after the separation of the volume of the group $SL(2, \mathbb{C})$. Here $k_i^2 = -8$, $i = 1, \ldots N$.

For four particles we have

$$B_4 = \int_{\mathbb{C}} d^2 z |z|^{1/2 k_1 \cdot k_4} |1 - z|^{1/2 k_2 \cdot k_4}. \qquad (1.16)$$

In string theory the amplitudes (1.1) correspond to open strings and (1.16) correspond to closed strings. Note that these expressions for scattering amplitudes are only the first terms in the so-called decomposition of the number of loops for string amplitudes. By analogy to (1.16) one can consider instead of (1.1) the expression

$$A_4 = \int_{\mathbb{R}} dx |x|^{k_1 \cdot k_4} |1 - x|^{k_2 \cdot k_4} \tag{1.17}$$

where the integral is over all real axis. Then we obtain the symmetric amplitude

$$A_4 = A(s,t) + A(t,s) + A(u,s). \tag{1.18}$$

2. *p-Adic Amplitudes*

As it was shown by the development of mathematical physics during the last twenty years, a rather substantial modern string theory stems from a development of the simple formula (1.1). Therefore it is very interesting to find a generalization of the expressions (1.16) and (1.17) using only pure mathematical resources. According to two representations (1.1) and (1.6) two possible generalizations exist. One can use either the integral representation or the Γ-function representation. Here we shall consider the integral representation. The function $|x|^a$ is a character, so one can interpret the integral in (1.1) or in (1.17) as a convolution of two multiplicative characters on the real axis. Therefore the following generalization is suggested.

Let K be a field, $\gamma_\alpha(x)$ and $\gamma_\beta(x)$ be multiplicative characters on K, where α and β are some parameters on which these characters depend. Let

$$A(\gamma_\alpha, \gamma_\beta) = \int_K \gamma_\alpha(x) \gamma_\beta(1-x) dx, \tag{2.1}$$

where dx is a measure on K. For different fields K and different characters γ_α this formula gives us the known as well as new amplitudes. In particular, for $\gamma_\alpha(x) = |x|^\alpha$ and $K = \mathbb{R}$ we obtain the expression (1.17), and for $K = \mathbb{C}$ we obtain the expression (1.16).

Let us consider the case of $K = \mathbb{Q}_p$, $\gamma_\alpha(x) = |x|_p^{\alpha-1}$, $\gamma_\beta(x) = |x|_p^{\beta-1}$, where α and β are complex parameters. The formula (2.1) then gives the $\mathcal{B}$-function considered in Sec. 8.3,

$$\mathcal{B}_p(\alpha, \beta) = \int_{\mathbb{Q}_p} |x|_p^{\alpha-1} |1 - x|_p^{\beta-1} dx \ . \tag{2.2}$$

The integral (2.2) absolutely converges in the region where $\mathrm{Re}\,\alpha > 0$, $\mathrm{Re}\,\beta > 0$, $\mathrm{Re}(\alpha+\beta) < 1$. In Sec. 8, see (3.9), the expression for B-function (2.2) in terms of Γ_p-function was given,

$$B_p(\alpha,\beta) = \frac{\Gamma_p(\alpha)\Gamma_p(\beta)}{\Gamma_p(\alpha+\beta)}, \tag{2.3}$$

where

$$\Gamma_p(\alpha) = \frac{1-p^{\alpha-1}}{1-p^{-\alpha}}, \tag{2.4}$$

or in more symmetric form

$$B_p(\alpha,\beta) = \Gamma_p(\alpha)\Gamma_p(\beta)\Gamma_p(\gamma)\,, \tag{2.5}$$

where

$$\alpha+\beta+\gamma = 1\,. \tag{2.6}$$

If we put in (2.3) $\alpha = -\alpha(s)$, $\beta = -\alpha(t)$, where $\alpha(s)$ is defined by (1.5), and s and t are defined by the formulas (1.3), then we get the simple p-adic dual amplitude $B_p(-\alpha(s),-\alpha(t))$. By virtue of the relation

$$\alpha(s)+\alpha(t)+\alpha(u) = -1,$$

and using (2.5) and (2.6), it can be written in the symmetric form

$$B_p(-\alpha(s),-\alpha(t)) = \prod_{x=s,t,u} \frac{1-p^{-\alpha(x)-1}}{1-p^{\alpha(x)}}\,. \tag{2.8}$$

The formula (1.13) admits a direct extension to p-adic case

$$A = \int_{\mathbb{Q}_p^N} \prod_{i=1}^{N} dx_i \prod_{1\le l<m\le N} |x_l - x_m|_p^{-k_l\cdot k_m}\,. \tag{2.9}$$

After the separation of volume of the group $SL(2,\mathbb{Q}_p)$ from (2.9) we obtain

$$A_{N,p} = \int_{\mathbb{Q}_p^{N-3}} dx_2 \ldots dx_{N-2} \prod_{j=1}^{N-2} |x_j|_p^{-k_1\cdot k_j} |1-x_j|_p^{-k_2\cdot k_j} \cdot \prod_{2\le l<m\le N-2} |x_l - x_m|_p^{-k_l\cdot k_m}. \tag{2.10}$$

The integral (2.10) can be calculated explicitly giving an answer similar to (2.8).

It is interesting to note that these p-adic dual amplitudes can be obtained by the rules of quantum field theory from the Lagrangian corresponding to the following nonlinear equation

$$p^{-\Box/2}\phi = \phi^p . \tag{2.11}$$

Here $\phi = \phi(x)$ is a real field in n-dimensional real space-time with coordinates $x = (x_0, \dots, x_{n-1})$, $\Box$ is the d'Alembert operator

$$\frac{\partial^2}{\partial x_0^2} - \frac{\partial^2}{\partial x_1^2} - \dots - \frac{\partial^2}{\partial x_{n-1}^2}. \tag{2.12}$$

The equation (2.12) has the static solitonic solution of very simple form

$$\phi = \frac{n-1}{p^{2(p-1)}} \exp\left[-\frac{1}{2}\frac{p-1}{p\ln p}\bar{x}^2\right], \tag{2.13}$$

where $\bar{x} = (x_1, \dots, x_{n-1})$.

We shall use the notations

$$B(\alpha,\beta) = \int_{\mathbb{R}} dx |x|^{\alpha-1} |1-x|^{\beta-1} , \tag{2.14}$$

$$\alpha + \beta + \gamma = 1 . \tag{2.15}$$

Lemma 2.1. *The following formula for the function* (2.14) *is valid*

$$B(\alpha,\beta) = \frac{\zeta(1-\alpha)}{\zeta(\alpha)} \frac{\zeta(1-\beta)}{\zeta(\beta)} \frac{\zeta(1-\gamma)}{\zeta(\gamma)}, \tag{2.16}$$

where ζ is the Riemann zeta function, and arguments α and β belong to the region $\alpha > 0$, $\beta > 0$, $\alpha + \beta < 1$, γ is defined by the formula (2.15).

■ Dividing the integration region in the integral (2.14) on three parts we transform it in the form

$$B(\alpha,\beta) = \frac{\Gamma(\alpha)\Gamma(\beta)}{\Gamma(\alpha+\beta)} + \frac{\Gamma(\beta)\Gamma(\gamma)}{\Gamma(\beta+\gamma)} + \frac{\Gamma(\gamma)\Gamma(\alpha)}{\Gamma(\gamma+\alpha)}. \tag{2.17}$$

Using the relation for the gamma function

$$\Gamma(x)\Gamma(1-x) = \frac{\pi}{\sin \pi x},$$

and doing simple trigonometrical calculations we reduce (2.17) to the form

$$B(\alpha, \beta) = \frac{4}{\pi} \cos \frac{\pi\alpha}{2} \cos \frac{\pi\beta}{2} \cos \frac{\pi\gamma}{2} \Gamma(\alpha)\Gamma(\beta)\Gamma(\gamma). \tag{2.18}$$

Taking into account the functional relation for the zeta function

$$(2\pi)^x \zeta(1-x) = 2 \cos \frac{\pi x}{2} \Gamma(x)\zeta(x)$$

and the relation (2.15), one get from (2.18)

$$B(\alpha, \beta) = \frac{\zeta(1-\alpha)}{\zeta(\alpha)} \frac{\zeta(1-\beta)}{\zeta(\beta)} \frac{\zeta(1-\gamma)}{\zeta(\gamma)} .$$

■

Remark. Let us note the following relation between the zeta and the gamma functions

$$\frac{\Gamma(\alpha)\Gamma(\beta)}{\Gamma(\alpha+\beta)} + \frac{\Gamma(\beta)\Gamma(\gamma)}{\Gamma(\beta+\gamma)} + \frac{\Gamma(\gamma)\Gamma(\alpha)}{\Gamma(\gamma+\alpha)} = \frac{\zeta(1-\alpha)}{\zeta(\alpha)} \frac{\zeta(1-\beta)}{\zeta(\beta)} \frac{\zeta(1-\gamma)}{\zeta(\gamma)} \tag{2.19}$$

where

$$\alpha + \beta + \gamma = 1.$$

3. *Adelic Products*

The following theorem is valid.

Theorem 3.1. *Let real parameters α and β take their values in the region*

$$\alpha < -1, \beta < -1 \tag{3.1}$$

Then we have the following relations for the functions (2.3) *and* (2.16)

$$\prod_p (-1) B_p(\alpha, \beta) = \frac{\zeta(-\alpha)}{\zeta(1-\alpha)} \frac{\zeta(-\beta)}{\zeta(1-\beta)} \frac{\zeta(1-\alpha-\beta)}{\zeta(\alpha+\beta)}, \tag{3.2}$$

$$B(\alpha,\beta)\prod_p |B_p(\alpha,\beta)| = \frac{\zeta(-\alpha)}{\zeta(\alpha)}\frac{\zeta(-\beta)}{\zeta(\beta)}\frac{\zeta(\alpha+\beta)}{\zeta(-\alpha-\beta)}, \tag{3.3}$$

with the products in (3.2) *and* (3.3) *on all prime p being absolutely convergent.*

■ Let us rewrite the expression (2.3) in the form

$$B_p(\alpha,\beta) = (-1)p^{\alpha}\frac{1-p^{\alpha-1}}{1-p^{\alpha}}p^{\beta}\frac{1-p^{\beta-1}}{1-p^{\beta}}p^{\gamma-1}\frac{1-p^{1-\gamma}}{1-p^{-\gamma}}$$

or taking into account (2.15) in the form

$$B_p(\alpha,\beta) = -\frac{1-p^{\alpha-1}}{1-p^{\alpha}}\frac{1-p^{\beta-1}}{1-p^{\beta}}\frac{1-p^{1-\gamma}}{1-p^{-\gamma}}. \tag{3.4}$$

In the range of values of the parameters (3.16) we have $\gamma > 2$ and

$$B_p(\alpha,\beta) < 0. \tag{3.5}$$

Using the Euler formula (0.1) from Sec. 7.7, we find from (3.4)

$$\begin{aligned}\prod_p(-1)B_p(\alpha,\beta) &= \frac{\zeta(-\alpha)}{\zeta(1-\alpha)}\frac{\zeta(-\beta)}{\zeta(1-\beta)}\frac{\zeta(-\gamma)}{\zeta(1-\gamma)}\\ &= \frac{\zeta(-\alpha)}{\zeta(1-\alpha)}\frac{\zeta(-\beta)}{\zeta(1-\beta)}\frac{\zeta(1-\alpha-\beta)}{\zeta(\alpha+\beta)}\end{aligned} \tag{3.6}$$

that is the relation (3.2). The product in (3.6) absolutely converges. In fact, as it is known for the absolute convergence of a product

$$\prod_p(1+x_p)$$

it is necessary and sufficient to have the absolute convergence of a series $\sum_p x_p$. We have

$$\frac{1-p^{\alpha-1}}{1-p^{\alpha}} = 1+(1-p^{-1})\frac{1}{p^{-\alpha}-1}$$

and for $\alpha < -1$

$$\sum_p \frac{1}{p^{-\alpha}-1} < \infty.$$

Analogously one proves convergence of the products on p in other terms in (3.4). The formula (3.3) follows from (3.2) taking into account the Lemma 2.1. ∎

Remark. In a paper by Freund and Witten the following remarkable formula was derived

$$B(\alpha,\beta)\prod_p B_p(\alpha,\beta)=1 \tag{3.7}$$

There are some subtleties with a convergence of the product and a result depends on the choice of a regularization.

We show that there exists a connection between the formulas (3.2), (3.3), (3.7) and the Tate formula. From the formula (2.2) of Sec. 7.7 for $\theta(\lambda)\equiv 1$ it follows the relation

$$\prod_p \int \varphi_p(x_p)|x_p|_p^\beta dx_p = \prod_p \int \tilde{\varphi}_p(y_p)|y_p|_p^{1-\beta} dy_p, \tag{3.8}$$

here the case $p=\infty$ is allowed, and β is a real parameter. The function $\varphi_p(x_p)$ here must belong to the Bruhat-Schwartz space. However we formally put

$$\varphi_p(x_p)=|1-x|_p^\alpha, \tag{3.9}$$

where α is a real parameter and then the relation (3.8) yields the formula (3.7). Indeed, one has

$$\tilde{\varphi}_p(y)=C_p(\alpha)\chi(y)|y_p|_p^{-1-\alpha}, \qquad p\neq\infty,$$
$$\tilde{\varphi}_\infty(y)=C_\infty(\alpha)e^{2\pi iy}|y|_p^{-1-\alpha}, \qquad p=\infty$$

where

$$C_p(\alpha)=\frac{1-p^\alpha}{1-p^{-1-\alpha}},p\neq\infty; \qquad C_\infty(\alpha)=-2(2\pi)^{-1-\alpha}\sin\frac{\pi\alpha}{2}\Gamma(1+\alpha)\,.$$

Therefore according to (3.8)

$$\prod_p \int |1-x_p|_p^\alpha |x|_p^\beta dx_p = \prod_p C_p(\alpha)C_p(-\alpha-b-1)=1,$$

as

$$\prod_p C_p(\alpha) = -2(2\pi)^{-1-\alpha} \sin\frac{\pi\alpha}{2}\Gamma(1+\alpha)\prod_p \frac{1-p^\alpha}{1-p^{-1-\alpha}}$$
$$= -2(2\pi)^{-1-\alpha} \sin\frac{\pi\alpha}{2}\Gamma(1+\alpha)\frac{\zeta(1+\alpha)}{\zeta(-\alpha)} = 1\,.$$

If we put

$$\varphi_p(x) = -|1-x|_p^\alpha$$

instead of (3.3), then we obtain the relation (3.7).

4. *String Action*

We have discussed the Veneziano amplitude and its generalizations. In fact these amplitudes are the only first terms in the so-called loop expansion. Full answer has the form

$$A(k_1,\dots k_N) = \sum_{h=0}^{\infty} A_h(k_1,\dots k_N),$$
$$A_h(k_1,\dots k_N)$$
$$= \int_{M_h} d\mu(\tau) \int_{\sum_h} \exp\left(-\frac{1}{2}\sum_{i<j} k_i\cdot k_j G(\xi_i,\xi_j;\tau)\right)\prod_{j=1}^N dS_j\,. \tag{4.1}$$

Here $\sum_h$ is a compact oriented surface of genus h, ξ_j, $j=1,\dots,N$ are coordinates of N points on it, $dS_j = \sqrt{g(\xi_j)}d^2\xi_j$, $g(\xi) = \det g_{\alpha\beta}(\xi)$, where $g_{\alpha\beta}$ is a metric on $\sum_h$, $\alpha,\beta = 1,2$. M_h is the moduli space of Riemann surfaces of genus h,τ are local coordinates on M_h, $G(\xi,\zeta;\tau)$ is a Green function of the scalar Laplacian on $\sum_h$, $d\mu(\tau)$ is a measure on $M_h, k_1,\dots k_N$ are momenta of particles (tachyons), $k_j \in \mathbb{R}^d$, $j=1,\dots N$. One has $k_j^2 = 2$ and $k_1+\dots+k_N = 0$.

We will discuss this expression in more details in the next subsection. Here we note that this formula is obtained from a formal functional integral

$$\delta(k)A_h(k_1,\dots k_N) = \int [Dg_{\alpha\beta}]\int [DX^\mu]\prod_{j=1}^N V(k_j)e^{-I}, \tag{4.2}$$

where $\delta(k) = \delta\left(\sum_{i=1}^{N} k_i\right)$ is the Dirac δ-function. Here the classical action is

$$I = \frac{1}{2}\int_{\Sigma_h} d^2\xi\sqrt{g}g^{\alpha\beta}\partial_\alpha X^\mu \partial_\beta X^\mu, \tag{4.3}$$

$X = X^\mu(\xi), \mu = 1, \ldots, d$ is a map from $\sum_h$ to $\mathbb{R}^d$, $\partial_\alpha X^\mu = \frac{\partial}{\partial\xi^\alpha}X^\mu(\xi)$, vertex operator $V(k)$ has the form

$$V(k) = \int_{\Sigma_h} d^2\xi\sqrt{g}\exp(ik\cdot X(\xi)) .$$

Symbols $[Dg_{\alpha\beta}]$ and $[DX^\mu]$ mean that one should integrate over all metrics on $\sum_h$ and over all maps X^μ.

5. *Moduli Space and Theta-Functions*

The action I possesses two infinite symmetry groups. One of them is the group of *Weyl* rescalings which acts by $g_{\alpha\beta} \to e^\sigma g_{\alpha\beta}$. It is called conformal or Weyl group. The other group of symmetries is the group of diffeomorphisms *Diff*, or reparametrizations of $\sum_h$. A diffeomorphism that is continuously connected to the identity is called a local diffeomorphism. The group of these diffeomorphisms is denoted by Diff_0. There are also global diffeomorphisms that are not in the connected component of the identity.

The space

$$M_h = \left\{ \text{Metrics on} \sum_h \right\} / Diff \cdot Weyl$$

is called moduli space. The moduli space M_h is a complex variety of complex dimension 0 for $h = 0$, 1 for $h = 1$ and $3h - 3$ for $h \geq 2$. The bigger space

$$T_h = \left\{ \text{Metrics on} \sum_h \right\} / Diff_0 \cdot Weyl$$

is called Teichmuller space. The group obtained by taking all of the diffeomorphism modulo the local diffeomorphism is the *mapping class group* (or modular group). Let us choose a slice S through the space of metrics which is transversal to the actions of Diff_0 and Weyl. A subdomain of S in which no two points are related by a global diffeomorphism is a *fundamental domain* for the mapping class group. The conformal class of a metric amounts to a complex structure on $\sum_h$.

The first homology group $H_1(\sum_h, Z)$ has $2h$ generators. Let us choose a symplectic basis $\{a_1, \ldots, a_h, b_1, \ldots b_h\}$ of 1-cycles on $\sum_h$ satisfying the following conditions on intersection numbers:

$$a_i \cdot a_j = b_i \cdot b_j = 0, \qquad a_i \cdot b_j = \delta_{ij}.$$

Any two symplectic bases are related by a transformation from the symplectic modular group $S_p(2h, Z)$. The space of holomorphic 1-forms on $\sum_h$ has the basis $\{\omega_1, \ldots, \omega_h\}$ uniquely determined by the conditions

$$\int_{a_i} \omega_j = \delta_{ij}, \quad i, j = 1, \ldots h.$$

A matrix $\tau = (\tau_{ij})$,

$$\tau_{ij} = \int_{b_i} \omega_j$$

is called the period matrix of $\sum_h$ and is a complex symmetric $h \times h$ matrix with positive defined imaginary part. Given a base point $Q_0 \in \sum_h$ one associates to every point Q on $\sum_h$ a complex h-component vector $f(Q)$ by the Jacobi map defined as

$$Q \longrightarrow f_i(Q) = \int_{Q_0}^{Q} \omega_i, \quad i = 1, \ldots h.$$

This vector is unique up to the periods τ. So f is an element of the complex torus

$$J(\textstyle\sum_h) = C^h / Z^h + \tau Z^h$$

called the Jacobian variety of $\sum_h$.

The theta function is defined for $z \in J(\sum_h)$ by the sum

$$\theta(z, \tau) = \sum_{N \in Z^h} \exp(i\pi n^t \tau n + 2\pi i n^t z).$$

A theta function with characteristics $\alpha, \beta \in 1/2\mathbb{Z}^h$ is defined by

$$\theta\begin{bmatrix}\alpha\\ \beta\end{bmatrix}(z,\tau) = \exp(i\pi n^t\tau\beta + 2\pi i\alpha^t(z+\beta))\theta(z+\tau\alpha+\beta,\tau)\,.$$

The parity of $4\alpha^t\beta$ is called the parity of characteristic $\{\alpha,\beta\}$.

One needs the prime form $E(w,\xi)$ which is defined as the holomorphic differential form on $\sum_h \times \sum_h$ of weight $(-1/2,0)\times(-1/2,0)$:

$$E(w,\xi) = \frac{\theta\begin{bmatrix}\alpha\\ \beta\end{bmatrix}\left(\int_\xi^w \omega, \tau\right)}{(h(w))^{1/2}(h(\xi))^{1/2}} \tag{5.1}$$

where

$$h(w) = \sum_{i=1}^h \frac{\partial}{\partial z_i}\theta\begin{bmatrix}\alpha\\ \beta\end{bmatrix}(0,\tau)\cdot\omega_i(w)$$

$E(w,\xi)$ is independent of the choice of the odd characteristic $\{\alpha,\beta\}$, it is only zero for $w=\xi$.

There exists a measure on M_h reflecting the holomorphic structure of moduli space. According to the Mumford theorem the line bundle

$$K\otimes\lambda^{-13} \tag{5.2}$$

over moduli space is holomorphically trivial. Here K is the canonical bundle of moduli space, i.e. the highest wedge power of its cotangent bundle. λ is the highest wedge power of the bundle of holomorphic 1-forms on the surface. There is a unique holomorphic nowhere vanishing section F of $K\otimes\lambda^{-13}$. Let $v_1,\ldots v_{3h-3}$ be analytic coordinates on $M_h(h\geq 2)$ then

$$d\mu = |F(v)|^2(\det\ \mathrm{Im}\ \tau)^{-13}dv\wedge d\bar{v} \tag{5.3}$$

is a measure on M_h, here

$$dv = dv_1\wedge\ldots\wedge dv_{3h-3}.$$

6. *Multiloop Amplitudes*

To define the functional integral (4.2) one uses the procedure of quantization of gauge theories. One can argue that the amplitude (4.2) has the form

$$A_h(k_1,\ldots k_N) = \int_{M_h} d\mu(v)T(k_1,\ldots k_N;v) \tag{6.1}$$

where $T(k_1,\dots k_N;v)$ comes from the Gaussian integral

$$\delta(k)T(k_1,\dots k_N;v) = \int [DX^\mu]e^{-I}\prod_{j=1}^{N} V(k_j)$$
$$= \delta(k)\int \exp\left(-\frac{1}{2}\sum_{i<j} k_i\cdot k_j G(\xi_i,\xi_j)\right)\prod_{j=1}^{N}\sqrt{g}(\xi_j)d^2\xi_j$$

This can be expressed in terms of prime form (5.1):

$$T(k_1,\dots k_N;v) = \int \prod_{i<j}|f(\xi_i,\xi_j)|^{2k_ik_j}\prod_{j=1}^{N} d^2\xi_j, \tag{6.2}$$
$$f(\xi_i,\xi_j) = E(\xi_i,\xi_j)\exp\{-\pi I_m z_{ij}^t(\text{ Im }\tau)^{-1}\text{ Im } z_{ij}\},$$
$$z_{ij} = \int_{\xi_i}^{\xi_j}\omega.$$

In particular for torus ($h = 1$) one has

$$A_1(k_1,\dots k_N) = \int_{M_h}\frac{d^2v}{(\text{Im } v)^{14}e^{-4\pi \text{Im} v}}|\eta(v)|^{-48} \tag{6.3}$$
$$\cdot\int \prod_{r=1}^{N-1} d^2\nu_r\prod_{r<s}(\chi_{rs})^{k_r\cdot k_s/2}$$

Here

$$\eta(v) = e^{\frac{i\pi v}{12}}\prod_{n=1}^{\infty}(1-e^{2\pi i n v})$$

is the Dedekind eta function,

$$\chi_{rs} = \chi(\nu_r-\nu_s, e^{2\pi i v}),$$
$$\chi(x,w) = \exp\left(\frac{\ln^2|x|}{2\ln|w|}\right)\left|\frac{(1-x)}{x^{1/2}}\prod_{n=1}^{\infty}\frac{(1-w^n x)(1-w^n/x)}{(1-w^n)^2}\right|.$$

The integral runs over the fundamental domain $M_1 = \{v\in\mathbb{C} : \text{Im}v > 0,\ |\text{Re}\, v|\le 1/2\}$ and

$$|e^{2\pi i v}|\le|e^{2\pi\nu_r}|\le 1,\qquad r = 1,\dots N-1.$$

The integral in (6.3) is invariant under modular $SL(2,\mathbb{Z})$ transformations. The integral (6.3) diverges and one needs to use a regularization. The problem of divergence is avoided in the superstring theory. The four massless particles scattering amplitude for superstring theory converges and 1-loop amplitude has a simple form

$$\int_{M_1} \frac{d^2\tau}{(\text{Im } \tau)^5} \int \prod_{r=1}^{3} d^2\nu_r \prod_{r<s} (\chi_{rs})^{k_r k_s/2}.$$

The requirement of modular invariance of the string measure $d\mu$ permits for low genus to express the measure in terms of the modular forms. For genus $h = 2$ one can use the Siegel upper half plane H_2 to describe the moduli space M_2. More precisely, M_2 is isomorphic to the space $H_2/Sp(4,\mathbb{Z})$ modulo the equivalence class of the space of diagonal period matrices. The measure $d\mu$ on H_2 should have the form

$$d\mu = |f(\tau)|^2 (\det \text{Im } \tau)^{-13} \prod_{i \le j} d\tau_{ij} \ . \tag{6.4}$$

A holomorphic function $f : H_h \to \mathbb{C}$ is called a genus h modular form of weight k if f satisfies

$$f((A\tau + B)(C\tau + D)^{-1}) = \det(C\tau + D)^k f(\tau)$$

for all $\begin{pmatrix} A & B \\ C & D \end{pmatrix} \in Sp(2h,\mathbb{Z})$. Under modular transformations one has

$$\prod_{i \le j} d\tau_{ij} \longrightarrow \det(C\tau + D)^{-3} \prod_{i \le j} d\tau_j$$

$$\det \text{Im } \tau \longrightarrow \frac{\det \text{Im } \tau}{|\det(C\tau + D)|^2}$$

therefore the function f^{-1} in (5.5) should be a modular form of weight 10. The ring of genus-two modular forms is known. There exist a unique modular form ψ_{10} of weight 10 vanishing on the diagonal period matrices,

$$\psi_{10} = \prod_{\alpha,\beta} \theta^2 \begin{bmatrix} \alpha \\ \beta \end{bmatrix} (0,\tau),$$

the product over even characteristics.

Thus we have

$$d\mu = |\psi_{10}|^{-2}(\det \mathrm{Im}\ \tau)^{-13} \prod_{i \leq j} d\tau_{ij}.$$

7. *Rigid Analytic Geometry and p-Adic Strings*

We see from the previous discussion that the theory of Riemann surfaces in particular moduli space and theta functions are used in string theory. To develop p-adic string theory one needs a p-adic analog of complex analysis including the basic principle of analytic continuation. The p-adic field $\mathbb{Q}_p$ is totally disconnected so one cannot use to this end the "naive" notion of analyticity considered in Sec. 2. Nevertheless an appropriate theory was constructed by Tate and others. It is called rigid analytic geometry.

The main idea is to perform analytic continuation only with respect to certain *admissible open coverings* of rigid space. Analytic functions are no longer considered on all open subsets of such a space; one has to restrict oneself to admissible open sets which form the so-called Grothendick topology. There exists a nontrivial notion of connectedness of such a space. This rigid analytic approach was used for the uniformization of algebraic curves and it is natural to use it for p-adic strings. We are going to describe shortly some points from rigid analytic geometry.

Let $T_n = K\langle z_1, \ldots z_n\rangle$ be a ring consisting of power series converging on

$$B_n = \{(x_1, \ldots x_n) \in K^n : |x_i| \leq 1, i = 1, \ldots, n\}.$$

Here K is a complete non-archimedean field for example $\mathbb{Q}_p$.

An *affinoid algebra* A over K (or Tate-algebra) is a K-algebra which is a finite extension of T_n. That means: there exists a K-algebra homomorphism $T_n \to A$ which makes A into a finitely generated T_n-module. The space $Sp(A)$ of maximal ideals of A is called an *affinoid space*. Every affinoid algebra A has the form T_n/I for some ideal $I \subset T_n$ and is a Banach space with respect to the quotient norm. One can identify $Sp(A)$ with the points x in B_n such that $f(x) = 0$ for all $f \in I$.

Let $X = Sp(A)$. Let $f_0, f_1, \ldots f_n \in A$ be such that they have no common zeros on X. Then the subspace $U \subset X$

$$U = \{x \in X : |f_i(x)| \leq |f_0(x)|, i = 1, \ldots n\}$$

is called a rational domain. We can define a Grothendick topology on X by taking as open sets all rational domains. A Grothendick topology on a topological space X is defined as follows. Let F be a set of open subsets of X and let $Cov(U)$ be a collection of coverings of $U \subset X$. Then (F, Cov) is called a *Grothendick topology* if it has the following properties:

1) $\phi, X \in F$ and if $U, V \in F$ then $U \cap V \in F$,
2) $\{U\} \in Cov(U)$ for all $U \in F$,
3) if $\mathcal{U} \in Cov(U)$ and $V \subset U$ with $U, V \in F$ then $\mathcal{U} \cap V \in Cov(V)$
4) if $\mathcal{U}_i \in Cov(U_i)$ and $(U_i) \in Cov(U)$ then $\bigcup_i \mathcal{U}_i \in Cov(U)$.

The elements of F are called admissible open sets and the elements of $Cov(U)$ admissible coverings of U.

With a rational domain U one associates the affinoid algebra

$$A_U = A\langle z_1, \dots z_n\rangle / (f_1 - z_1 f_0, \dots f_n - z_n f_0) \, .$$

Then we can define a presheaf O_X on X by associating A_U to each admissible open set U. A basic result is that O_X is in fact a sheaf for the Grothendick topology. The space $X = Sp(A)$ with its Grothendick topology and the structure sheaf O_X is called an affinoid space. An *analytic space* X is defined as follows. X has a Grothendick topology and a sheaf O_X such that there exists a $(X_i) \in Cov(X)$ with $(X_i, O_X|X_i)$ is an affinoid space.

Example. If X is a complete, non-singular and irreducible curve over K with function field $K(X)$, then for any $f \in K(X)$ with $f \neq 0$ the set $X_f = \{x \in X : |f(x)| \leq 1\}$ is an admissible affinoid subset of X. The affinoid algebra O_{X_f} is a completion of algebraic holomorphic functions on X_f.

A uniformization of X is its representation in the form Y/Γ where Y is an analytic space and Γ is an infinite group that acts discontinuously on Y. As it is known the upper half plane H uniformizes curves of genus $g \geq 2$ over $\mathbb{C}$. The problem of uniformization of curves over a non-archimedean field K is more complicated. First consider an elliptic curve X. If X is defined over $\mathbb{C}$ then X has uniformization $\mathbb{C}/\Gamma$ where Γ is a lattice. Such an uniformization does not exists over a non-archimedean field K. If we take $\Gamma = \mathbb{Z} + \mathbb{Z} \cdot \tau$ with Im $\tau > 0$ then we can construct another uniformization of X by using the exponent map. One gets the uniformization $\mathbb{C}^*/\langle q\rangle$ of X where $\langle q\rangle$ is the multiplicative group generated by $q = e^{2\pi i\tau}$. This uniformization does have an analogy over non-archimedean field K. For $q \in K^*$ with $|q| < 1$ an analytic torus $X_q = K^*/\langle q \rangle$ is an elliptic curve and is called Tate curve. The j-invariant $j(q)$ of X_q satisfies $|j(q)| > 1$. The reduction of X_q is a rational curve with singularities over $\bar{K}$ where $\bar{K}$ denotes the residue field over K.

To uniformize a complete non singular curve X of genus $g \geq 2$ one uses the Schottky uniformization. A Schottky group Γ is a subgroup of $PGL(2, K)$ which is finitely generated, discontinues and has no elements of finite order. It has a nice fundamental domain: $F = \mathbb{P} - (2g$ open disks). Let L be set of limit points for the action of Γ in $\mathbb{P}$. Then $(\mathbb{P} - L)/\Gamma$ is an analytic space and analytically isomorphic to a complete nonsingular and irreducible curve of genus g. Such curve is called a Mumford curve. For Mumford curve there exists a nice theory of automorphic forms, the Jacobian variety, period matrix, non-archimedean Siegel half-space and theta functions. In particular one can define the prime form

$$E(x, y) = (x - y) \prod_{\gamma \in \Gamma} \frac{x - \gamma(y)}{x - \gamma(x)} \cdot \frac{y - \gamma(x)}{y - \gamma(y)}$$

and try to use it in p-adic string theory simply by replacing the complex variables by p-adic ones. If we want to develop p-adic string theory in a more systematic way one should start with a classical p-adic action and then quantize it. Recall that closed strings correspond to compact Riemann surfaces without boundary and open strings correspond to Riemann surfaces with boundary.

One can think of the curve X, endowed with a rigid analytic structure, as the non-archimedean analogue of a compact Riemann surface without boundary. Then the definition of X_f suggests that an affinoid space is the non-archimedean analogue of a compact Riemann surface with boundary. One can try to define the naive boundary of X_f to be the set $B = \{x \in X_f : |f(x)| = 1\}$. However this boundary depends very much on the choice of f. Using rigid geometry we can define the canonical boundary ∂X_f in such a way that $B \subset \partial X_f$.

An approach to p-adic string theory based on a p-adic analogue of the classical string action is not available at the moment. One attempts to use the $SL(2, \mathbb{Q}_p)$-symmetry of the operator D and consider an action

$$S = \int_F \int_F \frac{(X^\mu(x) - X^\mu(y))^2}{|x - y|^2} dx dy$$

where F is a fundamental domain of a Schottky group. One really can obtain p-adic tree amplitudes from this action if $F = \mathbb{Q}_p$. It was also suggested to interpret the open p-adic string world sheet as a coset space T/Γ where T is the Bruhat-Tits tree, $T = PGL(2, \mathbb{Q}_p)/PGL(2, \mathbb{Z}_p)$ and Γ is a Schottky group. The tree T is the connected infinite graph with no loops. Each vertex of T is connected with $p+1$ neighbor vertices by edges. Note that there is a canonical tree $T(X)$ for any compact subset X of $\mathbb{P}$.

It is an open problem what would be a string measure on p-adic moduli space. Perhaps most principal approach to p-adic string theory would be to start with a theory over the global field of rational numbers $\mathbb{Q}$ in accordance with the discussion in the Introduction.

One can expect that the beautiful rigid analytic geometry will also find an application in p-adic gravity.

XV. q-Analysis (Quantum Groups) and p-Adic Analysis

We are going to discuss some remarkable relation between theory of quantum groups (q-analysis) and p-adic analysis.

1. *p-Adic and q-Integrals*

The quantum group $SU_q(2)$ is a Hopf algebra with generators a and c satisfying quadratic relations. It was shown by Woronovič that there exists a Haar measure m on $SU_q(2)$.

Proposition 1. *The Haar measure on $SU_q(2)$ coincides with the Haar measure on the field of p-adic numbers $\mathbb{Q}_p$ if $q = 1/p$.*

■ It is known that the cohomology group $H^3(SU_q(2)) = \mathbb{C}$, therefore there exist a unique linear functional

$$\int : \Gamma^3 \longrightarrow C,$$

such that $\int d\zeta = 0$ for $\zeta \in \Gamma^2$, where Γ^n is the module of n-differential forms. For coordinate ring of $SU_q(2)$ one has

$$\int f\omega_0\omega_1\omega_2 = m(f)$$

where m is the Haar measure, ω_i are differential forms on $SU_q(2)$. If f is a polynomial in cc^*, we have

$$m(f) = (1-q^2)\sum_{n=0}^{\infty} f(q^{2n})q^{2n}, \quad 0 < q < 1.$$

This is in fact the well known Jackson integral in q-analysis

$$\int_0^1 f(x)d_q x = (1-q)\sum_{n=0}^{\infty} f(q^n)q^n . \tag{1.1}$$

Now we recall that the integral over $\mathbb{Q}_p$ with respect to the Haar measure is

$$\int_{|x|_p\leq 1} f(|x|_p)dx = \left(1-\frac{1}{p}\right)\sum_{n=0}^{\infty} f(q^{-n})q^{-n} . \tag{1.2}$$

Hence we get that (1.1) is equal to (1.2) if

$$q = \frac{1}{p}$$

i.e.

$$\int_0^1 f(x)d_{1/p}x = \int_{|x|_p\leq 1} f(|x|_p)dx . \qquad \blacksquare$$

There is the following generalization of this observation. Let us consider the operator

$$\begin{aligned} M[f](|x|_p) &= \frac{1}{(1-p^{-1})|x|_p}\int_{|y|_p=|x|_p} f(y)dy \\ &= \int_{|u|_p=1} f(u/|x|_p)\frac{du}{1-p^{-1}}, \qquad x\in G . \end{aligned}$$

Here $f \in L^1_{\text{loc}}(G)$, $G = \underset{\gamma\in B}{\cup} S_\gamma$, $S_\gamma = [|x|_p = p^\gamma]$. Then

$$\int_G f(x)dx = \int_B M[f](r)d_{1/p}r,$$

where

$$\int_B \varphi(r) d_q r = (1-q) \sum_{\gamma \in B} q^\gamma \varphi(q^\gamma).$$

2. *Differential Operators*

Let us now consider a relation between differential operators in q-analysis and in p-adic analysis. In q-analysis one has the following operator of differentiation

$$\partial_q f(x) = \frac{f(x) - f(qx)}{(1-q)x}. \tag{2.1}$$

One uses also

$$\partial_q f(x) = \sum_i^{\infty} f(q^{i+1/2} x)(xq^i - xq^{i+1}) \, .$$

In p-adic analysis if we consider a real valued function $f(x)$ depending on a p-adic variable x we cannot use the standard definition of differention, because one cannot multiply real and p-adic numbers. In Sec. 9 it was considered the following operator

$$D^\alpha f(x) = \frac{p-1}{1-p^{-1-\alpha}} \int_{\mathbb{Q}_p} \frac{f(x) - f(y)}{|x-y|^{\alpha+1}} dy. \tag{2.2}$$

Restricting to rational x in (2.1) and (2.2) we see that the expressions (2.1) and (2.2) are very similar, in particular, both are non-local operators.

3. *Spectra of the q-Deformed Oscillator and the p-Adic Model*

Let us discuss the connections between the spectra of the q-deformed oscillator and of the p-adic model.

The spectrum of p-adic equation

$$D\psi(x) + V(|x|_p)\psi(x) = E\psi$$

for

$$V(|x|_p) = \frac{p}{1-p^2} \left[|x|_p - \frac{1+p+p^2}{p|x|_p} \right]$$

has eigenvalues

$$E_n = [n]_{1/p} = \frac{p^{-n} - p^n}{p^{-1} - p}, \qquad p \in Z. \tag{3.1}$$

For the q-deformed oscillator

$$aa^+ - qa^+a = q^{-N}, \quad [N,a] = -a, [N,a] = a^+$$

the spectrum was found by Biedenharn

$$H = aa^+, \quad H|n>= [n]_q|n> . \tag{3.2}$$

If

$$q = \frac{1}{p}$$

one gets the p-adic spectrum (3.1).

In Sec. 10 it was shown that in the spectrum of more general equation

$$D^\alpha \psi(x) + V(|x|_p)\psi(x) = E\psi, \qquad \alpha > 0$$

there is the following family of eigenvalues

$$E_{n,l} = p^{\alpha n} + V(p^{l-n}), \qquad l = 1,2,3,\dots; \quad n = 0, \pm 1, \dots$$

Similar spectrum there appeared for a q-deformed Schrodinger equation.

XVI. Stochastic Processes over the Field of p-adic Numbers

This section presents elements of a theory of stochastic processes over the field of p-adic numbers. Following a summary of basic definitions of the probability theory we consider a Brownian motion on the p-adic line as a Markov process. We next discuss generalized random fields and p-adic quantum field theory.

1. *Random Maps and Markov Processes*

A measure space is a pair $\{\Omega, \Sigma\}$, where Ω is a set and Σ is a σ-algebra of its subsets. A triple $\{\Omega, \Sigma, P\}$, where $\{\Omega, \Sigma\}$ is a measure space and P is a σ-additive nonnegative measure on Σ which satisfies the condition

$$P(\Omega) = 1, \tag{1.1}$$

is a *probability space*. Elements $A \in \Sigma$ are called events and $P(A)$ is the probability of an event A.

A *random variable* $\xi = \xi(\omega)$ is a Σ-measurable real-valued function on Ω, that is $\xi^{-1}(S) \in \Sigma$ for every Borel set S on the real line $\mathbb{R}^1$. The random variable ξ gives rise to a probability measure P_ξ on $\mathbb{R}^1$, defined by

$$P_\xi(S) = P(\xi^{-1}(S)) \tag{1.2}$$

for any Borel set S on $\mathbb{R}^1$. The measure P_ξ is said to be the distribution of the variable ξ. The mean of ξ, $E(\xi)$ is defined by

$$E(\xi) = \int_\Omega \xi(\omega) dP(\omega).$$

If $A, B \in \Sigma$ and $P(B) \neq 0$ we define

$$P(A|B) = \frac{P(A \cap B)}{P(B)}.$$

$P(A|B)$ is called the conditional probability of A given B.

Let F be an arbitrary σ-algebra, which is contained in Σ, ξ be a random variable on a probability space (Ω, Σ, P) with mean equal to zero. Conditional expectation of ξ with respect to the σ-algebra F is a random variable $E\{\xi|F\}$, which is F-measurable and satisfies the relation

$$\int_B E\{\xi|F\} dP = \int_B \xi dP$$

for an arbitrary $B \in F$.

Conditional probability $P\{A|F\}$ with respect to a σ-algebra F is defined as the special case of conditional expectation, setting $\xi = \chi_A(\omega)$, where $\chi_A(\omega)$ is the indicator of the set A. Namely, for fixed A the conditional probability $P\{A|F\}$ is a F-measurable random variable satisfies the relation

$$\int_B P\{A|F\} dP = P(A \cap B)$$

for every $B \in F$

Conditional probability $P\{A|\xi\}$ with respect to a random variable ξ is defined by the formula

$$P\{A|F\} = P\{A|F_\xi\},$$

where $F_\xi = \{B : B = \xi^{-1}(S), S-$ belongs to measurable Borel sets in $\mathbb{R}\}$ is σ-algebra generated by the map ξ.

Let X be a set and $\{Y, \mathfrak{B}\}$ be a measure space. *Random map* of a set X to measure space $\{Y, \mathfrak{B}\}$ is a map $\xi = \xi(x, \omega) : X \times \Omega \to Y$ which for an arbitrary fixed x is a measurable map from $\{\Omega, \Sigma\}$ to $\{Y, \mathfrak{B}\}$ that is for any $B \in \mathfrak{B}$

$$\{\omega \in \Omega : \xi(x, \omega) \in B\} \in \Sigma.$$

Below two examples of random maps are considered. The first when X is a subset (semi-axis) of real line, in this case we write t instead of x. This random map is called *random process* $\xi(t)$. Independent variable t is interpreted as time. In the second example X is a space of distributions $\mathcal{D}'(\mathbb{Q}_p^n)$. Here random map reduces to a *generalized random field.*

Let n be an arbitrary positive integer, x_k, $k = 1, 2, \ldots, n$ be arbitrary points from X. Measures $P_{x_1 x_2 \ldots x_n}(B)$ on $\mathfrak{B}^n$ of the form

$$P_{x_1 x_2 \ldots x_n}(B) = P\{\xi(x_1, \omega), \xi(x_2, \omega), \ldots, \xi(x_n, \omega) \in B\}, \quad B \in \mathfrak{B}^n$$

is said to be *joint distributions* of a random map $\xi(x)$. Under sufficiently wide assumptions a set of joint distributions defines a random map.

We consider Markov process $\xi = \xi(t)$ on the semi-axis $t \geq 0$. Let Y be a complete metric space and $\mathfrak{B}$ be the σ-algebra of Borel subsets of Y. A random process $\xi = \xi(t)$, $t \geq 0$ with values in Y is called a (homogeneous) *Markov* process, if the following conditions are valid:

1) for any $0 \leq t_1 < t_2 < \ldots < t_n < t$ and $B \in \mathfrak{B}$ the relations for conditional probabilities are fulfilled

$$P\{\xi(t) \in B|\xi(t_1), \xi(t_2), \ldots, \xi(t_n)\} = P\{\xi(t) \in B|\xi(t_n)\} \pmod{P};$$

2) there is a function $P(t, x, B)$ which satisfies the following conditions:
i) it is $\mathfrak{B}$-measurable on x for fixed t and B;
ii) it is a probability measure on the space (Y, B, P) for fixed t and x;
iii) it satisfies the Chapman-Kolmogorov equation

$$P(s+t, x, B) = \int_Y P(s, x, dy) P(t, y, B), \qquad s, t \geq 0;$$

iv) it coincides with probability 1 with conditional probability

$$P(t, x, B) = P\{\xi(s+t) \in B | \xi(s) = x\}, \qquad s, t > 0.$$

The function $P(t, x, B)$ is called the *transition function* of Markov process $\xi(t)$. If a probability measure μ on $\{Y, \mathfrak{B}\}$ and a function $P(t, x, B)$, which satisfies the conditions 2), i)–iii) are defined then there is a Markov process $\xi(t)$ with this transition function.

Often it is more convenient to work with *transition density*. Let μ be a nonnegative measure on $\{Y, \mathfrak{B}\}$. A function $p(t, x, y)$, $t > 0$, $x, y \in Y$, is called the transition density if the following conditions are valid:

1) $p(t, x, y) \geq 0$, $t > 0$, $x, y \in Y$;
2) $p(t, x, y)$ for fixed $t > 0$ is a $\mathfrak{B} \times \mathfrak{B}$-measurable function of x, y;
3) $\int\limits_Y p(t, x, y)\mu(dy) \leq 1$, $t \geq 0$, $x \in Y$;
4) $p(s+t, x, z) = \int\limits_Y p(s, x, y)p(t, y, z)\mu(dy)$, $s, t \geq 0$; $x, z \in Y$.

If a transition density $p(t, x, y)$ is given then the function

$$P(t, x, B) = \begin{cases} \int\limits_B p(t, x, y)\mu(dy), & t > 0, \ x \in Y, \ B \in \mathfrak{B} \\ \chi_B(x), & t = 0 \end{cases}$$

defines a transition function which corresponds to some Markov process.

A transition function defines a set of linear operators on the Banach space E of all bounded measurable complex valued functions $f(x)$, $x \in Y$ with the norm

$$||f|| = \sup_{x \in Y} |f(x)|$$

by the formula

$$T_t f(x) = \int\limits_Y P(t, x, dy) f(y), \quad t \geq 0.$$

The set $T_t, t \geq 0$ is a contraction semigroup of operators, that is bounded operators T_t which satisfy the following condition

$$T_{s+t} = T_s T_t, \quad s, t \geq 0,$$
$$||T_t|| \leq 1, \quad t \geq 0.$$

An infinitesimal generator A of this semigroup is also called an infinitesimal operator of a function $P(t,x,B)$. The domain D_A consists of all functions f for which the limit in the relation

$$Af(x) = \lim_{t\to+0} \frac{1}{t}\left[\int_Y P(t,x,dy)f(y) - f(x)\right]$$

does exist uniformly with respect to $x \in Y$.

Let $C(Y)$ be a space of continuous bounded complex valued functions on Y. A Markov process is *stochastically continuous* if

$$\lim_{t\to+0} T_t f = f, \qquad \forall f \in C(Y)\ ,$$

and is called *Feller's* process, if

$$T_t C(Y) \subset C(Y), \qquad \forall t > 0\ .$$

There is the following criterion. If for a transition function $P(t,x,B)$ for any compact B the following conditions are valid:
$\forall s \geq 0$

$$\lim_{x\to\infty} \sup_{t\leq s} P(t,x,B) = 0, \tag{A}$$

and $\forall \varepsilon > 0$

$$\lim_{t\to+\infty} \sup_{x\in B} P(t,x,\overline{U_\varepsilon(x)}) = 0, \tag{B}$$

where $U_\varepsilon(x)$ is a disk of radius ε with the center at x, and $\overline{U_\varepsilon(x)} = Y\backslash U_\varepsilon(x)$, then the corresponding Markov process is *bounded, continuous from the right and has no discontinuities of the second kind with probability 1.*

2. *Brownian Motion on the p-Adic Line*

It is known that a Brownian motion on the real line can be described by means of the diffusion equation

$$\frac{\partial U}{\partial t} = a\frac{\partial^2 U}{\partial x^2}\ . \tag{2.1}$$

A solution of the equation (2.1) defines a transition probability of Brownian motion considered as a Markov process. Here we consider the equation

$$\frac{\partial U}{\partial t} = -D_x^2 U \tag{2.2}$$

and associate with it a Markov process, which we shall call a Brownian motion on p-adic line. Here real valued function $U = U(t, x)$ in (2.2) depends on variables $t \in \mathbb{R}$ and $x \in \mathbb{Q}_p$.

Let us consider a fundamental solution of the Cauchy problem

$$U(x, 0) = \varphi(x) \tag{2.3}$$

for the equation (2.2). It can be given by using the following functions:

$$p(t, x, y) = K(t, x - y) = \int_{\mathbb{Q}_p} \chi(\xi(x - y)) \exp(-t|\xi|_p^2) d\xi, \tag{2.4}$$

where $t > 0$, $x, y \in \mathbb{Q}_p$.

Theorem 1. *The function $p(t, x, y)$ (2.4) satisfies the conditions 1)–4) for a transition density if $Y = \mathbb{Q}_p$, $\mu(dx) = dx$ is a Haar measure on $\mathbb{Q}_p$, and hence defines a Markov process which we call the Brownian motion on the p-adic line.*

Remark. In contrast to the Winer process, trajectories of which are continuous, there are no nonconstant continuous functions from a real segment to $\mathbb{Q}_p$.

■ Let us consider the function

$$K_t(x) = \int_{\mathbb{Q}_p} e^{-t|\xi|_p^2} \chi_p(x\xi) d\xi, \qquad t > 0, \ x \in \mathbb{Q}_p. \tag{2.5}$$

We prove the following properties:

(i) $K_t(x) > 0$,
(ii) $\int_{\mathbb{Q}_p} K_t(x) dx = 1$,
(iii) $K_t(x) \to \delta(x)$, $t \to +0$ in $\mathcal{D}'$,
(iv) $K_t * K_{t'} = K_{t+t'}$, $t, t' > 0$.

Using the formula (3.3) of Sec. 4.3 for $f(\alpha) = e^{-t\alpha^2}$ we have

$$K_t(x) = \left(1 - \frac{1}{p}\right) |x|_p^{-1} \sum_{\gamma=0}^{\infty} p^{-\gamma} e^{-tp^{-2\gamma}|x|_p^{-2}} - |x|_p^{-1} e^{-tp^2|x|_p^{-2}}$$

$$= \left(1 - \frac{1}{p}\right) |x|_p^{-1} \sum_{\gamma=0}^{\infty} p^{-\gamma} e^{-tp^2|x|_p^{-2}} \left[e^{tp^2|x|_p^{-2}(1-p^{-2\gamma-2})} - 1\right] > 0.$$

This gives the proof of (i). Then

$$\int_{B_k} K_t(x)dx = (K_t, \Delta_k) = (F[e^{-t|\xi|_p^2}], \Delta_k) = (e^{-t|\xi|_p^2}, \tilde{\Delta}_k)$$

$$= (e^{-t|\xi|_p^2}, \delta_k) = \int_{\mathbb{Q}_p} \delta_k(\xi) e^{-t|\xi|_p^2} d\xi \longrightarrow 1, \qquad k \to \infty$$

and thus we have the property (ii). One has

$$K_t(x) = F[e^{-t|\xi|_p^2}] \longrightarrow \tilde{1} = \delta(x), \qquad t \to +0 \text{ in } \mathcal{D}'$$

and (iii) is proved.

Taking into account the Theorem of Sec. 7.5 we get the property (iv):

$$K_t * K_{t'} = F[e^{-t|\xi|_p^2} \cdot e^{-t'|\xi|_p^2}] = F[e^{-(t+t')|\xi|_p^2}] = K_{t+t'} \,. \qquad \blacksquare$$

The function $K_t(x)$ is the transition function of a Markov process which satisfies the conditions (A) and (B).

3. *Generalized Stochastic Processes*

Let L be a locally convex space, L' dual space and $\{\Omega, \Sigma, P\}$ a probability space. A linear map ϕ from L to the random variables on $\{\Omega, \Sigma, P\}$ is called a *generalized stochastic processes* over L. A set

$$C(\varphi_1, \ldots, \varphi_n; A) = \{\phi \in L : (\phi(\varphi_1), \ldots, \phi(\varphi_n)) \in A \subset \mathbb{R}^n\}$$

is called a *cylinder set*. Here $\varphi_i \in L$, $i = 1, \ldots, n$ and A is a Borel set in $\mathbb{R}^n$. The family of cylinder sets Σ_0 is an algebra. Any subset N from the factor space L'/Ψ where Ψ is the annulator is called a base. A *cylinder measure* μ is a real valued function $\mu(c)$ on Σ_0 with the following properties:

i) $0 \le \mu(c) \le 1$ for every $c \in \Sigma_0$

ii) $\mu(\Sigma_0) = 1$

iii) if a set c is a disjoint union of cylinder sets $c_1, \ldots, c_n, \ldots$ based on N then $\mu(c) = \sum_{n=1}^{\infty} \mu(c_n)$

iv) $\mu(c) = \inf \mu(U)$

for every cylinder set C where U runs over all cylinder sets containing C. If L is a nuclear space then the cylinder measure μ can be extended to a countably additive measure on Σ where Σ is the Borel σ-algebra generated by Σ_0. According to the Minlos theorem if $J : L \to \mathbb{C}$ is a continuous function of positive type and $J(0) = 1$ then there is a probability measure μ on $\{L', \Sigma\}$ such that

$$J(\varphi) = \int_{L'} \exp(i\phi(\varphi)) d\mu(\phi).$$

J is called the characteristic function of μ. Let $B(\varphi, \psi)$ be a continuous inner product on a nuclear space L. Then

$$J(\varphi) = \exp\left(-\frac{1}{2}B(\varphi,\varphi)\right)$$

is a continuous characteristic function and thus there is a probability measure μ on $\{L', \Sigma\}$ so that

$$\exp\left(-\frac{1}{2}B(\varphi,\varphi)\right) = \int_{L'} \exp(i\phi(\varphi)) d\mu(\phi).$$

The measure μ is called a Gaussian measure on L' and $\phi(\varphi)$ is called the Gaussian process with mean zero and covariance $B(\varphi, \varphi)$.

4. *Quantum Field Theory*

Let us take $L = \mathcal{D}(\mathbb{Q}_p^n)$ as the space of real test functions which was considered in Sec. 6. It is a locally convex nuclear space. Consider the following inner product

$$B(\varphi, \psi) = \int_{\mathbb{Q}_p^n} \frac{\tilde{\varphi}(k)\tilde{\psi}(-k)}{a(k)} dk$$

where $a(k) \geq c > 0$ is some function of k. This inner product gives rise to a Gaussian generalized process. If we take

$$a(k) = \left|\sum_{i=1}^{n} k_i^2\right|_p + m^2, \quad m > 0$$

then by analogy with the Euclidean formulation of quantum field theory one can call the corresponding Gaussian process the free scalar p-adic quantum field. This field is invariant under group $SO(n, \mathbb{Q}_p)$. Another natural choice of the function $a(k)$ is

$$a(k) = \|k\| + m^2, \quad \|k\| = \max_i |k_i|_p.$$

p-adic white noise is obtained if $a(k) = 1$. The measure $d\mu$ can be formally written in the form

$$d\mu \sim e^{-S_0(\phi)} \mathcal{D}\phi$$

where the free action

$$S_0(\phi) = \int\limits_{\mathbb{Q}_p^n} \phi(x) a(D) \phi(x) dx. \tag{3.1}$$

Here $a(D)$ is a pseudodifferential operator. The expression (3.1) is not defined for an arbitrary generalized functions $\phi(x)$ and it requires a regularization or restriction to a subspace. In p-adic string theory one takes in (3.1) $a(D) = D$.

To describe an interaction one needs to construct a non-Gaussian measure corresponding to an action

$$S(\phi) = S_0(\phi) + \int\limits_{\mathbb{Q}_p^n} V(\phi(x)) dx$$

with some function $V(\phi)$.

Bibliography

In the following we present a brief chapter by chapter bibliographic discussion of main ideas and results in p-adic mathematical physics. This branch of mathematical physics being rather new is related with different branches of mathematics. Namely, it is closely connected with very traditional mathematics having a long history, such as number theory as well as with modern abstract algebraic geometry and representation theory. The list of bibliography contains references to books on p-adic analysis. This list is not complete and the selected books have proved as more suitable for current mathematical physics applications. There are references to pure p-adic topics as well as to applications. Physical papers motivated p-adic considerations are also collected in the list. We apologize in advance for any errors or oversights and possible omissions.

Introduction

We are not going to review enormous literature on the space-time geometry, we point out only a few references on this subject. A general mathematical discussion of the notion of the physical space has been performed by many authors, notably by Poincare [172] and H.Weyl [228]. Riemann [177] considered a continuous (renowned Riemannian geometry) as well as discrete models of the space. Our intuitive understanding of properties of space is expressed in the axioms of elementary geometry. A complete list of geometrical axioms was presented in the Hilbert famous "Grundlagen der Geometry" [104]. The role of the Archimedean axiom and a possibility of construction of a non-Archimedean geometry was pointed out by Veroneze and Hilbert.

An absolute limitation on length measurements in quantum gravity and string theory is discussed in [176, 171, 147, 229, 200, 109, 220] and [7, 94]. As Witten summarized, "There are many reasons to believe that in the string theory there is no such thing as distances less than the fundamental length $\sqrt{\alpha'}$.... At distances below $\sqrt{\alpha'}$, not just physics as we know it but local physics altogether has disappeared. There will be no distances, no times, no energies, no particles, no local signals — only differential topology, or its string theoretic successor." [232]*

A hypothesis on a possible non-Archimedean p-adic structure of space-time at the Planck scale was suggested and considered by Volovich [221]. The basic role of rational numbers was stressed and the idea of fluctuating number fields was suggested [219-221].

p-Adic numbers were introduced in 1899 by K.Hensel. The books by Koblitz [121], Mahler [138] and Schikhof [186] can serve as an introduction to p-adic numbers and p-adic analysis. As a background reading on algebra and number theory one can use Borevich and Shafarevich [37], Kostrikin [126], Leng [128], Serre [189], Vinogradov [204] and Weil [225].

A possible role of p-adic analysis in mathematical physics in the context of superanalysis and supersymmetry was noted by Vladimirov and Volovich [211].

An application of ultrametricity in solid state physics was discussed by Mezard, Parisi, Rammal, Sourlas, Toulouse and Virasoro [157, 178]. The Parisi matrix [169] describing a hierarchical structure leads naturally to ultrametricity.

A discussion of a possible role of number theory in physics one can find in Manin [142, 145], Atiyah [25] and Bott [39].

Comments on references related with p-adic analysis , p-adic quantum mechanics and strings will be done under discussion of corresponding chapters.

Chapter I

p-Adic analysis is treated in [8, 38, 61, 82, 85, 121, 122, 179, 186].

A homeomorphism (see subsect.1.6) of p-adic numbers $\mathbb{Q}_p$ to some Cantor subset of the field of real numbers $\mathbb{R}$ was constructed by Zelenov [240]. In the presentation of the theory of additive characters we follow the approach of the Pontryagin book [172]. Gaussian integrals on an arbitrary locally compact abelian group were considered by Weil [225]. The explicit calculations for a special case of the field $\mathbb{Q}_p$ were performed by authors

* Here $\sqrt{\alpha'}$ is the string theory notation for the Planck length.

[212, 215, 218] and independently by Alacoque, Ruelle, Thiran, Verstegen and Weyers [3, 182] and by Meurice [151]. The number theory function $\lambda_p(a)$ was introduced and investigated by Vladimirov and Volovich [212], and by Alacoque at al. [3]. The proof of the adelic formula for $\lambda_p(a)$ is presented here for the first time.

A theory of distributions on an arbitrary locally compact group has been developed by Bruhat [43] and on a locally compact disconnected field by Gel'fand, Graev and Piyatieckii-Shapiro [82]. We present elements of this theory for the field $\mathbb{Q}_p$ along the line of the paper [206]. A distinguished feature of the theory of p-adic distributions as compare with Sobolev-Schwartz distributions over real numbers [205] is the fact that in the p-adic case from the linearity of a functional follows its continuity. A theory of convolution and multiplication of p-adic distributions using the Fourier transformation was firstly developed by Vladimirov [206]. Dealing with the theory of one-dimensional homogeneous distributions we follow the book by Gelfand, Graev and Pijatetsskii-Shapiro [82] and for multidimensional case the paper by Smirnov [193]. Properties of the two-dimensional Green function were considered by Bikulov [35].

Chapter II

A notion of a pseudo-differential operator on the space $\mathbb{Q}_p^n$ has been introduced by Vladimirov [208]. A reader can find the notions of the spectral theory, used in this chapter, in Reed and Simon [175], Dunford and Schwartz [60] and Yosida [237].

The non-local operator of fraction differentiation and integration D^α was introduced and investigated by Vladimirov [206] (see also [15, 168, 196, 245]).

The spectral theory of the operator D^α, $\alpha > 0$, acting in $\mathbb{Q}_p$ has been performed by Vladimirov [207]. This treatment includes a calculation of an explicit form of eigenfunctions (see also [218]). The orthonormal basis of eigenfunctions of operator D^α, (including $p = 2$) in $\mathbb{Q}_p$ is firstly presented in subsection 9.5.

Spectral theory of a pseudodifferential operator of the form $a^* + V(x)$ on an open-closed set G (bounded or unbounded) with functions $\tilde{a}(\xi)$ and $V(x)$ being bounded from below and going to $+\infty$ at the infinity has been constructed in [208]. There one can also find an explicit form of the eigenfunctions and eigenvalues of the operator $D^\alpha, \alpha > 0$ in the disk B_γ and on the circle S_γ (if $p \neq 2$). Here all these results are reproduced, the case of $p = 2$ is also considered, and the inversion of the main theorem is done in the stronger form.

The method of finding of invariant eigenfunctions $\psi(|x|_p)$ of the p-adic Schrodinger operator $D^\alpha + V(|x|_p)$ has been developed by Vladimirov and is given here for the first time. Further developments of spectral theory of the operator $D^\alpha + V(|x|_p), \alpha > 0$ in $\mathbb{Q}_p$ without a condition at the infinity have been performed by Kochubej [123].

A decomposition of the field $\mathbb{Q}_p$ on sectors has been suggested by Vladimirov [207].

Stationary and non-stationary p-adic Schrodinger equations were suggested by Vladimirov and Volovich [214].

Chapter III

About foundations of quantum mechanics see for example Dirac [52], Holevo [107], Mackey [137], Maslov [149].

A formalism of p-adic quantum mechanics based on a triplet $(L_p(\mathbb{Q}_p), W(z), U(t))$ was suggested by Vladimirov and Volovich [212, 213]. This formalism there appeared as a quantization of p-adic classical mechanics. Slightly different approaches were considered by Freund and Olson [74], Alacoque et al. [3] and Meurice [154]. Instead of the dynamical operator $U(t)$ these approaches use a unitary representation of a non-abelian group and therefore are restricted by quadratic Hamiltonians. Several ways of construction of p-adic quantum mechanics were discussed by Parisi [168]. A Lagrangian formalism and Feynman path integral were briefly discussed in [213] and have been developed by Zelenov [240]. Another approach to p-adic quantum mechanics based on the probability measures on the space of distributions was also pointed out in [213]. Quantum mechanics with p-adic valued functions is also sensible [212]. A general approach to quantum mechanics with p-adic valued functions based on the theory of Gaussian distributions was elaborated by Khrennikov [114, 116, 120]. p-Adic Hilbert spaces are discussed by Bayod [29]. All the above-mentioned approaches are equivalent over real numbers but in the case of p-adic numbers their equivalence is questionable and only fragmentary information concerning relations between different approaches is available at the moment. Note that in the construction of p-adic quantum mechanics ideas and results by H.Weyl [227], Mackey [136], Segal [187] and A.Weil [225] are essentially used. The theory of Jacquet and Langlends [81] should be helpful in further development of p-adic quantum mechanics.

Much recent effort have been concentrated on studying spectral theory in p-adic quantum mechanics. An appealing property of this theory is the appearance of a rich spectrum for a rather simple system (p-adic harmonic oscillator). Spectral theory in p-adic quantum mechanics was carried out by authors [218], similar results were obtained by Ruelle et al. [182] and Meurice [151].

An investigation of p-adic Weyl systems in finite and infinite dimensional cases and study of coherent states and eigenfunctions for $p = 3 \pmod 4$ were performed by Zelenov [241-243].

Formulations of p-adic string theory with p-adic valued and complex valued amplitudes as convolutions of characters were suggested by Volovich [219-221]. Freund and Olson [73] made an important proposal to consider the p-adic B-functions as string amplitudes. Freund has stressed an idea of using p-adic calculus in order to simplify calculations for real systems. A connection of p-adic string theory and the Weil conjecture in the number theory was discussed by Grossman [95]. An idea of adelic approach was suggested by Manin [145]. An important adelic formula was pointed out by Freund and Witten [75]. Problems of regularization of this formula and an alternative adelic formula were considered by Aref'eva, Dragovic and Volovich [16].

Five-point p-adic amplitudes were considered by Marinary and Parisi [146] and by Frampton and Okada [65]. N-point p-adic amplitudes were elaborated by Frampton and Okada [65,68] and by Brekke et al. [40, 42]. They derived an effective field theory with the remarkable soliton solution. Interesting papers by Frampton, Nishino, Okada and Ubriaco [69, 164, 165, 167] are devoted to p-adic σ-models.

The action from Sec. 16.4 was considered in quantum field theory by Parisi [168], Aref'eva and Volovich [15], Lerner and Missarov [131] and in string theory by Spokoiny [196] and Zhang [245]. This action uses the operator D considered by Vladimirov. p-Adic multiloop amplitudes on the Bruhat-Tits tree were considered by Chekhov, Mironov and Zabrodin [238, 44, 45].

Subsections 14.4–14.6, which deal with multiloop amplitudes, is a short extraction of enormous current literature on string theory. We shall indicate here the sources of most of the material presented in these subsections, and attempt to direct the reader to additional papers. A nice short introduction to string theory is the old Scherk review [184]. Excellent reference on string theory is Green, Schwartz and Witten [92]. Multiloop calculations based on the path-integral method suggested by Polyakov [173] have been recently developed into a field of its own, and references on more detailed aspects of multiloop calculations are [6, 31, 87, 108, 143, 203]. Multiloop calculations are closely related with geometry of complex manifolds [93, 162]. Properties of moduli space described in subsect. 14.5 are due to Mumford [160, 161] and Deligne and Mumford [50]. About p-adic Schottky groups and p-adic uniformization see Manin [141] and Gerritzen and van der Put [85].

Rigid analytic geometry was introduced by Tate [198]. It is considered in Gerritzen and van der Put [85], Bosch, Guntzer and Remmert [28] and Fresnel and van der Put [72]. The theory of p-adic differential equations is presented in Dwork [61].

Gervais [86] suggested an idea of using p-adic variables to extend conformal symmetry to higher-dimensional case.

An approach to p-adic string theory with p-adic valued amplitudes using the Morita gamma function was suggested by Volovich [221] and Grossman [95]. By means of Gross-Koblitz formula it leads to string theory on the Galois fields. The Galois-Veneziano-Jacobi string amplitudes [221] can be expressed in terms of Frobenius action on the space of etale cohomology. This approach is connected with the theory of motives and L-functions [158, 100, 48, 49, 28].

Witten proposed that at short distances there is a phase in which general covariance is unbroken. This phase is described by topological quantum field theories which are systems without local, propagating degrees of freedom. These theories are associated with the cohomology of various moduli spaces [233, 51]. It seems it would be a natural next step to consider analogous theories over p-adic and other number fields. Note that recently methods of quantum field theory proved useful for investigation of properties of moduli spaces of Riemann surfaces [99, 170, 234, 124].

Section 15, which deals with relations between q-analysis and quantum groups is mainly taken from [22]. Other connections between them were pointed out by Macdonald [134] and Freund [78,79]. In the last years there has been considerable work done on the theory of quantum groups. Some references are Drinfeld [58], Jimbo [80], Faddeev, Reshetichin and Tachtadjan [62], Woronowicz [235] and Manin [144]. The q-deformed oscillator was considered by Biedenharn [34] and Macfarlane [135]. A general approach to non-commutative geometry has been developed by Connes [47]. Non-commutative differential calculus was considered by Wess and Zumino [224]. About possible applications of quantum groups in field theory see Aref'eva and Volovich [23]. q-Analysis is discussed in Andrews [10] and Askey [24]. Koornwinder [125] discussed a relation between q-analysis and quantum groups.

Stochastic processes and generalized random fields are considered by many authors. Note in particular Gelfand and Vilenkin [83], Gihman and Skorokhod [88], Accardi, Frigeiro and Lewis [2], Hida [103] and Heyer [102]. The probability theory approach to quantum field theory is considered by Glimm and Jaffe [89] and Simon [91]. Diffusion on p-adic numbers is considered by Albeverio and Karwowski [4].

Recently new interesting approaches to the description of the space-time structure at small distances have been suggested by Bennet, Nielsen and Picek [33], Alvarez, Cespedes and Verdaguer [4] and Isham, Kubyshin and Renteln [111]. Basic quantum variable is the two-point distance on a metric space. Such consideration naturally includes also the ultrametric.

The first steps in investigation of p-adic Einstein equation and quantum p-adic gravity were performed by Areféva, Dragovic, Frampton and Volovich in [20, 71, 21, 54, 57]

References

1. S.Abyankar: Local analytic geometry, N.Y.: Academic Press, 1964.
2. L.C.Accardi, A.Frigerio and J.T.Lewis: Quantum stochastic processes. Publications RIMS **18**, 97 (1982).
3. L.C.Alacoque, P.Ruelle, E.Thiran, D.Verstegen and J.Weyers: Quantum amplitudes on p-adic fields. *Phys. Lett.* **211B**, 59–62 (1988).
4. S.Albeverio, and W.Karwowski: Diffusion on p-adic numbers, in "Gaussian Random Fields", The third Nagoya Lévy Seminar, 15–20 Aug. 1990, eds. K.Itô, T.Hida, World Scientific Publ. 1991.
5. E.Alvarez: *Phys. Lett.* **210B**, 73–78 (1988); E.Alvarez, J.Cespedesm and E.Verdaguer: Dynamical generation of space time dimensions. Preprint CERN-TH. 5764/90 (1990).
6. L.Alvarez-Gaume, J.-B.Bost, G.Moore, P.Nelson and C.Vafa: Bosonization on Higher Genus Riemann Surfaces. *Comm. Math. Phys.* **112**, 503–569 (1987).
7. D.Amati, M.Ciafaloni and G.Veneziano: *Phys. Lett.* **197**, 81 (1987); *Int. Mod. Phys.* **3A**, 1615 (1988); *Phys. Lett.* **216**, 41 (1989).
8. Y.Amis: Les nombres p-adiques. Presses Universitaires de France, 1975.
9. Y.Andre: G-functions and Geometry, Aspects of Mathematics **E13**, Vieweg, 1988.
10. G.E.Andrews: q-Series: Their develpment and application in analysis, number theory, combinatorics, physics, and computer algebra, in "Regional Conference Series in Math." v.66, *Amer. Math. Soc.*, Providence, RI, 1986.
11. S.J.Arakelov: An intersection theory for divisors on an arithmeic surface. Izv. Akad Nauk SSSR, *Ser. Math.* **38**, 1179–1192 (1974).
12. V.I. Arnold: Mathematical Methods of Classical Mechanics. Moscow: Nauka, 1974.
13. I.Ya.Aref'eva: String Field Theory, in "Conformal Invariace and String Theory", eds. P.Dita and V.Gergescu, Boston: Acadic Press, 1989.
14. I.Ya.Aref'eva: Physics at the Planck Length and p-adic Field Theory, in "Differential Geometry Methods in Theoretical Physics", eds. L.L. Chau and W.Nahm, N.Y.:Plenum Press, 1990.
15. I.Ya.Aref'eva and I.V.Volovich: "String, gravity and p-adic space-time", in: Quantum Gravity, eds. M.A.Markov, V.A.Berezin and V.P.Frolov, World Scientific Publishing, 1988.
16. I.Ya.Aref'eva, B.Dragovič and I.V.Volovich: On the adelic string amplitudes. *Phys. Lett.* **B209**, 445–450 (1988).
17. I.Ya.Aref'eva, B.Dragovič and I.V.Volovich: On the p-adic summability of the anharmonic oscillator. *Phys. Lett.* **B200**, 512–514 (1988).
18. I.Ya.Aref'eva, B.Dragovič and I.V.Volovich: Open and closed p-adic strings and quadratic extensions of number fields. *Phys. Lett.* **B212**, 283–289 (1988).
19. I.Ya.Aref'eva, B.Dragovič and I.V.Volovich: P-adic superstrings. *Phys. Lett.* **B214**, 339–346 (1988).
20. I.Ya.Aref'eva, B.Dragovic, P.Frampton and I.V.Volovich: Wave function of the universe and p-adic gravity. *Mod. Phys. Lett.* **A6**, 4341–4358 (1991).

21. I.Ya.Aref'eva and P.H.Frampton: Beyond Planck energy to non-Archimedean geometry. *Mod. Phys. Lett.* **A6**, 313–316 (1991).
22. I.Ya.Aref'eva and I.V.Volovich: Quantum groups particles and non-Archimedean geometry. *Phys. Lett.* **B268**, 179–193 (1991).
23. I.Ya.Aref'eva and I.V.Volovich: Quantum group gauge fields. *Mod. Phys. Lett.* **A6**, 893–906 (1991).
24. R.Askey: The q-gamma and q-beta functions, *Appl. Anal.* **8**, 125–146 (1978).
25. M.F.Atiyah: Commentary on the article of Manin. *Lect. Notes in Math.* **1111**, pp.103–109, Springer, 1985
26. M.F.Atiyah and R.Bott: The Yang-Mills equation over Riemann surfaces. *Phil. Trans. R. Soc. London* **A308**, 523–615 (1982).
27. J.Atick and E.Witten: The Hagedorn transition and the number of degrees of freedom of string theory, Preprint IASSNS-HEP-88/4 (1988).
28. F.Baldassarri: p-Adic interpolation of Evans sums and deformation of the Selberg integrals: an application of Dwork's theory. In: Special Differencial Equations. Proc. of the Taniguchi Workshop, p.7–35, 1991.
29. J.M.B. Bayod: Productos Internos en Espacios Normados no Arquimedianos. Memoria presentada para optar al grado de Doctor en Ciencias (Seccion de Matematicas). Univ. de Bilbao, 1976.
30. J.M. Bayod, N. De Grande-DeKimpe and J.Martinez-Maurica: p-Adic functional analysis. N.Y.: Marcel Dekker, Inc., 1992.
31. A.A.Belavin and V.G.Knizhnik: Algebraic geometry and the geometry of quantum strings. *Phys. Lett.* **B168**, 201 (1986).
32. S. Ben-Menahem: p-adic iteractions. Preprint TAUP-1627-88, 1–17 (1988).
33. D.L.Bennett, H.B.Nielsen and I.Picek: *Phys. Lett.* **B208**, 275–284 (1988).
34. L.C.Biedenharn: *J. Phys. A. Math. Gen.* **22**, L873 (1989).
35. A.H.Bikulov: Investigation of the p-adic Green function. *Theor. Math. Phys.* **87**, 376–390 (1991).
36. J.-M. Bismut, H.Gillet and C.Soule: Analytic torsion and holomorphic determinant bundles I, II, III. *Comm. Math. Phys.* **115**, 301–350 (1988).
37. Z.I.Borevich and I.R.Shafarevich: Number theory. N.Y.: Academic Press, 1966.
38. S.Bosch, U.Guntzer and R.Remmert: Non-Archimedean Analysis,Berlin: Springer-Verlag, 1984.
39. R.Bott: On the shape of the curve. Advances in Mathematics, **16**, 144–159 (1975).
40. L.Brekke, P.G.O.Freund, E.Melzer and M.Olson: Adelic N-point amplitudes. *Phys. Lett.* **B216**, (1989).
41. L.Brekke and M.Olson: p-Adic diffusion and relaxation in glasses. Preprint UTTG-16-89, EFI-89-23 (1989).
42. L.Brekke, P.G.O.Freund, M.Olson and E.Witten: Non-archmedean string dynamics. *Nucl. Phys.* **B302**, 365–402 (1988).
43. F.Bruhat: Distributions sur un groupe localemont compact et applications a l'etude des representations des groupes p-adiques. *Bulletin Soc. Mathem. de France* **89**, 43–75 (1961).

44. L.Chekhov: A note on multiloop calculus in p-adic string theory. *Mod. Phys. Lett.* **A4**, 1151–1158 (1989).
45. L.O.Chekhov, A.D.Mironov and A.Zabrodin: Multiloop calculations in p-adic string theory and Bruhat-Tits trees I,II. Preprints ITP-89-41E, 42E (1989); *Mod. Phys. Lett.* **A4**, 1127–1235 (1989).
46. G.Christol: Solutious algebriques des equatious differenielles p-adique. *Prog. Math.* **38**, 51–58 (1983).
47. A.Connes: Non Commutative geometry. *Publ. Math.* IHES, **62** (1986).
48. P.Deligne: Hodge Cycles on Abelian Varieties. In: P.Deligne, J.S.Milne, A.Ogus and Shih Kuang-yen. Hodge Cycles, Motives and Shimura Varieties. *Lect. Notes in Math.*, No.900, pp.9–100, Springer-Verlag, 1982.
49. P.Deligne: Valeurs de functions L et periodes g'integrales. *Proc. Symp. Pure Math.* v.XXXIII, 313–336, *Amer. Math. Soc.* Providence, 1979.
50. P.Delinge and D.Mumford: The irreducibility of the space of curves of a given genus, Publ.I .H.E.S. 36, 1969.
51. R.Dijkgraaf, H.Verlinde and E.Verlinde: Notes on Topological String Theory and 2D Quantum Gravity. Preprint PUPT-1217, IASSNS-HEP-90/80 (1990).
52. P.A.M.Dirac: The principles of Quantum Mechanics. Oxford: At the Clarendon Press, 1958.
53. B.G. Dragovič: Adelic summability of perturbation series. Preprint IF-19/89, 1–15 (1989).
54. B.G. Dragovič, P.H.Frampton and B.V.Urosevič: Classical p-adic space-time. *Mod. Phys. Lett.* **A5**, 1521–1528 (1990).
55. B.G. Dragovič: p-Adic perturbation series and adelic summability. *Phys. Lett.* **256B**, 392–396 (1991).
56. B.G.Dragovič: On factorial perturbation series. Preprint IF-91-11, 1–8 (1991).
57. B.G.Dragovič: On signature change in p-adic space-times. *Mod. Phys. Lett.* **A6**, 2301–2307 (1991).
58. V.G.Drinfeld: Quantum groups. In: Proc. Intern. Congr. Math., Berkeley, 1986, pp.798–820.
59. B.A.Dubrovin, S.P.Novikov and A.T.Fomenko, Modern Geometry,Moscow: Nauka, 1979.
60. N.Dunford and J.T.Schwartz: Linear Operators, part 2. Spectral Theory. Self-adjoint Operators in Hilbert Space. N.Y.-London: Interscience Publishers, 1963.
61. B.M.Dwork: Lectures on p-adic differential equations, N.Y.: Springer-Verlag, 1977.
62. L.D.Faddeev, N.Reshetikhin and L.A.Tachtadjan: *Alg. Anal.* **1**, 129 (1988).
63. G.Faltings: Calculus on arithmetic surfaces. *Ann. of Math.* **119**, 387–424 (1984).
64. G.M.Fichtengolz: Differential and integral Calculus. Vol.3, Gostechizdat, 1949.
65. P.H.Frampton and Y.Okada: The p-adic string N-point function. *Phys. Rev. Lett.* **60**, 484–488 (1988).

66. P.H.Frampton, Y.Okada and M.R.Ubriaco: On adelic formulas for the p-adic string. *Phys. Lett.* **B213**, 260–264 (1988).
67. P.H.Frampton, Y.Okada and M.R.Ubriaco: New p-adic strings from old dual models. *Phys. Rev.* **D39**, 1152–1156 (1989).
68. P.H.Frampton and Y.Okada: Effective scalar field theory of p-adic string. *Phys. Rev.* **D37**, 3077–3079 (1989).
69. P.H.Frampton and H.Nishino: Theory of p-adic closed strings. *Phys. Rev. Lett.* **62**, 1960–1964 (1989).
70. P.H.Frampton: p-Adic number fields and string tree amplitudes. In: Problems in high energy physics and field theory. Proc. of the XI workshop, Protvino 1988, Nauka, 1989.
71. P.H.Frampton and I.V.Volovich: Cosmogenesis and primary quantization. *Mod. Phys. Lett.* **A4**, 1825–1832 (1990).
72. J.Fresnel and M.van der Put: Geometrie Analytique Rigide et Applicatiions, Boston: Birkhauser, 1981.
73. P.G.O.Freund and M.Olson: Non-archimedean strings. *Phys. Lett.* **B199**, 186–190 (1987).
74. P.G.O.Freund and M.Olson: *Nucl. Phys.* **B297**, 86–102 (1988).
75. P.G.O.Freund and E.Witten: Adelic string amplitudes. *Phys. Lett.* **B199**, 191–194 (1987).
76. P.G.O.Freund: Real, p-adic and adelic strings. Proceedings of IXth Int. Congress on Mathematical physics, Swansea, 1988, Adam Hilder, 282–285 (1989).
77. P.G.O.Freund: Scattering on p-adic and on adelic symmetric spaces. Preprint EFI-90-87, 13–17 (1990).
78. P.G.O.Freund: in Superstrings and Particle Theory, L.Clavelle and B.Harms eds., World Scientific Publishing, Singapore, 251 (1990).
79. P.G.O.Freund: On the quantum group – p-adic connection. Chicago preprint (1991).
80. M.Jimbo: A q-difference analogue of $U(g)$ and the Yang-Baxter equation. *Lett. Math. Phys.* **10**, 63–69 (1985).
81. H.Jacquet and R.P.Langlands: Automorphic Forms on $GL(2)$, Lecture Notes in Mathematics, v.114, Berlin: Springer-Verlag, 1970.
82. I.M.Gelfand, M.I.Graev and I.I.Pjatetskii-Shapiro: Representation Theory and Automorphic Functions. London: Saunders, 1966.
83. I.M.Gelfand and N.Ya.Vilenkin: Generalized functions, vol.IV, Moscow, Nauka, 1961
84. L.Gerritzen: p-Adic Siegel halfspace. Groupe Etude Anal. Ultrametrique **3**, J9, 1981/82.
85. L.Gerritzen and M. van der Put: Schottky Groups and Mumford Curves. Lecture Notes in Mathematics, v.817. Berlin: Springer-Verlag, 1980.
86. J.L.Gervais: p-Adic analyticity and Virasoro algebras for conformal theories in more than two dimensions. *Phys. Lett.* **B201**, 306–310 (1988).
87. S.W. Giddings: Fundamental strings Preprint HUTP-88/A061 (1988).

88. I.I.Gihman and A.V.Skorokhod: Introduction to the theory of random processes. Moscow: Nauka, 1965.
89. J.Glimm and A.Jaffe: Quantum Physics, A Functional Intyegral Point of View, N.Y.: Springer-Verlag, 1981.
90. J.A.Gracey and D.Verstegen: The $O(N)$ Gross-Neveu and supersymmetric σ-models on p-adic fields. *Mod. Phys. Lett.* **A5**, 243–254 (1990).
91. N.De Grande-De Kimpe and L.Van Hamme (eds.) Proceedings of the Conference on p-adic analysis, Houthalen, Brussel:Vrije Universiteit Brussel, 1986.
92. M.B. Green, J.H.Scwarz and E.Witten: Superstring Theory. London: Cambridge University Press, 1987.
93. Ph.Griffiths and J.Harris: Principles of Algebraic Geometry, N.Y.: J.Wiley and Sons, 1978
94. D.Gross and P.F.Mende: String theory beyond the Planck scale. *Nucl. Phys.* **303**, 407 (1988).
95. B.Grossman: p-Adic strings, the Weyl conjectures and anomalies. *Phys. Lett.* **B197** 101–105 (1987).
96. B.Grossman: The adelic components of the Dirac operator. *J. Phys.* **A21**, L1051–L1060 (1988).
97. B.Grossman: Adelic conformal field theory. *Phys. Lett.* **B215**, 14–19 (1988).
98. B.Grossman: Arithmetic Directions in Topologycal Quantum Field Theory and Strings. Preprint Rockefeller Univ. DOE/ER40 325-52-Task.B. (1988).
99. J.Harer and D.Zagier: The Euler characteristic of the moduli space of curves. *Invent. Math.* **185**, 457–486 (1986).
100. R.Hartshorne: Algebraic Geometry, N.Y.: Springer-Verlag, 1977.
101. F.Herrlich: Moduli and Teichmuller space for degenerating curves. Proc. Conf. p-adic anal. Hendelhoef, 1986, pp.83–95.
102. H.Heyer: Probability measures on locally compact groups. Berlin: Springer, 1977.
103. T.Hida: Brownian motion. Springer-Verlag, 1980.
104. D.Hilbert: Grundlagen der Geometrie, Leipzig, 1930.
105. Z.Hlousek and D.Spector: Scattering amplitudes in p-adic string theory. *Phys. Lett.* **B214**, 19–25 (1988).
106. Z.Hlousek and D.Spector: p-Adic string theory. Preprint CLNS 88/832, 1988.
107. A.S.Holevo: Probabilistic and statistical aspects of quantum theory. Moscow: Nauka, 1980.
108. E.D'Hoker and D.Phong: The geometry of string perturbation theory, Preprint PUPT-103, 1988.
109. G.'t Hooft: Nucl.Phys, B256, 1985: "Gravitational Collapse and Quantum Mechanics":, Lectures given at the 5th Adriatic Meeting on Particle Physisc, Dubrovnik, 1986.
110. C.J.Isham, Topological and global aspects of quantum theory, in:Relativity, groups and topology II, eds. B.S.DeWitt and R.Stora , 1983.
111. C.J.Isham, Y.Kubyshin and P.Renteln: Quantum norm theory and the quantization of metric topology. *Class. Quant. Grav.* **7**, 1053–1074 (1990).

112. E.Kani: Potential theory on curves, NSERC preprint,1987.
113. G.Kato: Frobenius map of Fermat curves for p-adic strings. Princeton preprint, 1988.
114. A.Yu.Khrennikov: Quantum mechanics over non Archimedean number field. *Theor. Math. Phys.* **83**, 406–418 (1990).
115. A.Yu.Khrennikov: Quantum mechanics over Galois extantions of number field. *Dokl. Acad. Nauk USSR* **315**, 840–864 (1990).
116. A.Yu.Khrennikov: Mathematical methods of non-Archimedean physics. *Usp. Math. Nauk.* **45**, 79–110 (1990).
117. A.Yu.Khrennikov: Pseudodifferential operators over non Archimedean spaces. *Diff. Equat.* **26**, 1044–1053 (1990).
118. A.Yu.Khrennikov: Real and non-Archimedean structure of the space-time. *Theor. Math. Phys.* **86**, 177–190 (1991).
119. A.Yu.Khrennikov: Generalized functions and Gauss path integrals over non-Archimedean functional spaces. *Izv. Acad. Nauk USSR* **55**, 780–814 (1991).
120. A.Yu.Khrennikov: p-Adic quantum mechanics with p-adic valued functions. *J. Math. Phys.* **32**, 932–936 (1991).
121. N.Koblitz: p-Adic numbers, p-adic analysis, and zeta-functions. Berlin, Heidelberg, New York: Springer, 1984.
122. N.Koblitz: p-Adic analysis: a short course on recent work. London: Cambridge University Press, Mathematical Society Lecture Notes, series 46, 1980.
123. A.N.Kochubej: Schrodinger-type operator over the field of p-adic numbers. *Theor. Math. Phys.* **86**, 323–333 (1991).
124. M.Kontsevich: Intersection theory on the moduli space of curves and the matrix Airy functions. Max-Planck-Institute preprint MPI/91-77 (1991).
125. T.H.Koornwinder: Orthogonal polynomials in connection with quantum groups. In "Orthogonal Polynomials:Theory and Practice", ed. P.Nevai, NATO ASI Series C, v.294, Kluwer Academic Publ.,1990.
126. A.I.Kostrikin: Introduction to algebra, M.:Nauka, 1977.
127. S.Lang: Fundamentals of Diophantine Geometry. N.Y.: Springer-Verlag, 1983.
128. S. Lang: Algebra. Addison-Wesley, 1965.
129. D.R.Lebedev and A.Yu.Morosov: An attempt of p-adic one-loop computations. Preprint ITEP 163-88 (1988)
130. E.Y.Lerner and M.D.Missarov: p-Adic Feynman string amplitudes. *Commun. Math. Phys.* **121**, 35–48 (1989).
131. E.Yu.Lerner and M.D.Missarov: Scalar models in p-adic quantum field theory and hierarchical models. *Theor. Math. Phys.* **78**, 248–257 (1989).
132. L.Leroy: Particle in a p-adic box. *Mod. Phys. Lett.* **A5**, 1359–1364 (1990).
133. J.L.Lucio and Y. Meurice: Asymptotic properties of vandom walks on p-adic spaces. Preprint U. of Iowa 90-33, 1–5 (1990).
134. I.G.Macdonald: Orthogonal polynomias associated with root systems, preprint.
135. A.J.Macfarlane: *J. Phys. A. Math. Gen.* **22**, 4581 (1989).

136. G.W.Mackey: Unitary Group Representations in Physics, Probability, and Number Theory, N.Y.: Addison-Wesley Publ. Comp.,, Inc, 1989.
137. G.W.Mackey: Mathemetical foundation of quantum mechanics, W.A. Benjamin, N.Y., 1963.
138. K. Mahler: p-Adic numbers and their functions, London: Cambridge University Press, Cambridge tracts in mathematics 76,1980.
139. B.Mandelbrot: The Fractal Geometry of Nature. Freeman, 1982.
140. Yu.I.Manin and V.G.Drinfeld: Periods of p-adic Schotktky groups, *J.reine angew.Math.*, **262/263**, 239 (1973).
141. Yu.I.Manin. p-Adic Automorphic Functions, Itogi Naukii Tekhniki, Sovremennie Problemy Mathematiki, **3**, 5–92, 1974.
142. Yu.I.Manin: New dimensions in geometry. *Lect. Notes in Math.* **1111**, 59–101, Springer, 1985.
143. Yu.I.Manin: Quantum strings and algebraic curves, *Proc. Int. Cong. Math.*, Berkeley, 1986, p.1286–1296.
144. Yu.I.Manin: Multiparametric deformations of the general linear supergroup. *Comm. Math. Phys.* **123**, 163 (1989).
145. Yu.I.Manin: Reflections on Arithmetical Physics. In "Conformal Invariace and String Theory", eds. P.Dita and V.Gergescu, p.293–303. Boston: Acadic Press, 1989.
146. E.Marinari and G.Parisi: On the p-adic five point function. *Phys. Lett.* **203B**, 52–56 (1988).
147. M.A.Markov: *Progr.Theor. Phys. Suppl.* **85**, (1965).
148. A.Marshakov and A.Zabrodin: New p-adic string amplitudes. *Phys. Lett.* (1991).
149. V.P.Maslov: Perturbation theory and asymptotic methods. Moscow: Moscow University Press, 1965.
150. V.P.Maslov: Operator Methods. Moscow: Nauka, 1973.
151. Y.Meurice: Quantum mechanics with p-adic numbers. *Int. J. Mod. Phys.* **A4**, 5133–5137 (1989).
152. Y.Meurice: A path integral formulation of p-adic quantum mechanics. *Phys. Lett.* **245B**, 99–104 (1990).
153. Y.Meurice: A discretization of p-adic quantum mechanics. *Comm. Math. Phys.* **135**, 303–312 (1991).
154. Y.Meurice: The classical harmonic oscillator on Galois and p-adic fields. *Intern. Journ. of Mod. Phys.* **A4**, 2211–2233 (1989).
155. Y.Meurice: Quantum mechanics with p-adic time. Preprint CINVESTAV-FIS-12-89, 1–14 (1989).
156. Y.Meurice: Symanzik's Field Theory on p-adic Spaces, Iowa Preprint, (1991).
157. M.Mezard, G.Parisi, N.Sourlas, G.Toulouse and M.A.Virasoro: *J. Phys.* (Paris) **45**, 843 (1984).
158. J.S.Milne: Etale Cohomology. Princeton: Princeton University Press, 1980.
159. M.D.Missarov: Random fields on the adele ring and Wilson's renormalization group. *Ann. Inst. H. Poincare* **49**, 357–367 (1989).

160. D.Mumford: An analytic construction of degenerating curves over complete local fields. *Composito Math.* **24**, 129 (1972).
161. D.Mumford: Stability of projective varieties. *Enseign. Math.* **23**, 39–100 (1977).
162. D.Mumford: Tata lectures on Tetha I,II. Boston: Birkhauser, 1983,1984.
163. M.A.Naimark: Normed rings. moscow: Nauka, 1968.
164. H.Nishino, Y.Okada and M.R.Ubriaco: Effective field theory from a p-adic string. *Phys. Rev.* **D40**, 1153–1157 (1989).
165. H.Nishino and Y.Okada: Beta function approach to p-adic string. *Phys Lett.* **B219**, 258–262 (1989).
166. Y.Okada: p-Adic string amplitude. *Nucl.Phys.* **B6**, 177–182 (1989).
167. Y.Okada and M.R.Ubriaco: Renormalization of $O(N)$ Nonlinear σ-Model on a p-adic Field. *Phys. Rev. Lett.* **61**, 1910–1913 (1988); **A3**, 639–643 (1988).
168. G.Parisi: On p-adic functional integrals. *Mod. Phys. Lett.* **A4**, 369–374 (1988).
169. G.Parisi: *J. Phys.* **A13**, 1887–1892 (1980).
170. R.C.Penner: Perturbation series and the moduli space of Riemann surface. *J. Diff. Geom.* **27**, 35–59 (1988).
171. A.Peres and N.Rosen: Quantum limitation to the measurement of gravitational field. *Phys. Rev.* **118**, 335–376 (1960).
172. H.Poincare: La Science et l'Hypothese, Paris: Flammarion, 1923.
173. A.M.Polyakov: Quantum geometry of bosonic strings. *Phys. Lett.* **B103**, 207–212 (1981).
174. L.S.Pontrjagin: Continous groups. Moscow:Nauka, 1972.
175. M.Reed and B.Simon: Methods of Modern Mathematical Physics. 1: Functional Analysis, 1972; 4: Analysis of Operators, 1978.-Academic Press.
176. T.Regge: Nuovo Cimento **7**, 215 (1958).
177. B.Riemann: Uber die Hypothesen die der Geometrie Zugrunde leigen. Nachr. Ges. Gottingen, Bd.**13**, S.133 (1868).
178. R.Rammal, G.Toulouse and M.A.Virasoro: Ultrametricity for physicist. *Rev. Mod. Phys.* **58**, 765–821 (1986).
179. A. van Rooij: Non-archimedean functional analysis. N.Y.: Marcel Dekker Inc., 1978.
180. B.O.B.Roth: A general approach to quantum fields and strings on adeles. *Phys. Lett.* **B213**, 263–268 (1988).
181. P.Ruelle, E.Thiran, D.Verstegen and J.Weyers: Adelic string and superstring amplitudes. *Mod. Phys. Lett.* **A4**, 1745–1753 (1989).
182. P.Ruelle, E.Thiran, D.Verstegen and J.Weyers: Quantum mechanics on p-adic fields. *J. Math. Phys.* **30**, 2854–2859 (1989).
183. Z.Ryzak: Scattering amplitudes from higner dimensional p-adic world sheets. *Phys. Lett.* **B208**, 411–416 (1988).
184. J.Scherk: An introduction to the theory of dual models and strings. *Rev. Mod. Phys.* **47**, 123–165 (1975).

185. W.H.Schikhof: Non-Archimedean Harmonic Analysis, Ph.D. Thesis, Rotterdam University, 1967.
186. W.H.Schikhof: Ultrametric calculus. An introduction to p-adic analysis. Cambridge University Press, 1984.
187. I.E.Segal: Mathematical Problems of Relativistic Physics. *Amer. Math. Soc. Providence*, R.I., 1963.
188. I.R.Shafarevich: Foundations of Algebraic geometry. Moscow: Nauka, 1980.
189. J.P.Serre: Cours d'Arithmetique. Presses Universitaires de France, Paris, 1970.
190. J.P.Serre: Lie Algebras and Lie groups. - N.Y.: Benjamin, 1985.
191. B.Simon: The $P(\phi)$ Euclidean (Quantum) Field Theory. Princeton, N.J. 1974.
192. V.A.Smirnov: Renormalization in p-adic quantum field theory. *Mod. Phys. Lett.* **A6**, 1421–1427 (1991).
193. V.A. Smirnov: p-Adic Feynman amplitudes. Preprint MPI-Ph/91-48, 1–13 (1991).
194. O.G.Smolianov: Analysis on linear topological spaces and its applications. Moscow. Moscow University Press, 1979.
195. C.Soule: Geometrie d'Arakelov des surfaces arithmetiques. Seminare Bourbaki, 1988–89, No.713, Asterisque 177–178 (1989), p.327–343.
196. B.L.Spokoiny: Quantum geometry of non-Archimedean particles and strings. *Phys. Lett.* **207B**, 401–406 (1988).
197. B.L.Spokoiny: Non-Archmedean geometry and quantum mechanics. *Phys. Lett.* **B211**, 120–125 (1989).
198. J.Tate: Rigid Analytic spaces. *Invent. Math.* **12**, 257–293 (1971).
199. E.Thiran, D.Verstegen and J.Weyers: p-Adic Dynamics. *J. Stat. Phys.* **54**, 893–913 (1989).
200. H.-J.Treder: in "Relativity, Quants and Cosmology", ed. F.de Finis, N.Y., 1979.
201. M.R.Ubriaco: Fermions on the field of p-adic numbers. *Phys.Rev.* **D41**, 2631–2636 (1990).
202. M.R.Ubriaco: Field quantization in nonarchimedean field theory. Preprint LTP-012-UPR.
203. E.Verlinde and H.Verlinde: Chiral bosonization, determinats and string partition functions. Preprint Utrecht, 1986.
204. I.M.Vinogradov: Elements of Number theory. Moscow: Nauka, 1972.
205. V.S.Vladimirov: Generalized functions in mathematical physics. Moscow: Nauka, 1979.
206. V.S.Vladimirov: Generalized functions over p-adic number field. *Usp. Mat. Nauk* **43**, 17–53 (1988).
207. V.S.Vladimirov: p-Adic analysis and p-adic quantum mechanics. *Ann. of the NY Ac. of Sci: Symposium in Frontiers of Math.* (1988).
208. V.S.Vladimirov: On the spectrum of some pseudo-differential operators over p-adic number field. Algebra and analysis **2**, 107–124 (1990).

209. V.S.Vladimirov: An application of p-adic numbers in quantum mechanics. In: Selected topics in Statistical mechanics, 22–24 August 1989, Dubna, USSR, World Scientific Publishing, 1990, p.282–297.
210. V.S.Vladimirov: Applications of p-adic numbers in mathematical physics. Delph: Culturel Center, 1989.
211. V.S.Vladimirov and I.V.Volovich: Superanalysis. Differential calculus. *Theor. Math. Phys.* **59**, 3–27 (1984).
212. V.S.Vladimirov and I.V.Volovich: P-Adic Quantum Mechanics. *Soviet. Math,. Dokl.* **303**, No.2, 320–323 (1988).
213. V.S.Vladimirov and I.V.Volovich: P-Adic Quantum Mechanics. *Commun. Math. Phys.* **123**, 659–676 (1989).
214. V.S.Vladimirov and I.V.Volcvich: P-adic Schrodinger-type equation. *Lett. in Math. Phys.* **18**, 43–53 (1989).
215. V.S.Vladimirov and I.V.Volovich: A vacuum state in p-adic quantum mechanics. *Phys. Lett.* **B217**, 411–414 (1989).
216. V.S.Vladimirov and I.V.Volovich: Applications of p-adic numbers in Mathematical physics. Trudi MIAN **200**, 88–99 (1991).
217. V.S.Vladimirov, I.V..Volovich and E.I.Zelenov: Spectral theory in p-adic quantum mechanics and representation theory. *Soviet. Math. Dokl.* **41**, 40–44 (1990).
218. V.S.Vladimirov, I.V.Volovich and E.I.Zelenov: Spectral theory in p-adic quantum mechanics and representation theory. *Izv. Acad. Nauk USSR* **54**, 275–302 (1990); *Math. USSR Izvestya* **36**, 281–309 (1991).
219. I.V.Volovich: p-Adic space-time and string theory. Theor. *Math. Phys.* **71**, 337–340 (1987).
220. I.V.Volovich: Number theory as the ultimate physical theory. Preprint CERN-TH. **87**, 4781–4786 (1987).
221. I.V.Volovich: p-Adic string. *Class. Quantum Grav.* **4**, L83–L87 (1987).
222. I.V.Volovich: Harmonic analysis and p-adic strings. *Lett. Math. Phys.* **16**, 61–66 (1988).
223. I.V.Volovich: p-Adic quantum theory and srtings. In: Proceedings of IXth Int. Congress on Mathematical physics, Sweansea, 1988, Adam Hilger, 286–293 (1989).
224. J.Wess and B.Zumino: Covariant differential calculus on the quantum hyperplane. Preprint CERN-TH-5697/90 (1990).
225. A.Weil: Sur certains groupes d'operateurs unitaries. *Acta Math.* **111**, 143–211 (1964).
226. A.Weil: Basic Number Theory. Berlin: Springer, 1985
227. H.Weyl: The theory of groups and quantum mechanics. New York: Dover, 1931.
228. H.Weyl: Philosophy of Mathematics and Natural Science. Princeton University Press, 1949.
229. J.A.Wheeler: "Superspace and the Nature of Quantum Geometrodynamics". In: C.M.DeWitt and J.A.Wheeler, eds., Battell Rencontres: 1967 Lectures in

Mathematical Physics, pp.242–307, N.Y.: Benjamin, 1968.
230. B.De Witt: The Quantization of Geometry. In: "Gravitation: an Introduction to Current Research", ed. L.Witten, N.Y.:J.Wiley and Sons, 1962.
231. E.Witten: Free fermions on an algebraic curve. In: Proc. of the Symposium on the Mathematical Heritage of Herman Weyl, Durham, North Carolina, 1987.
232. E.Witten: The search for higher Symmetry in String Theory, Preprint IASSNS-HEP-88/55 (1988).
233. E.Witten: Topological Quantum Field Theories. *Comm. Math. Phys.* **117**, 353–391 (1988).
234. E.Witten: Two dimensional gravity and intersection theory on moduli space. *Surveys in Diff. Geom.* **1**, 243–289 (1991).
235. S.Woronovicz: *Comm. Math. Phys.* **111**, 613 (1987); **122**, 125 (1989); Publ. RIMS Kyoto Univ. **23**, 117 (1987).
236. H.Yamakoshi: Arithmetic of strings. *Phys. Lett.* **B207**, 426–428 (1988).
237. K.Yosida: Functional analysis. Springer-Verlag, 1965.
238. A.V.Zabrodin: Non-archimedean strings and Bruhat-Tits trees. *Mod. Phys. Lett.* **A4**, 367–376 (1989).
239. E.I.Zelenov: p-Adic quantum mechanics for $p = 2$. *Theor. Math. Phys.* **80**, 253–263 (1989).
240. E.I.Zelenov: p-Adic path integrals. *J. Math. Phys.* **32**, 147–152 (1991) 12.
241. E.I.Zelenov: p-Adic quantum mechanics and coherent states. 1. Weyl systems. *Theor. Math. Phys.* **86**, 210–220 (1991).
242. E.I.Zelenov: p-Adic quantum mechanics and cohorernt states 2. Oscillator eigenfunctions. *Theor. Math. phys.* **86**, 375–384 (1991).
243. E.I.Zelenov: Representations of commutation relations of p-adic systems of infinitely many degrees of freedom. *J. Math. Phys.* **33**, 178–188 (1992).
244. E.I.Zelenov: p-Adic Heisenberg group, Maslov index and metaplectic representation. Preprint Print-91-0439, 1–57 (1991).
245. R.B.Zhang: Lagrangian formulation of open and closed p-adic strings. *Phys. Lett.* **B209**, 229–232 (1988).

www.ingramcontent.com/pod-product-compliance
Lightning Source LLC
Chambersburg PA
CBHW050634190326
41458CB00008B/2268

* 9 7 8 9 8 1 0 2 0 8 8 0 6 *